U.S. AIR FORCE U-2 FLIGHT MANUAL

U.S. AIR FORCE

U-2
FLIGHT MANUAL

GOVERNMENT REPRINTS PRESS
Washington, D.C.

© Ross & Perry, Inc. 2001 All rights reserved.

No claim to U.S. government work contained throughout this book.

Protected under the Berne Convention. Published 2001

Printed in The United States of America
Ross & Perry, Inc. Publishers
717 Second St., N.E., Suite 200
Washington, D.C. 20002
Telephone (202) 675-8300
Facsimile (202) 675-8400
info@RossPerry.com

SAN 253-8555

Government Reprints Press Edition 2001

Government Reprints Press is an Imprint of Ross & Perry, Inc.

Library of Congress Control Number: 2001093418

http://www.GPOreprints.com

ISBN 1-931641-65-X

♾ The paper used in this publication meets the requirements for permanence established by the American National Standard for Information Sciences "Permanence of Paper for Printed Library Materials" (ANSI Z39.48-1984).

All rights reserved. No copyrighted part of this publication may be reproduced, stored in a retrieval system, or transmitted, in any form or by any means, electronic, photocopying, recording, or otherwise, without the prior written permission of the publisher.

AF(C)-1-1

COPY NO. 15

FLIGHT MANUAL

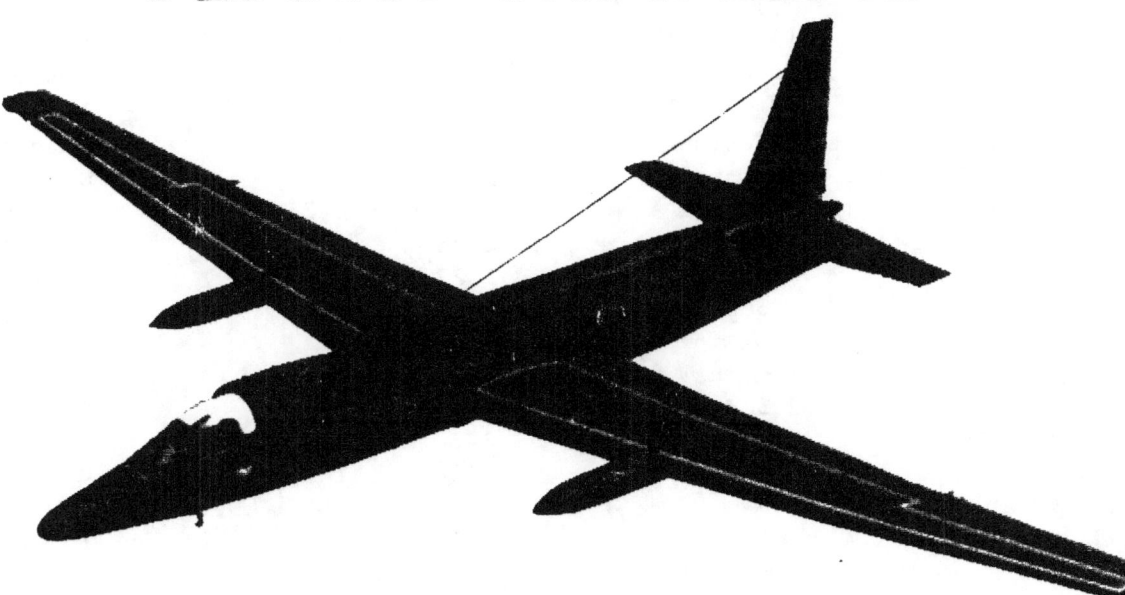

MODELS U-2C and U-2F AIRCRAFT

NOTICE

THIS REVISION, DATED 10 MAY 1967, SUPERSEDES FLIGHT MANUAL AF (C)-1-1 DATED 15 DECEMBER 1966, CHANGED 15 FEBRUARY 1967.

THIS CHANGE, DATED 15 OCTOBER 1968, AFFECTS INFORMATION IN SECTION II.

DESTROY SUPERSEDED DATA IN ACCORDANCE WITH AFR-205-1.

LATEST CHANGED PAGES SUPERSEDE
THE SAME PAGES OF PREVIOUS DATE

Insert changed pages into basic publication. Destroy superseded pages.

10 MAY 1967
Changed 15 October 1968

AF (C)-1-1

NOTE: The portion of the text affected by the changes is indicated by a vertical line in the outer margins of the page.

LIST OF EFFECTIVE PAGES

Insert Latest Changed Pages, Destroy Superseded Pages

TOTAL NUMBER OF PAGES IN THIS PUBLICATION IS 310 CONSISTING OF THE FOLLOWING:

Page No.	Issue	Page No.	Issue
*Title	15 Oct 68	2-14 thru 2-17	1 Sept 68
*A	15 Oct 68	2-18 thru 2-19	6 Nov 67
*B (Added)	15 Oct 68	2-20	1 Sept 68
C (Blank)	1 Sept 68	2-21	Original
i thru v	Original	2-22	6 Nov 67
vi (Blank)	Original	2-23 thru 2-25	1 Sept 68
1-1	1 Sept 68	2-26	Original
1-2 thru 1-3	Original	2-27	1 Sept 68
1-4	1 Sept 68	2-28 thru 2-29	Original
1-5	6 Nov 67	*2-30	15 Oct 68
1-6 thru 1-7	Original	2-31 thru 2-32	Original
1-8	1 Sept 68	2-33	1 Sept 68
1-9 thru 1-10	6 Nov 67	2-34 thru 2-35	6 Nov 67
1-10A (Added)	6 Nov 67	2-36 (Blank)	Original
1-10B (Blank)	6 Nov 67	2-37 thru 2-65 (Deleted)	6 Nov 67
1-11	6 Nov 67	3-1 thru 3-4	1 Sept 68
1-12	Original	3-4A (Added)	1 Sept 68
1-13	6 Nov 67	3-4B (Blank)	1 Sept 68
1-14	Original	3-5	1 Sept 68
1-15	1 Sept 68	3-6 thru 3-7	Original
1-16	Original	3-8	6 Nov 67
1-17	6 Nov 67	3-9 thru 3-10	Original
1-18 thru 1-20	Original	3-11	6 Nov 67
1-21 thru 1-22	1 Sept 68	3-12 thru 3-16	Original
1-23 thru 1-26	Original	3-17 thru 3-21	1 Sept 68
1-27	1 Sept 68	3-22 thru 3-26	Original
1-28 thru 1-29	Original	3-27	1 Sept 68
1-30	1 Sept 68	3-28	Original
1-31 thru 1-32	Original	3-29	1 Sept 68
1-33 thru 1-34	1 Sept 68	3-30 thru 3-34	Original
1-35	Original	3-35	15 Aug 68
1-36	1 Sept 68	3-36	1 Sept 68
1-36A thru 1-36B (Added)	1 Sept 68	3-37	15 Aug 68
1-37	1 Sept 68	3-38 thru 3-39	Original
1-38 thru 1-49	Original	3-40	6 Nov 67
1-50 (Blank)	Original	3-41 thru 3-66 (Deleted)	6 Nov 67
2-1	6 Nov 67	4-1	Original
2-2	Original	4-2	6 Nov 67
2-3	1 Sept 68	4-3	5 July 67
2-4 thru 2-5	Original	4-4 thru 4-6	1 Sept 68
2-6 thru 2-10	1 Sept 68	4-6A thru 4-6B (Added)	1 Sept 68
2-11	6 Nov 67	4-7 thru 4-20	Original
2-12 thru 2-13	Original	4-21	6 Nov 67
		4-22 thru 4-28	Original

* The asterisk indicates pages changed, added, or deleted by the current change.

Changed 15 October 1968

AF (C)-1-1

LIST OF EFFECTIVE PAGES

NOTE: The portion of the text affected by the changes is indicated by a vertical line in the outer margins of the page.

Insert Latest Changed Pages, Destroy Superseded Pages

Page No.	Issue	Page No.	Issue
4-29	1 Sept 68	A5-1	6 Nov 67
4-30 thru 4-33	Original	A5-2	Original
4-34	6 Nov 67	A5-3 thru A5-8	6 Nov 67
4-35 thru 4-48	Original	A5-9 (Added)	6 Nov 67
4-49	1 Sept 68	A5-10 (Blank)	6 Nov 67
4-50 thru 4-52	Original	A6-1 thru A6-3	6 Nov 67
4-53	6 Nov 67	A6-4 (Blank)	Original
4-54 thru 4-55	Original	A7-1 thru A7-2	6 Nov 67
4-56 thru 4-57	12 Dec 67	A8-1 thru A8-4	Original
4-58 thru 4-59	Original	A9-1 thru A9-2	Original
4-60 (Blank)	Original	X-1	1 Sept 68
5-1	6 Nov 67	X-2 thru X-3	6 Nov 67
5-2	1 Sept 68	X-4	1 Sept 68
5-3	6 Nov 67	X-5	6 Nov 67
5-4	Original	X-6	15 Aug 68
5-5 thru 5-6	1 Sept 68	X-7	1 Sept 68
5-7 thru 5-16	Original	X-8 thru X-9	6 Nov 67
6-1 thru 6-15	Original	X-10	1 Sept 68
6-16	1 Sept 68	X-11	15 Aug 68
7-1	Original	X-12	6 Nov 67
7-2 thru 7-3	6 Nov 67	X-13 thru X-14	15 Aug 68
7-4 thru 7-5	Original		
7-6 (Blank)	Original		
8-1	Original		
8-2 (Blank)	Original		
9-1 thru 9-9	Original		
9-10 (Blank)	Original		
A-1	Original		
A-2 (Blank)	Original		
A1-1 thru A1-2	Original		
A1-3	6 Nov 67		
A1-4 thru A1-7	Original		
A1-8 (Blank)	Original		
A2-1	6 Nov 67		
A2-2	Original		
A3-1	6 Nov 67		
A3-2 thru A3-5	Original		
A3-6 (Blank)	Original		
A4-1	6 Nov 67		
A4-2	Original		
A4-3 thru A4-17	6 Nov 67		
A4-18 (Blank)	Original		

* The asterisk indicates pages changed, added, or deleted by the current change.

Changed 15 October 1968

B/C

AF (C)-1-1

SECURITY   CLASSIFICATION

SPECIFIC INSTRUCTIONS FOR SAFEGUARDING

THIS

MILITARY INFORMATION

This document is UNCLASSIFIED. Its dissemination and handling, however, will be on an established "need-to-know" basis. By direction of the Chief of Staff, USAF, the following policies will govern its use, dissemination and handling:

This document may be issued to persons possessing an established need-to-know.

Strict accountability will be maintained of all copies issued.

This document will be controlled in a manner that will prevent its loss, destruction, or its falling into the hands of unauthorized persons.

In the event this document is lost or destroyed, this fact will be reported to the office of issue and/or to the Commander responsible for the custody of the material.

AF (C)-1-1

WARNINGS, CAUTIONS, AND NOTES

The following definitions apply to the Warning, Cautions, and Notes found throughout the handbook.

 Operating procedures, practices, etc., which will result in personal injury or loss of life if not carefully followed.

 Operating procedures, practices, etc., which if not strictly observed will result in damage to equipment.

NOTE An operating procedure, condition, etc., which it is essential to emphasize.

CODING

This Flight Manual covers U-2C and U-2F model series aircraft. A coding system to identify that part of the manual pertaining to individual aircraft models is presented below:

Text and illustrations applicable to all aircraft models are not coded.

Text and illustrations applicable to individual aircraft models are identified as U-2C or U-2F, respectively, or coded with the letters Ⓒ or Ⓕ.

AF (C)-1-1

U-2C U-2F

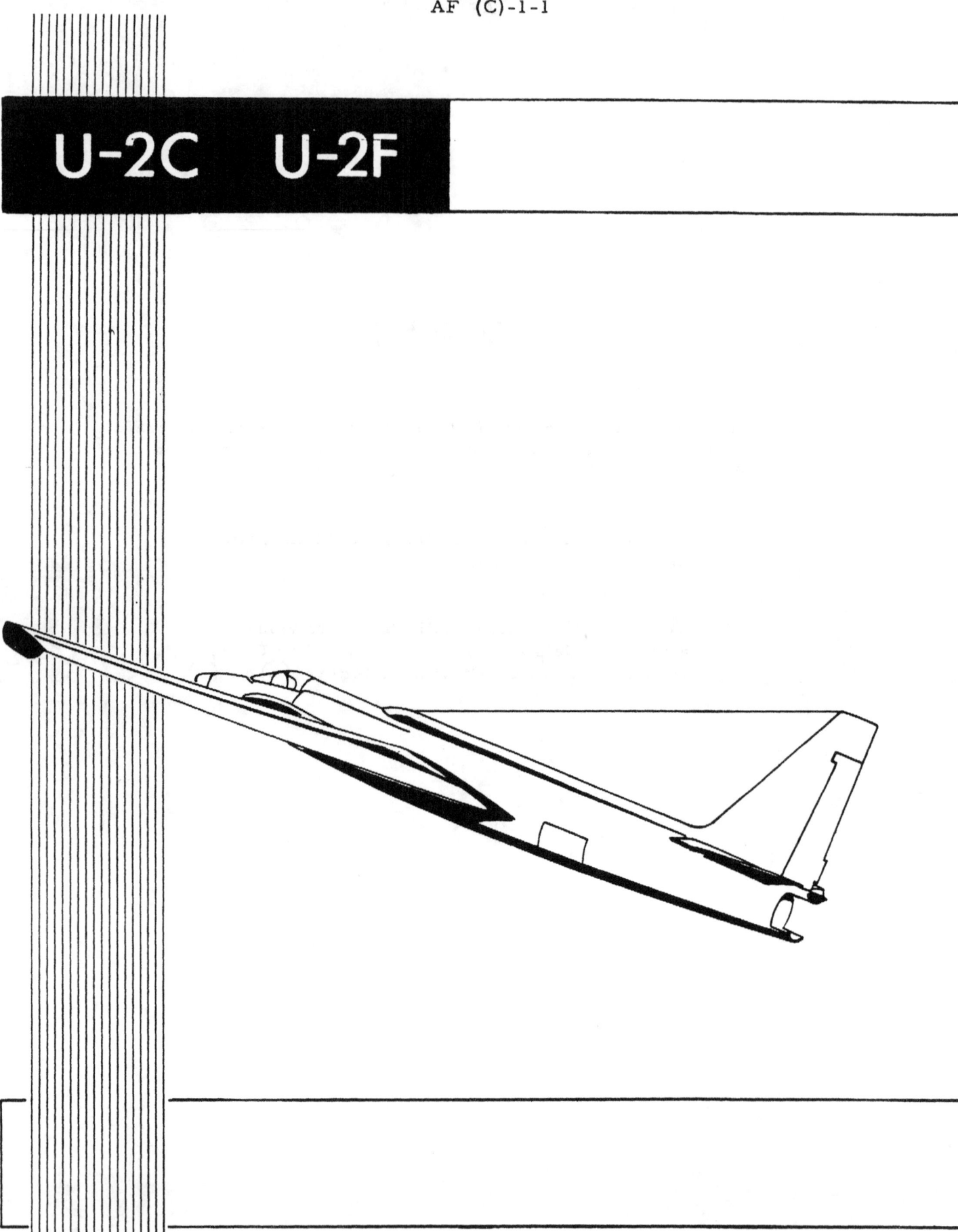

TABLE OF CONTENTS

		PAGE
SECTION I	DESCRIPTION	1-1
SECTION II	NORMAL PROCEDURES	2-1
SECTION III	EMERGENCY PROCEDURES	3-1
SECTION IV A	DESCRIPTION AND OPERATION OF AUXILIARY EQUIPMENT	4-1
SECTION IV B	DESCRIPTION AND OPERATION OF PHOTOGRAPHIC AND ELINT EQUIPMENT (UNDER SEPARATE COVER)	4-1
SECTION V	OPERATING LIMITATIONS	5-1
SECTION VI	FLIGHT CHARACTERISTICS	6-1
SECTION VII	SYSTEMS OPERATION	7-1
SECTION VIII	CREW DUTIES (NOT APPLICABLE)	8-1
SECTION IX	ALL WEATHER OPERATION	9-1
APPENDIX I	PERFORMANCE DATA	A-1
INDEX	ALPHABETICAL	X-1

AF(C)-1-1

DESCRIPTION SECTION 1

TABLE OF CONTENTS

	PAGE		PAGE
THE AIRCRAFT	1-1	WING FLAPS	1-28
ENGINE	1-4	RETRACTABLE STALL STRIPS	1-28
ENGINE FUEL SYSTEM	1-4	SPEED BRAKES	1-29
ENGINE OIL SYSTEM	1-7	LANDING GEAR	1-29
NORMAL IGNITION SYSTEM	1-9	STEERING SYSTEM	1-30
CONTINUOUS IGNITION	1-9	WHEEL BRAKES	1-30
STARTER SYSTEM	1-9	DRAG CHUTE	1-30
INTERCOMPRESSOR BLEED SYS	1-10	INSTRUMENTS	1-32
ENGINE INSTRUMENTS	1-10	FIRE WARNING SYSTEM	1-45
AIRPLANE FUEL SYSTEM	1-11	SPEED WARNING SYSTEM	1-46
ELECTRICAL SYSTEM	1-19	CANOPY	1-46
HYDRAULIC SYSTEM	1-24	AUXILIARY EQUIPMENT	1-47
FLIGHT CONTROL SYSTEM	1-26		

THE AIRCRAFT

The U-2C and U-2F are single place jet aircraft designed for high altitude, long range operation. The U-2F is air refuelable. Both models can be equipped with "slipper" and drop tanks.

The basic configuration is fitted for photo reconnaissance. Other equipment can also be installed for special purposes. Refer to figures 1-1 and 1-2 for general configuration information.

BASIC DIMENSIONS

Wing Area		600 sq ft
Wing Span		80 ft
Aspect Ratio		10.67
Wing Average Thickness		7.8%
Fuselage Length		49.7 ft

APPROXIMATE GROSS WEIGHTS

Condition	Gross Weight	
	U-2C	U-2F
Full main and empty aux tanks (1020 gal)	20,680 lb	20,980 lb
Full main and full aux tanks (1320 gal)	22,650 lb	22,950 lb
Full main, aux and slipper tanks (1520 gal)	23,970 lb	24,270 lb
Full main, aux, and drop tanks (1520 gal)	23,970 lb	24,270 lb
Zero fuel weight	13,870 lb	14,170 lb

Changed 1 September 1968

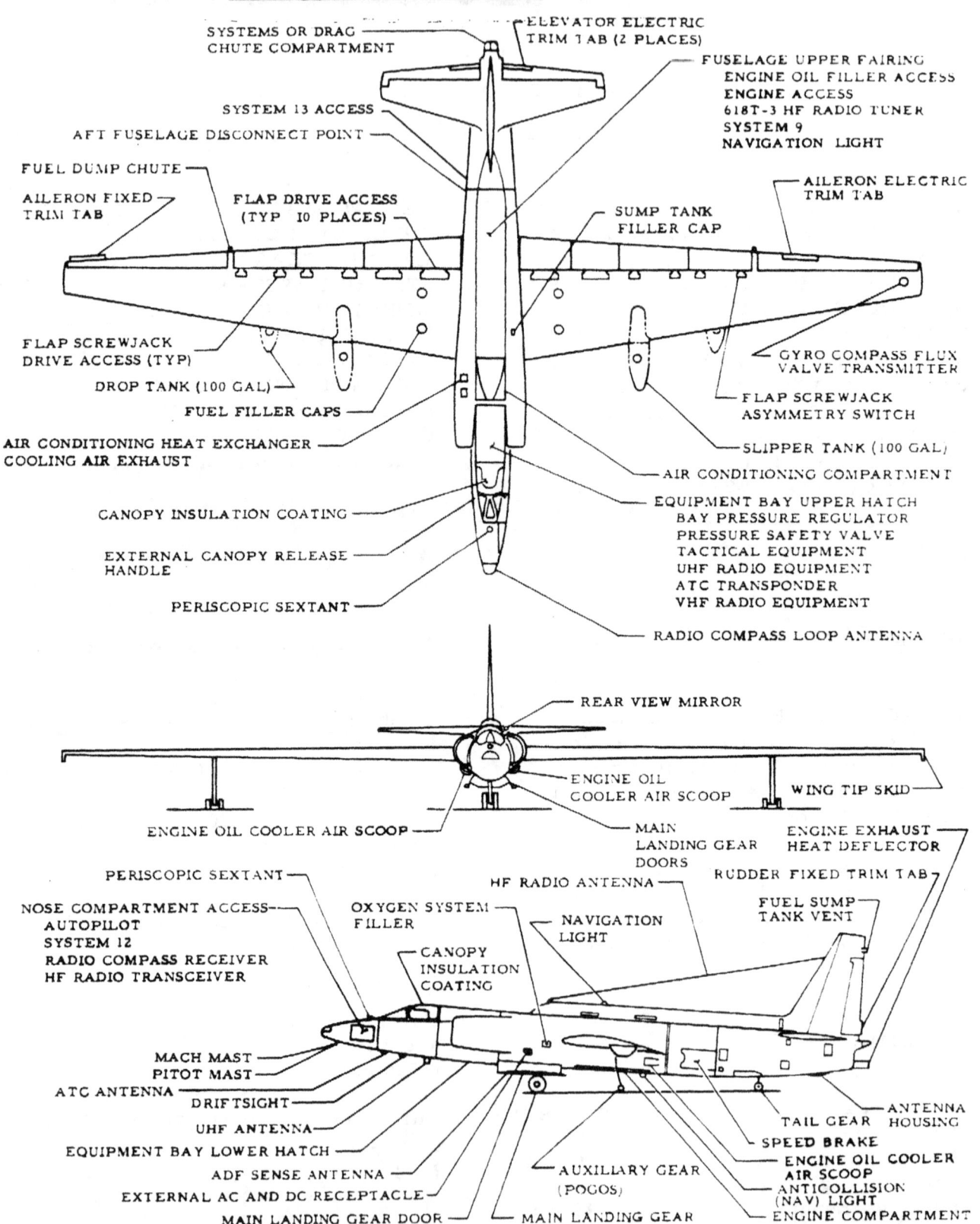

Figure 1-1

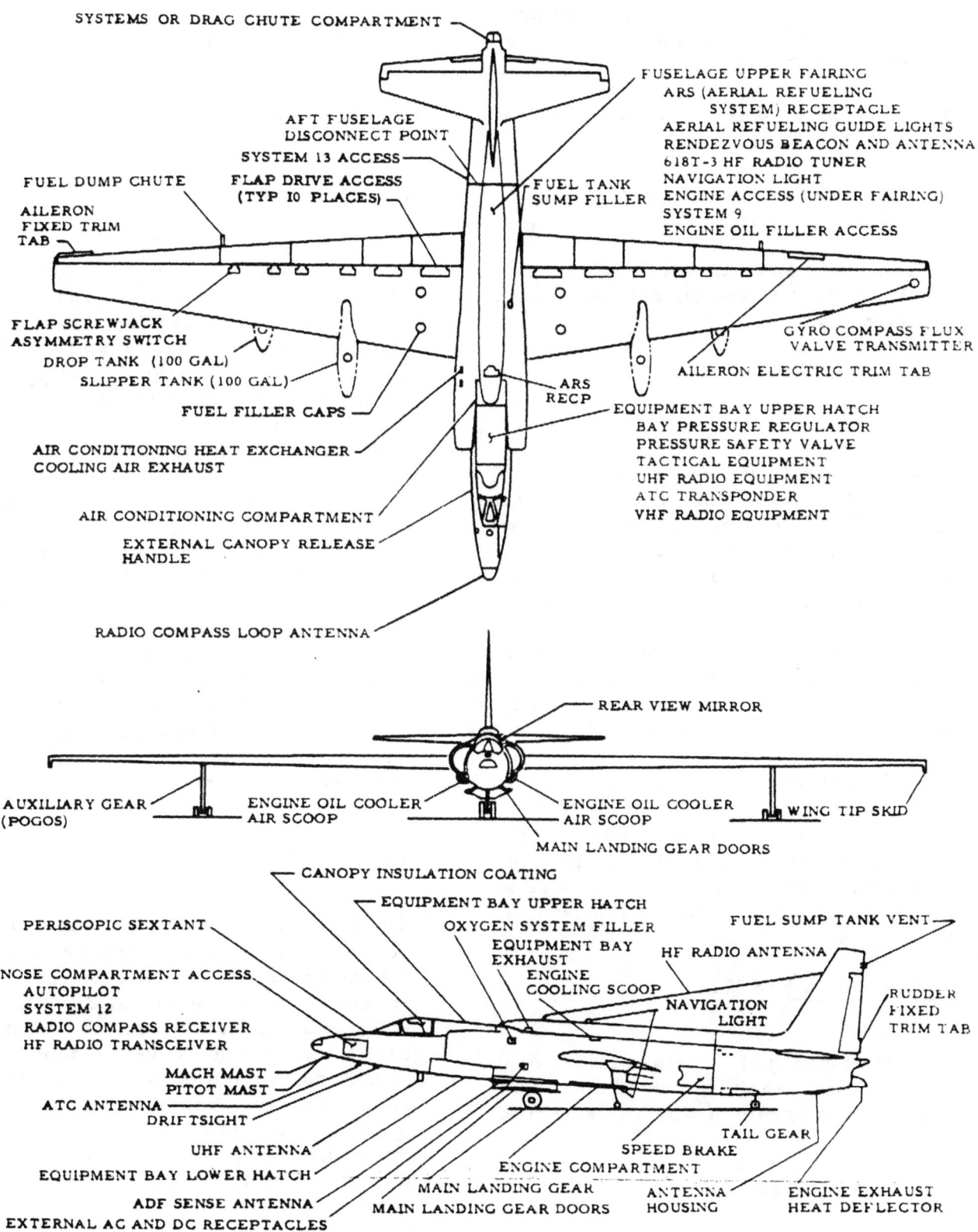

Figure 1-2

Notes:

1. Operational equipment load assumed to be equivalent to "B" and "T" with systems 1, 9, 12, and 13.

2. These weights are based on a specific fuel weight of 6.58 lb/gal.

3. The U-2F airplane is approximately 300 pounds heavier than the U-2C.

4. These are approximate weights and should not be used for detailed mission planning.

ENGINE

The aircraft is powered by a single, 15-stage, J75P-13B type nonafterburning engine (figure 1-3), which employs an axial flow, twin spool compressor, and a split three-stage turbine. The front compressor is an 8-stage low pressure unit which is connected by a through shaft to the second and third stage turbine wheels. The high compressor has seven stages of compression independent of and following the low compressor, and is connected by a hollow shaft to the first stage turbine wheel. This arrangement permits the low pressure rotor to turn at its best speed and allows higher compressor ratios. A low pressure overboard bleed valve is provided on each side of the High pressure compressor case. The engine is rated at approximately 17,000 pounds sea level static thrust.

ENGINE FUEL SYSTEM

The J-75 engine fuel system consists of the following major engine mounted components, a schematic diagram of which is shown in figure 1-4.

1. Two-stage gear type pump.

2. Hamilton Standard hydromechanical fuel control.

3. Fuel manifold pressurizing valve and automatic fuel manifold drain valve.

4. Forty eight dual-orifice fuel nozzles.

ENGINE FUEL PUMP

The engine-driven fuel pump has a low pressure stage and a high pressure stage and supplies pressure for both the normal and the emergency fuel systems. During operation, the low pressure stage acts as an engine-driven boost pump and increases fuel pressure to the inlet of the high pressure stage. The engine fuel pump will supply sufficient fuel to maintain flight if the airplane normal and auxiliary boost pumps fail, providing the airplane has been at altitude for approximately one hour, which is sufficient time to be cold soaked or weathered. In the event of a failure of the low pressure stage, adequate fuel pressure at the inlet to the high pressure (main) stage of the engine fuel pump can be maintained by the airplane boost through a by-pass valve. If the high pressure stage of the engine pump fails, flameout will result and a restart cannot be made.

MAIN FUEL CONTROL

The fuel control regulates engine speed by means of a fly-ball governor for all operating conditions, including starting, acceleration and deceleration. During starting and acceleration, the control limits fuel flow to prevent compressor surge and overtemperature. During deceleration, the control schedules a minimum fuel flow to prevent flameout. This minimum flow schedule also determines the idle RPM at altitude. In addition, the control incorporates a maximum burner pressure limiter which limits the maximum output of the engine to a safe level so as to prevent damage to the combustion chamber case due to overpressure.

Changed 1 September 1968

AF (C)-1-1　　　　　　　　　　　　　　　　　　　　　SECTION I

J75 ENGINE CUTAWAY

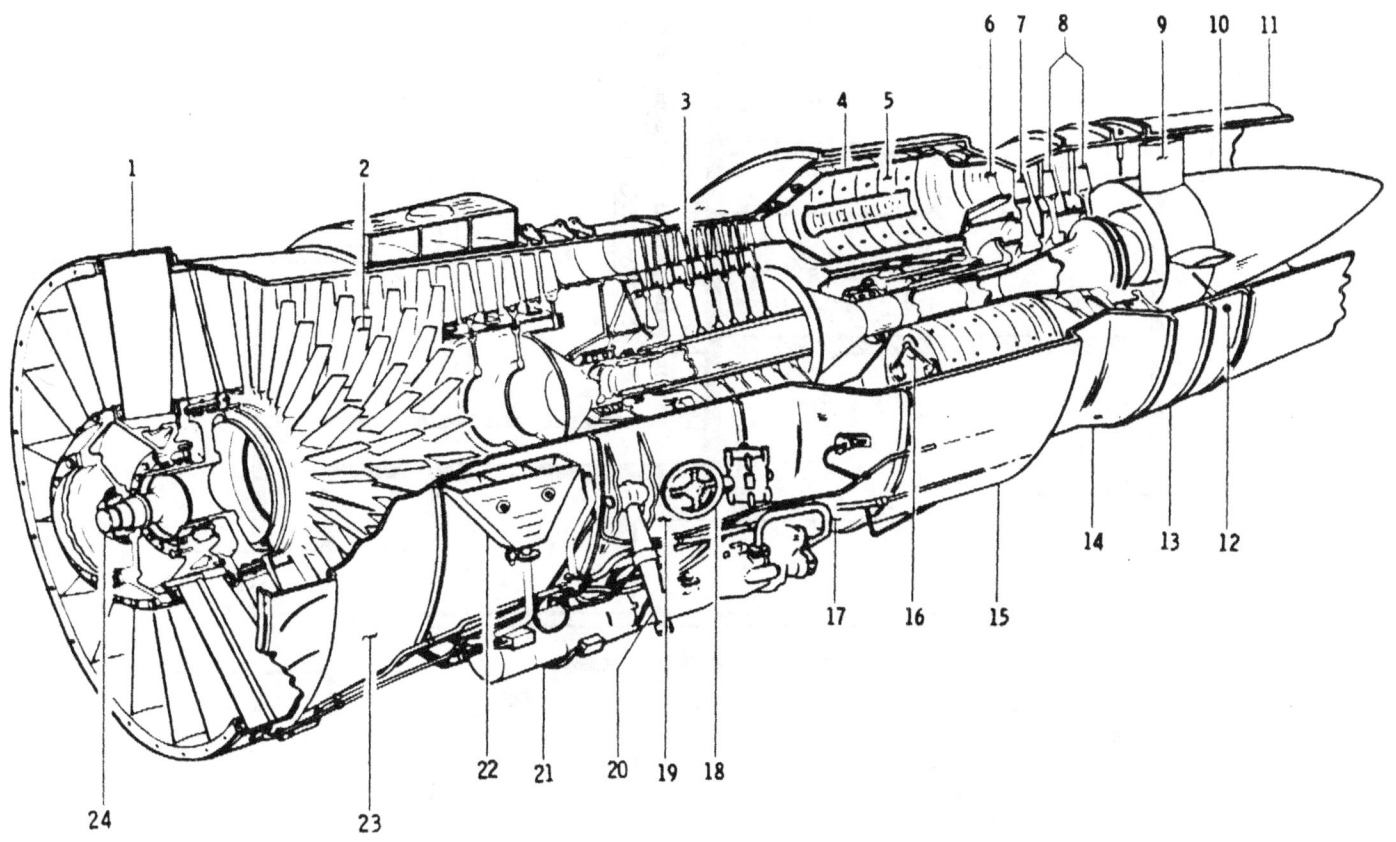

1. COMPRESSOR INLET GUIDE VANES AND SHROUD
2. LOW-PRESSURE COMPRESSOR (EIGHT-STAGE)
3. HIGH-PRESSURE COMPRESSOR (SEVEN-STAGE)
4. COMBUSTION CHAMBER (8)
5. FUEL NOZZLE (6 IN EACH COMUBSTION CHAMBER)
6. TURBINE NOZZLE
7. TURBINE WHEEL, FRONT (ONE, DRIVES HIGH PRESSURE COMPRESSOR)
8. TURBINE WHEEL, REAR (TWO, DRIVE LOW PRESSURE COMPRESSOR)
9. SWIRL STRAIGHTENER VANE (6)
10. EXHAUST CONE
11. EXHAUST TAILPIPE (NOT PART OF ENGINE)
12. EXHAUST GAS TEMPERATURE PROBE (6)
13. TURBINE EXHAUST CASE
14. TURBINE NOZZLE CASE
15. COMBUSTION CHAMBER CASE
16. FUEL MANIFOLD AND NOZZLES
17. DIFFUSER CASE
18. BLEED VALVE
19. COMPRESSOR INTERMEDIATE CASE
20. ENGINE MOUNT (BALL BAT)
21. ACCESSORY CASE (N_2)
22. OIL TANK
23. FRONT COMPRESSOR CASE
24. ACCESSORY CASE (N_1)

Figure 1-3

Changed 6 November 1967

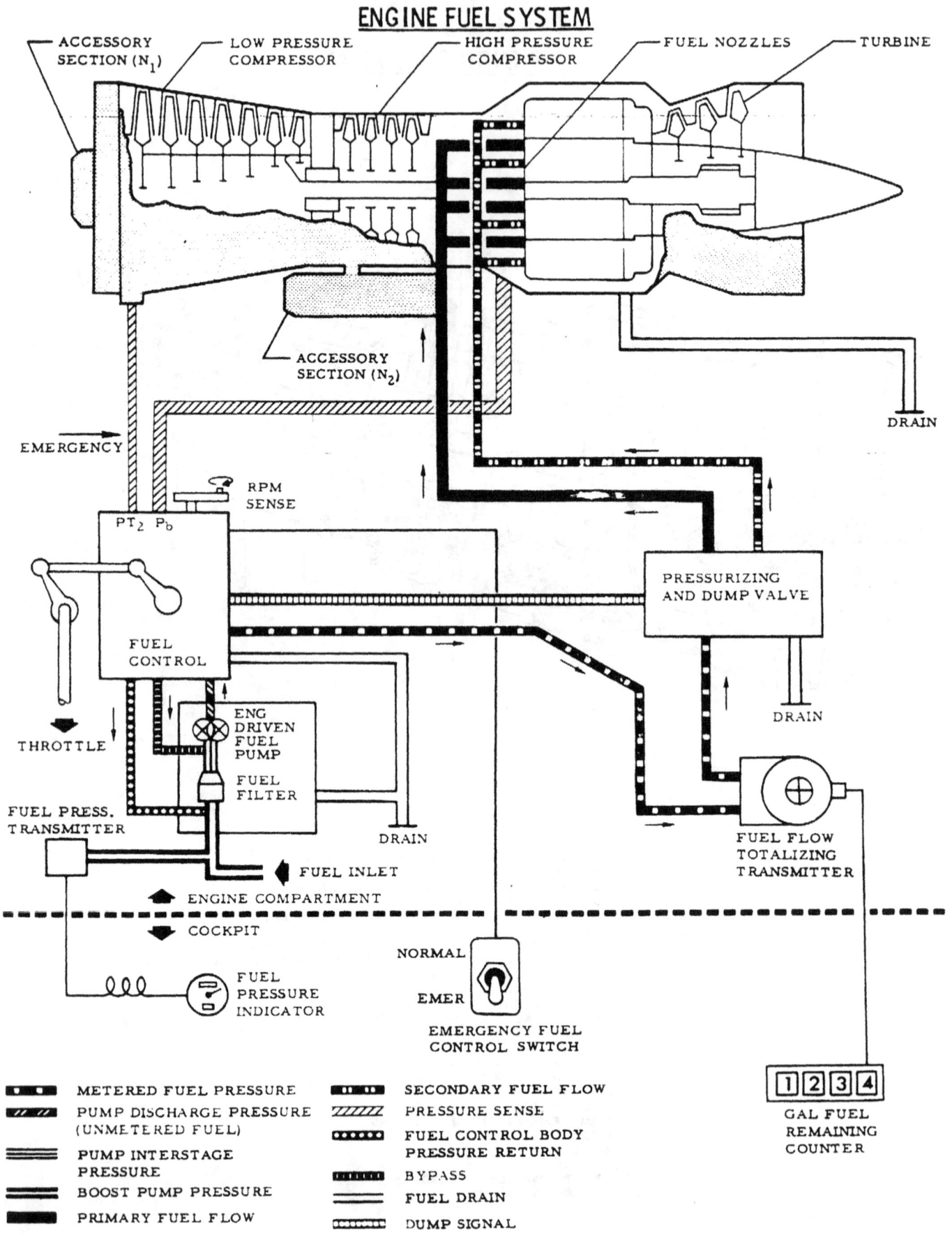

Figure 1-4

During climb at a fixed throttle setting, the fuel control in general will reduce engine speed. However, exhaust gas temperature (EGT) will increase during a fixed throttle climb. Some engines will require throttling to observe EGT limits and further throttling may be required at altitude to observe EPR limits.

EMERGENCY FUEL CONTROL

The emergency fuel system bypasses the compensating sections of the main fuel control so that fuel flow is manually selected and controlled. There is altitude compensation to only 30,000 feet; however, it is necessary to make extremely careful throttle movements during emergency operation in order to avoid overspeeding and overtemperature.

A fuel system selector switch is on the left side of the instrument panel. This is a two-position switch and is normally up for NORMAL fuel system operation. Placing the switch down activates the EMERG fuel system. An emergency fuel system light is on the annunciator panel. This is an amber light which is ON when operating on the emergency fuel system. The light also glows during the initial portion of a normal engine start but should be out before the throttle is moved from the OFF position.

NOTE

The emergency fuel system is not automatic and must be manually selected when desired.

THROTTLE

Engine power is controlled by the throttle. Throttle movement causes the fuel control to meter fuel to the engine with automatic compensation for RPM and burner pressure. The throttle has OFF, IDLE, GATE, and FULL positions. In OFF position no fuel reaches the burners. Due to possible engine overspeed or overtemp, a gate type throttle stop is provided to limit takeoff power to 93 to 95% RPM. To pass this gate, the throttle is moved outboard, then forward, or the gate stop can be removed by moving a small lever on the forward throttle quadrant to the INBOARD position. The gate automatically resets when the throttle is retarded unless the gate has been manually opened.

Avoid manually opening the gate stop if practical. This will preclude the stop being left open inadvertently. The gate stop should be in position during landing go-arounds or touch and go's.

Throttle friction is regulated by the small knob on the inboard side of the lower throttle console. A throttle vernier wheel on the same console is used for very small throttle adjustments. A microphone switch and the speed brake switch are on the throttle.

ENGINE OIL SYSTEM

The engine oil system is automatic and requires no controls. The system is comprised of a pressure oil system, scavenge oil system, breather pressurizing system and oil cooling system as shown in figure 1-5.

The pressure system is supplied with oil from a tank mounted on top of the engine compressor section. This tank contains 5.5 gallons.

Oil from the tank is fed to an engine driven boost pump which forces the oil through the oil cooling system consisting of a 14-inch air-oil cooler, a 9-inch air-oil cooler, and a fuel-oil cooler. From the fuel-oil cooler, the oil is forced by the engine driven main engine oil pressure pump into the engine components which require lubrication. The main engine oil pressure pump discharge pressure is regulated by a pressure relief valve.

Five engine driven scavenge pumps withdraw oil from the bearing compartments and accessories and return it to the supply tank

ENGINE OIL SYSTEM

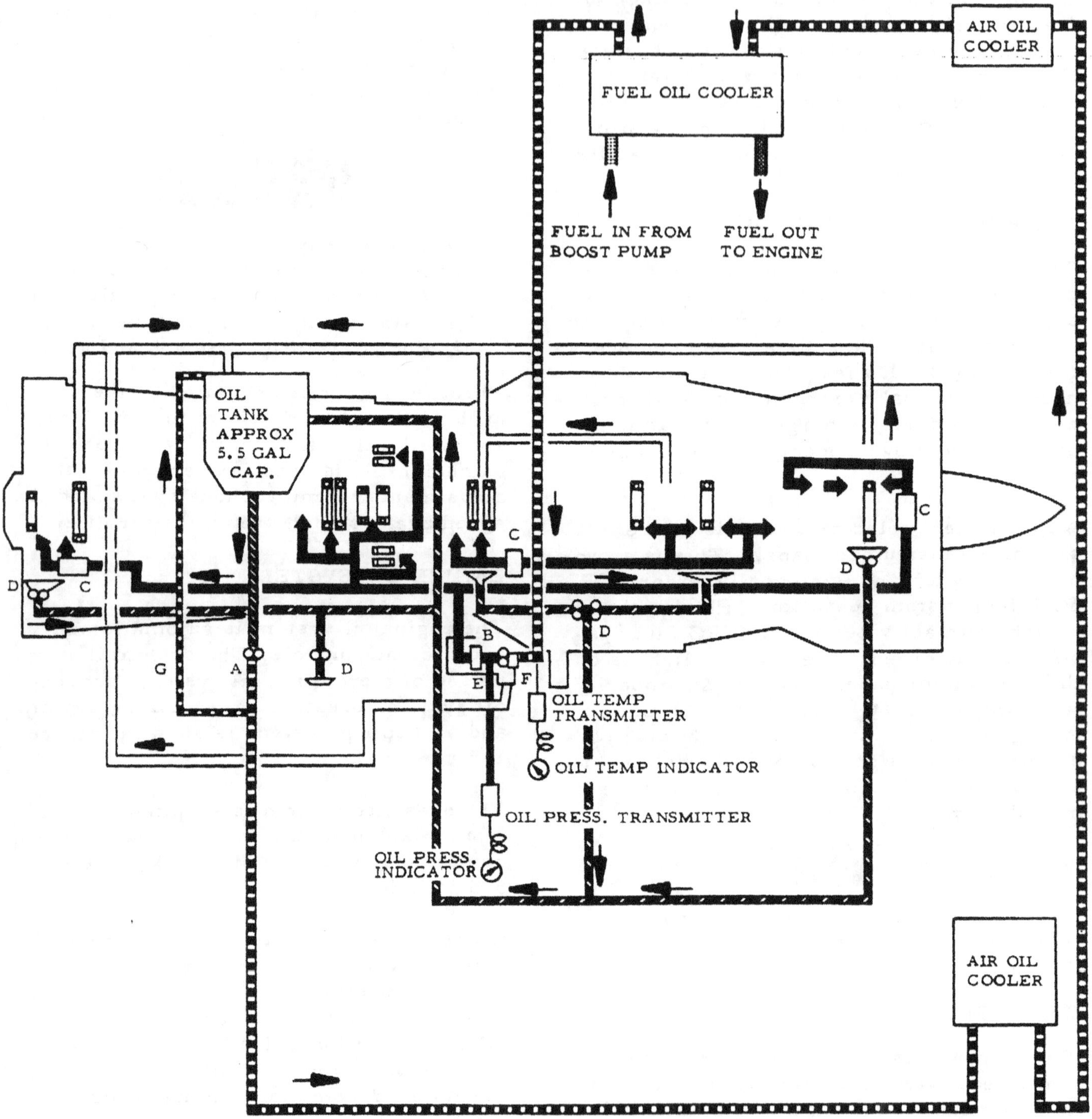

A - BOOST PUMP
B - MAIN ENG OIL PRESS. PUMP
C - LAST CHANCE OIL SCREEN
D - SCAVENGE PUMPS
E - MAIN ENG OIL SCREEN
F - BREATHER PRESSURIZING VALVE
    OVERBOARD CONNECTION
G - ANTISIPHON LINE

- GRAVITY FEED
- PRESSURE OIL
- SCAVENGE OIL
- BOOST PRESSURE
- BREATHER LINES

Figure 1-5

The oil temperature should be monitored during flight, particularly at high power.

The breather pressurizing system is provided to improve oil pump performance at altitude. The pressure is automatically controlled by an aneroid operated valve.

NORMAL IGNITION SYSTEM

The normal ignition system is a high energy capacitor-discharge 20-joule type, consisting of two identical, independent exciter units; one for each of two igniter plugs. One plug is in combustion chamber number 4 and the other is in number 5. The two exciter units are powered from separate electrical circuit breakers so that failure of one will not cause complete system failure. A single exciter is sufficient for making an air or ground start.

Ignition is controlled by a single spring-loaded switch on the lower left instrument panel. The ignition switch is wired directly to the 28-volt DC bus. Placing the ignition switch in start position energizes the igniters, regardless of BAT-GEN switch position.

CONTINUOUS IGNITION

The airplanes are equipped with a continuous ignition system to minimize the possibility of flameout due to transient conditions at high altitude.

The system is energized through the BOOST PUMP-CONT IGN - ON switch. The continuous ignition system and the fuel boost pump are energized simultaneously when the switch is in the ON position and the AC Generator is on the line.

This system consists of a separate 4-joule exciter and utilizes the ignition plug in combustion chamber number 5.

Operation

1. Battery switch - BAT & GEN.

2. AC generator - ON.

3. Auxiliary boost pump switch - ON.

NOTE

In the event the AC generator fails, power for the continuous ignition is supplied by the inverter. In order to prevent inverter overload under this condition, the APN 135 beacon and continuous ignition should not be operated simultaneously. Either the boost pump and continuous ignition or the beacon should be turned off as operating conditions dictate.

The AC generator must be on the bus for operation of the auxiliary boost pump. However, if the AC generator or the auxiliary boost pump fails, the auxiliary BOOST PUMP CONT IGN switch should remain in the ON position as required for operation of continuous ignition.

STARTER SYSTEM

An air turbine starter is provided for ground starts. An external air supply furnishes the necessary power. There are no airplane controls for this system. It is turned on and off by the ground crew according to signals given to the pilot. Airstarts do not require a starter and are made by windmilling the engine.

INTERCOMPRESSOR BLEED SYSTEM

During reduced power and idle descent conditions the P-13B engine requires intercompressor bleed since engine operation under these conditions is near compressor stall.

One low pressure overboard bleed valve is provided on each side of the high pressure compressor case. These automatic or manually actuated valves serve to prevent compressor surge at high altitude by ducting low pressure compressor air overboard during low thrust operation. High pressure compressor air is used to actuate the bleed valves in both the open and closed directions. A cam operated switch mounted on the throttle linkage actuates the solenoid valve which controls the high pressure air. In automatic operation, the bleed valves will open or close between 84 and 87% RPM. The valves operate as a function of throttle position and will actuate at a higher RPM at altitude when idle RPM is higher due to minimum fuel flow limitations.

BLEED VALVE CONTROL

The BLEED VALVE control switch is located on the right side of the lower instrument panel (see cockpit arrangement illustrations, this section). It is a 3-position lever lock switch. In the up or OPEN position the bleed valves are open, in the center or CLOSE position the bleed valves are closed and in the down or AUTO position the valves will operate automatically as a function of throttle position. DC power is required to automatically or manually open the bleed valves. Loss of DC power will close the bleed valves if they are open, or keep the bleed valves closed if they are closed.

BLEED VALVE OPEN LIGHT

Two amber bleed valve open lights, BLEED AIR LEFT and BLEED AIR RIGHT, are located in the annunciator panel (see cockpit arrangement illustrations, this section). These lights are actuated by switches mounted on the bleed valve actuating shafts and illuminate when the valves are open.

ENGINE INSTRUMENTS

The basic engine instruments are shown on figure 1-6.

OIL TEMPERATURE

The temperature of the oil entering the engine is monitored in the cockpit through a DC powered gage.

ENGINE PRESSURE RATIO (EPR) SYSTEM

The pressure ratio indicator gives the pilot an indication of thrust for all throttle settings. The transmitter senses engine inlet and exhaust pressures, and electrically transmits the ratio of these pressures to the indicator on the instrument panel.

Power for the operation of this system is supplied by either the number one or number two inverter.

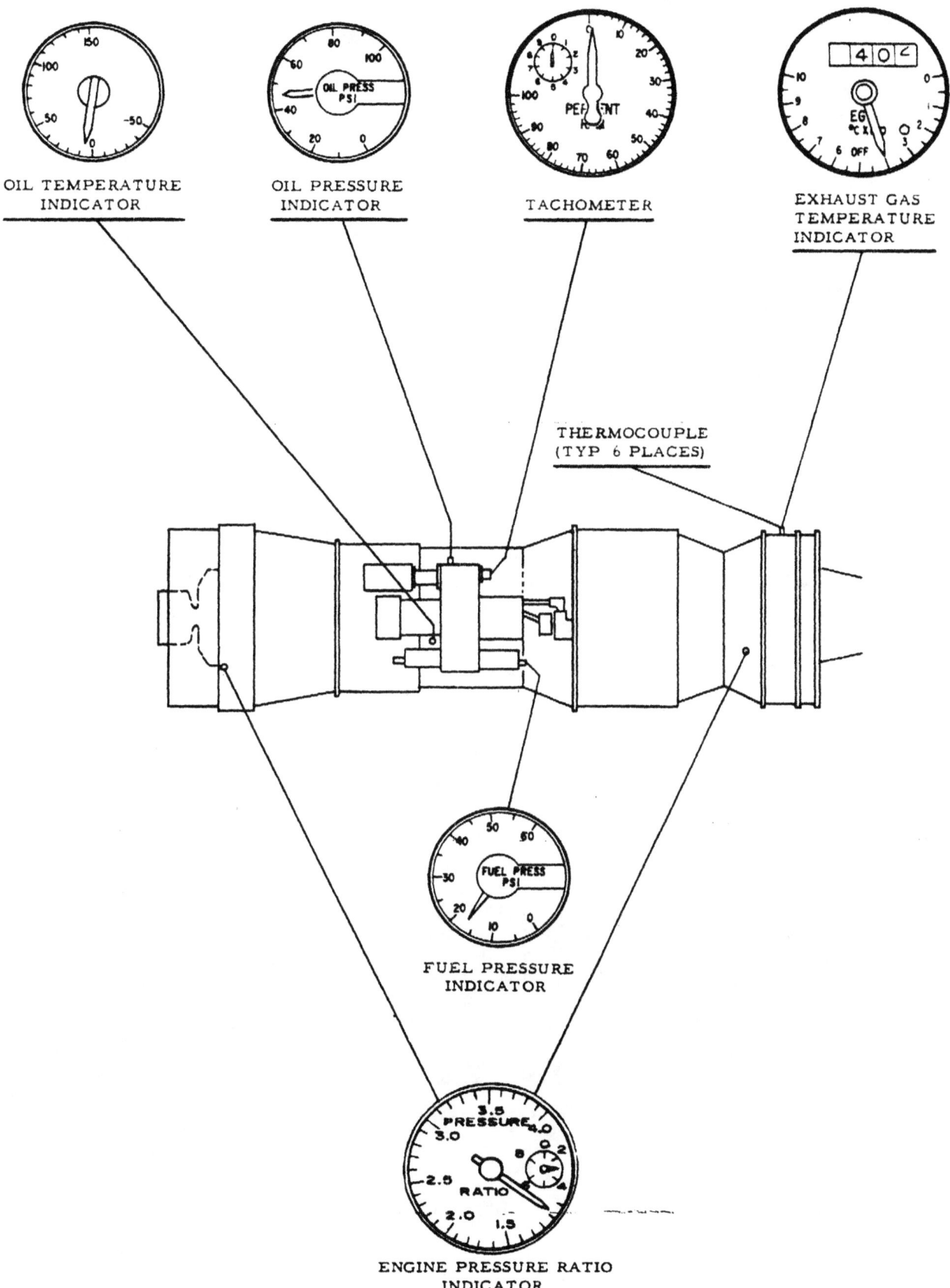

Figure 1-6

TACHOMETER

The tachometer indicates percent of rated high pressure rotor speed. 100% engine speed is nominally 8730 RPM. However, the J-75 engine is thrust rated, and "Military" thrust is usually obtained at different RPM for different engines.

The RPM should normally be used as a secondary indication of engine output. It also should be monitored for overspeeding (103.0% maximum allowable) and for starting.

The tachometer is a self-generating instrument having its own engine-driven generator.

EXHAUST GAS TEMPERATURE

This instrument indicates the turbine discharge temperature in degrees centigrade.

Temperature indication is presented on a digital counter for readings of high accuracy, while a needle pointer indicates the temperature on the dial for a quick or coarse reading. Graduations on the dial are in increments of 50°C over a range of 1000°C.

An integral red warning light on the dial glows when the exhaust gas temperature exceeds 671°C. A red OFF flag shows whenever the inverter power is off or fails.

Either normal or emergency inverter power will operate the EGT indicating system. No warmup time is required. Although the instrument is capable of following temperature changes of 150°C per second, the indication will be slower because of the time required for the probes in the engine turbine section to sense the temperature changes.

FUEL PRESSURE

This instrument shows the fuel pressure into the low pressure engine fuel pump and provides an indication of the output from the airplane boost pump.

The pressure will vary from 14 to 25 psi. At full throttle takeoff power, pressure may drop to zero. This system is remote indicating and is powered by either the number one or number two inverter.

OIL PRESSURE

This instrument indicates the engine oil pump discharge pressure. The normal range is 40 to 55 psi. This system is remote indicating and is powered by either the number one or number two inverter.

AIRPLANE FUEL SYSTEM

The airplane fuel system is very simple and requires little attention from the pilot other than monitoring of the boost pump fuel pressure, the fuel quantity indicator, and the fuel warning lights.

FUEL SEQUENCING SYSTEM

The airplane has four integral wing tanks, a fuselage sump tank and provisions for slipper and drop tanks. All tanks feed into the fuselage sump tank as shown in figure 1-7. The fuel from the sump tank feeds the engine as shown in figure 1-9.

SECTION I AF (C)-1-1

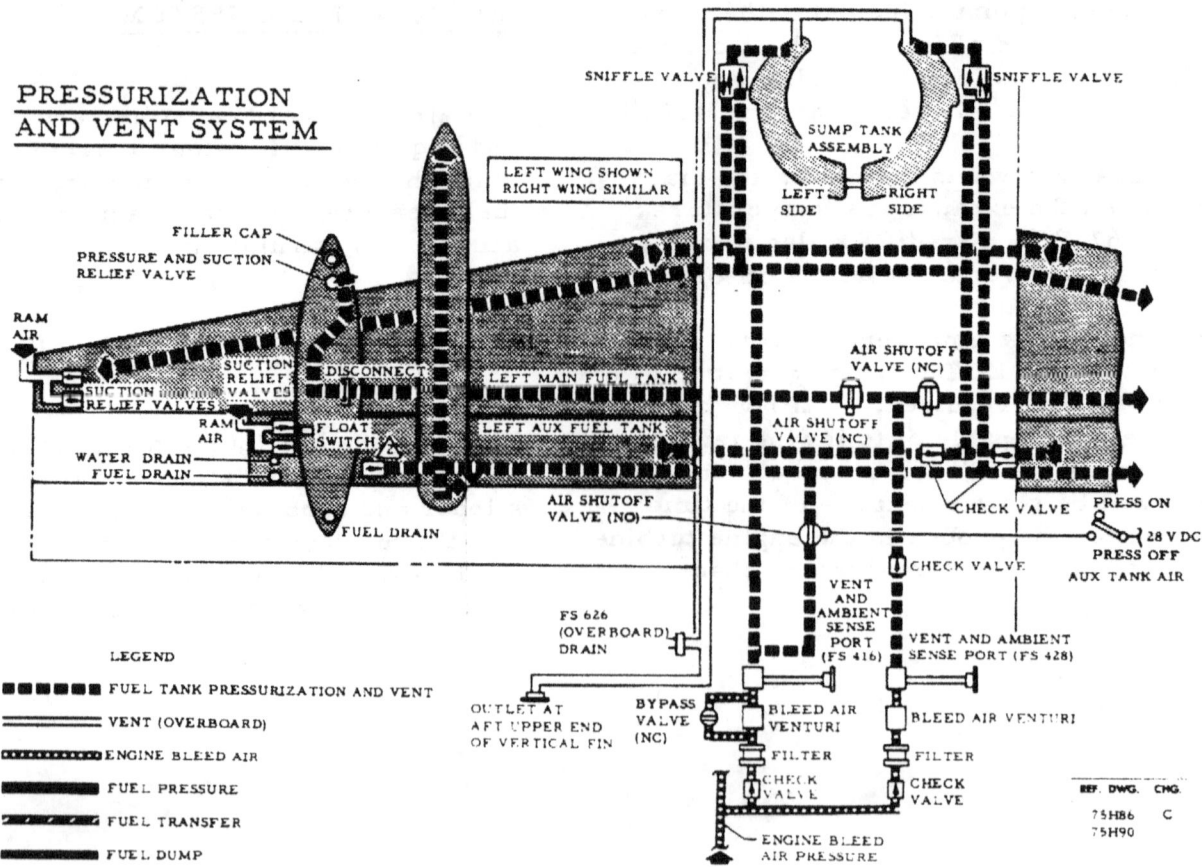

Figure 1-7

FUEL QUANTITY TABLE
(APPROX)

Tanks	Pounds	Gallons
Sump	625	95
Main	6,086	925
Auxiliary	1,974	300
Full Internal	8,685	1,320
Slippers	1,316	200
Drop Tanks	1,316	200
Full with Slippers Or Drop Tanks	10,001	1,520

NOTE

Each aircraft is individually placarded in the cockpit.

Fuel from all tanks flows into the sump tank by gravity and/or pressure feed. Engine compressor bleed air is regulated to 2 psi to pressurize the main, auxiliary, and slipper tanks, and to 7 psi to pressurize the drop tanks. The sump tank is not pressurized. The main wing tanks feed into the right-hand side of the sump, and the auxiliary tanks feed into the left-hand side of the sump tank. The fuel level in the left and right sides is equalized by a large crossover line.

Two sets of float valves control transfer sequence and maintain proper sump tank fuel level. These dual float valves prevent blocking of fuel flow due to a single stuck valve.

The upper set of float valves controls the fuel flow from the drop tanks or slipper tanks, and auxiliary tanks, in that sequence, so as to maintain the sump tank fuel at the float valve level until these tanks are empty. When the above tanks are empty, the sump tank fuel level drops to the lower set of float valves. These lower float valves then control fuel flow from the main tanks to maintain this fuel level in the sump tank until the main tanks are empty.

A secondary float valve is provided for each set of internal wing tanks at the 25-gallon level as a backup in case the primary float valves should be blocked and to ensure gravity feeding.

NOTE

Drop tanks and slipper tanks should not be carried at the same time.

FUEL SEQUENCING SYSTEM (F)

The fuel transfer system on the U-2F is the same as on the U-2C except that the drop tank fuel feeds into the main tank feed lines which are plumbed into the left-hand side of the sump tank. The fuel usage sequence is as follows: drop tanks, main tanks, slipper tanks, auxiliary tanks and sump tank. The drop tanks, and main tanks feed first because the main tanks are the only ones which can be refueled in flight. The main tank float valves in the sump therefore, are higher than the auxiliary float valves. See figure 1-8.

AUXILIARY TANK PRESSURIZATION SWITCH

Pressurization air to the auxiliary fuel tanks can be turned on or off by a lift lock switch on the right-hand cockpit sill. (Upper right-hand instrument panel Model U-2F.) Before takeoff, the AUX TANK AIR switch should be OFF if the auxiliary tanks are empty and ON if they contain fuel.

MAIN TANK PRESSURIZATION SWITCH (F)

A guarded switch identified MAIN TANK PRESS, used to override the main tank air shutoff float switch and provide main tank pressurization, is on the left side console forward of the lateral fuel transfer switches. The switch has a NORM (guard down) position, and a REPRESS position. This switch must be moved from the NORM position to the REPRESS position after refueling to allow the fuel in the sump tank to rise above the float valve and restore normal air pressurization. The switch is returned to the NORM position before the main tanks are again emptied to ensure slipper tank pressurization and complete fuel transfer.

FUEL SYSTEM MODEL U-2F

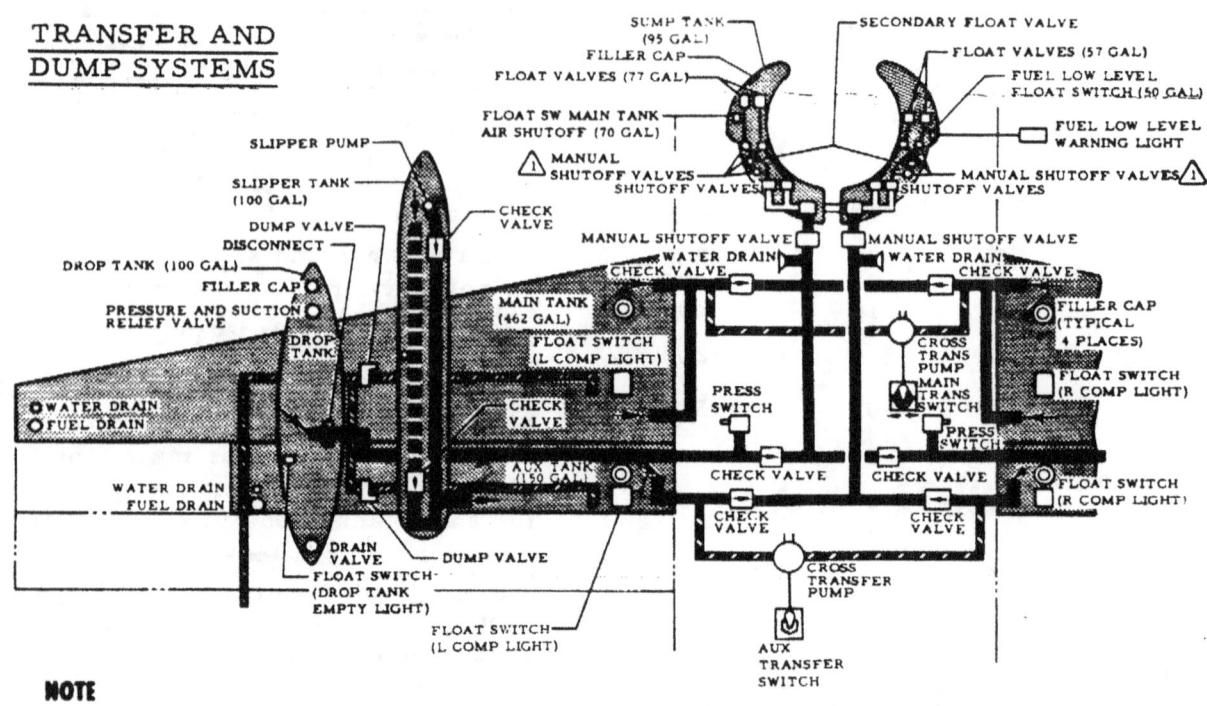

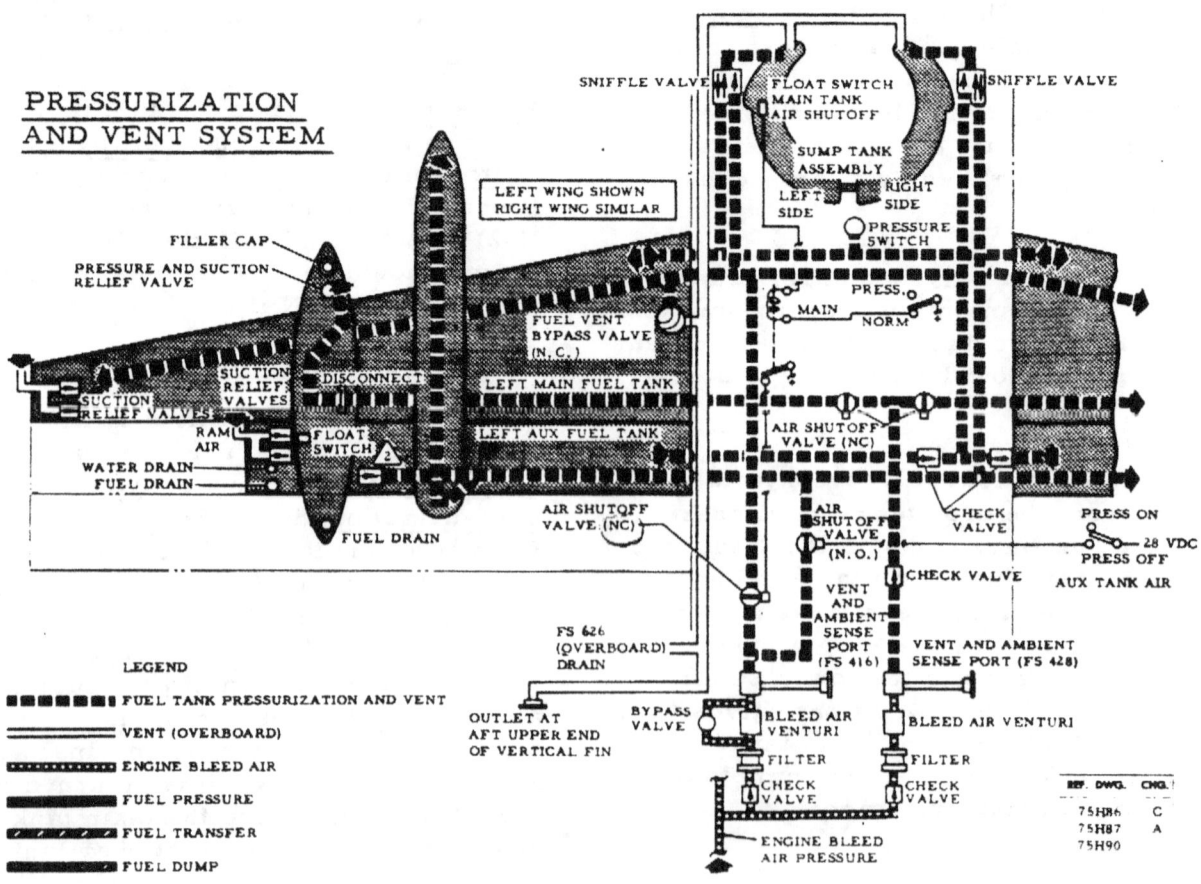

Figure 1-8

SLIPPER TANK INSTALLATION

This airplane has provisions for slipper tanks. These tanks are external pods that slip on over the leading edge of the wing approximately ten feet from the fuselage. Fuel from the slipper tanks is forced by air pressure into the auxiliary (aft) wing tanks. From here it feeds into the sump tank through the normal auxiliary tank transfer system. A small electric pump is installed in each slipper tank to ensure complete utilization of available fuel. The pumps are powered by the AC generator.

NOTE

Fuel from the slipper tanks will not gravity and/or pressure feed until empty. It is possible that 20 to 40 gallons will remain in each tank unless the slipper tank stripper pumps are used. Refer to Section VII, Systems Operations, for slipper tank operation.

CROSS TRANSFER PUMP

In order to provide a means of correcting wing heaviness due to uneven fuel feeding, a small 5-gallon-per-minute electric fuel pump is installed. The reversible pump is controlled by a three-position switch on the lower left instrument subpanel. It transfers fuel from either main wing tank to the other. Pushing the switch to the left transfers fuel into the left main tank, and vice versa. An annunciator panel light will be on when the transfer pump is operating.

CROSS TRANSFER PUMP

The system is identical to the U-2C except that on the U-2F, there are two fuel cross transfer systems. One is between the two main tanks and the other is between the two auxiliary tanks. The auxiliary tank transfer switch is above the left-hand console on the ARS panel.

DROP TANK INSTALLATION

Provisions are made for drop tanks with integral pylons which attach on the lower wing surface outboard of the slipper tanks. These tanks should not be used unless they are to be dropped when their fuel is expended as retention of dry tanks decreases range. The bottom surface of the wing is left clean when the drop tanks are released. Fuel from the drop tanks is forced by air pressure into the left-hand sump tank by way of the auxiliary tank transfer lines. Check valves in the auxiliary transfer lines prevent drop tank fuel from entering the auxiliary tanks. Two circuit breakers and a control panel are located in the cockpit. The circuit breakers are in the left-and right-hand drop tank air shutoff valve circuits. Drop tank fuel will not feed if these circuit breakers are pulled.

The cockpit control panel is on the left-hand step console. It contains two amber low pressure warning lights (one for each drop tank), two green drop tank empty lights (one for each drop tank), a two-position indicator lights ON-OFF switch, a drop tank release arming switch, and a guarded drop tank release switch.

DROP TANK INSTALLATION

The installation is the same except for the transfer system changes described in FUEL TRANSFER SYSTEM .

FUEL GRADE

The airplane fuel system is designed for low vapor pressure type fuel, and fuel loss through boiloff or ventline slugging is negligible. Fuel corresponding to MIL-F-25524B is presently designated as the primary fuel. In addition, fuel specified as LF1A or MIL-F-25524A, may be used without any restrictions as primary alternate fuels. JP-4 and JP-5 fuels may be used; however, check Section V for limitations.

FUEL DUMP SYSTEM

A fuel dump system allows reduction to landing weight without the necessity of extended flight at low altitude to burn out fuel. The system consists of a dump valve and float switch in each of the four wing tanks and an overboard line on the trailing edge of each wing between the flap and aileron. Each dump valve is controlled independently by switches marked DUMP and CLOSE on the lower left-hand console. System indicator lights are located on the top left side of the instrument panel.

When fuel is to be dumped, positioning the respective switch or switches to DUMP will actuate the valves to release fuel. When a dump switch is placed in the DUMP position, an OPEN light will illuminate to indicate the respective valve is open. Fuel dumping may be terminated at any time by returning the dump switches to CLOSE. The float switches in each tank actuate completion lights. Either main or auxiliary tank fuel dump float switch will actuate the respective L COMP or R COMP light if the dump switch is in DUMP position. Fuel in the slipper tanks will dump through the auxiliary tank dump valves but drop tank fuel cannot be dumped (only dropped).

Normal fuel tank pressurization is used to provide dump pressure. The pressurization system capacity has been increased for this purpose. If the dump switches are not returned to the closed position upon completion of dumping, tank pressurization will be lost and the sump tank will have to rely on gravity feed for replenishment.

The auxiliary tank dump lines will empty the auxiliary tanks, but the inlet end of the main tank dump lines are located above the tank bottom to prevent dumping fuel below approximately 130 gallons in each tank. However, if the valves are left open the amount will be reduced by turning, slipping, etc. If the dump switches are closed when the COMP light illuminates approximately 325 gallons of fuel will remain. To maintain a slightly conservative record of fuel aboard, the fuel counter should be reset to 300 upon completion of dumping. Fuel will dump at a rate of approximately 60 gallons per minute when dumping from both wings.

FUEL DUMP SYSTEM

The fuel dump system is identical to that in the U-2C except that the main tank will empty completely and each auxiliary tank will retain approximately 130 gallons of fuel. The system indicator lights are in the same panel as the ARS lights.

FUEL SYSTEM WARNING LIGHTS

Two fuel system warning lights are on the annunciator panel. The amber AUX BOOST PRESSURE light comes on whenever the auxiliary boost pump switch is on and the pressure is below 10 psi.

The FUEL LOW LEVEL red warning light comes on when the level in the sump tank has dropped to 50 gallons of usable fuel.

Drop tank system lights are described in DROP TANK INSTALLATION description, this section.

200-MESH FUEL STRAINER

Fuel from both sides of the sump feeds into a common line and into the 200-mesh strainer as shown in figure 1-9. This unit is equipped with a screen which prevents foreign material from entering the airplane main boost pump and the fuel system. The flow pattern inside the strainer is such that all fuel is strained through the screen in normal operation. However, if the screen should be blocked, the fuel will be bypassed with no interruption in flow.

NORMAL FUEL BOOST PUMP

The normal boost pump is hydraulically-driven and is not controllable from the cockpit. It furnishes fuel to the engine at a regulated pressure of 14 to 25 psi. A bypass around the boost pump provides an alternate path for fuel-flow in the event of pump failure or hydraulic system failure as shown in figure 1-9. The pump cannot supply sufficient fuel for full power at low altitudes. Under this condition the engine pump draws the additional fuel required for full power through the bypass. The engine-driven fuel pump will draw enough fuel through the bypass to run the engine at any altitude in case of failure of both the normal and auxiliary boost pumps. An exception to this could occur at the very beginning of a mission, before the fuel has had sufficient time to "weather". Approximately one hour is required to weather the fuel.

AUXILIARY FUEL BOOST PUMP

An electrically-driven submerged boost pump is installed in the right side of the sump tank. The system plumbing connects this pump and the normal fuel boost pump in parallel as shown in figure 1-9. Control is by a switch on the left-hand instrument subpanel. The switch is identified BOOST PUMP AND CONT IGN - ON. (The continuous ignition system and the auxiliary boost pump are energized simultaneously.) A pressure switch set for 10 psi is installed downstream of the electric pump. It is connected to an AUX BOOST PRESS light in the annunciator panel. The light is off when the pump is running and producing pressure.

In flight with the electric pump running, its output is less than the normal pump. If the normal boost pump fails or hydraulic pressure is lost, the electric boost pump will take over. The only noticeable change is a drop in fuel pressure. The output of the electric pump is purposely set lower than the normal boost pump but is adequate to keep the engine running.

In the event both fuel pumps fail, the engine will operate properly at any altitude on the engine-driven fuel pump only, providing that the fuel has been weathered. Electrical power for the boost pump is supplied only by the AC generator.

NOTE

In the event the AC generator drops out or fails, power for continuous ignition is supplied by either the No. 1 or No. 2 inverter. In order to prevent inverter overload under this condition, the AN/APN-135 rendezvous beacon and continuous ignition should not be operated simultaneously. Either the BOOST PUMP AND CONT IGN switch or the RENDZ BCN should be turned off, as operating conditions dictate.

FUEL SHUTOFF VALVE

A DC operated shutoff valve is in the main feed line downstream of the boost pumps. This valve is for emergency use in case of fire or for engine shutdown in case of an inoperative throttle. A guarded switch is provided on the left-hand console. The valve is open when the guard is down.

FUEL FEED SYSTEM

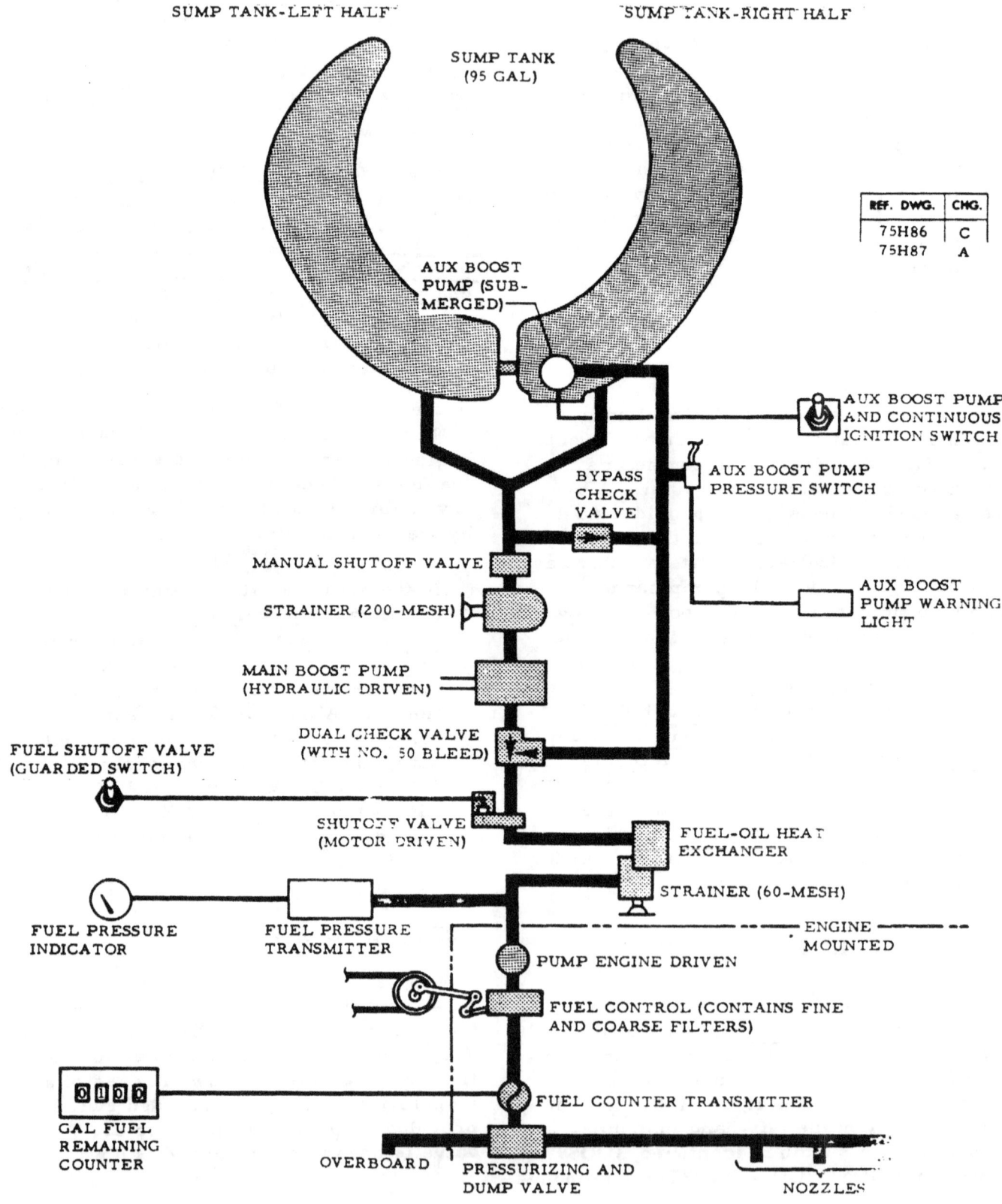

Figure 1-9

FUEL-OIL HEAT EXCHANGER

In order to prevent any tendency toward ice formation in the 60-mesh strainer, the fuel is warmed by a fuel-oil heat exchanger. This unit is thermostatically controlled and the amount of oil passing through is varied as necessary. Some oil cooling also takes place in the heat exchanger since the fuel is warmed by heat rejection from the oil. From the heat exchanger the fuel passes through a wire mesh strainer and into the engine fuel pump.

FUEL QUANTITY COUNTER

A positive displacement type flowmeter is in the engine high pressure fuel system. This unit counts the gallons of fuel used by the engine whether operating on the normal or emergency fuel system. A subtractive counter mounted on the instrument panel shows the gallons remaining in the aircraft. The fuel quantity counter and the low level warning lights are the only fuel quantity indicators.

TANK VENT SYSTEM

All wing tanks are vented into the sump tank through combination type suction and pressure relief valves. The top of the sump tank is vented to the outside air at the top trailing edge of the vertical fin. Dual suction relief valves are in the outboard end of each wing tank to prevent negative pressures from developing during rapid descents. The drop tanks have integral suction and pressure relief valves.

ELECTRICAL SYSTEM

DC SYSTEM

The direct current electrical system is a regulated 28-volt, single conductor type which utilizes the airframe structure for the ground return. DC power is furnished by one 400-ampere, 28-volt, engine-driven generator, which in this installation is derated to 225 amperes. A 35-ampere-hour, nickel-cadmium battery is installed to supply emergency power to the DC system.

AC SYSTEMS

The alternating current system power is furnished by a 750-VA inverter for the normal system. A backup 750-VA inverter is also provided. In addition, a 100-VA emergency inverter and a 10-KVA engine-driven AC generator are provided. The inverters are rated at 115-volt, 400-cycle, three-phase output, with an input of 26 volts DC. The AC generator is rated at 120/208 volts, 320-450 cycle, three-phase AC.

NOTE

Whenever the engine is operated below approximately 80% RPM the AC generator will trip off due to underfrequency (320 cps). The AC generator will come back on whenever the frequency exceeds 320 cps as long as the switch is in the AC GENERATOR position.

The AC generator system provides a source of 28-volt DC power which is used in the system control circuits and therefore does not require ships 28-V DC power for normal operation. However, ships 28-V DC is supplied through a 5-ampere circuit breaker in the equipment bay as backup in case of failure of the DC power generating circuits in the AC generator system.

A schematic of the direct current and inverter power distribution is shown in figure 1-10.

A schematic of the AC power distribution system is shown in figure 1-11.

EXTERNAL POWER RECEPTACLE

For ground operation, two external power receptacles are provided on the left side of the fuselage near the leading edge of the wing. One receptacle is for 28-volt DC power and the other is for 120/208-volt, three-phase, AC power. When external DC power is connected, the ships battery is automatically cut off the battery-generator switch. However, this is not true of the generator which may come on if the external voltage supply is low and the switch is in the generator and battery position.

BATTERY AND GENERATOR SWITCH

The battery-generator control is a three position switch on the lower left instrument subpanel. Center position is OFF. The up position, BAT & GEN, is the normal operating position. The down position is BAT EMERG. When the generator is not on the line due to low engine speed, or other cause, the battery is still on with the switch in the BAT & GEN position.

GENERATOR AMMETER

A DC ammeter is provided on the lower left instrument subpanel to indicate DC generator load. The maximum permissible load is 225 amperes.

ANNUNCIATOR WARNING PANELS

A master annunciator warning panel, see figure 1-12, and master caution light are installed to consolidate most warning and advisory lights in one location. A secondary panel is installed for fuel dump lights in the U-2C and for both fuel dump and ARS in the U-2F. The master caution light is on the top of the instrument panel to the right of the driftsight. The master annunciator panel is on the right-hand console forward panel. The secondary panel in the U-2C is on the top of the instrument panel to the left of the driftsight.

Those lights on the left and center rows of the master annunciator panel designated as warning lights will actuate the master caution light each time they illuminate. The master caution light will then remain on until reset by pressing on the master caution light capsule, or until the condition is corrected. The remaining lights are advisory lights and do not actuate the master caution light. Pressing the master caution light does not extinguish the lights on the annunciator panel.

DC AND INVERTER POWER DISTRIBUTION

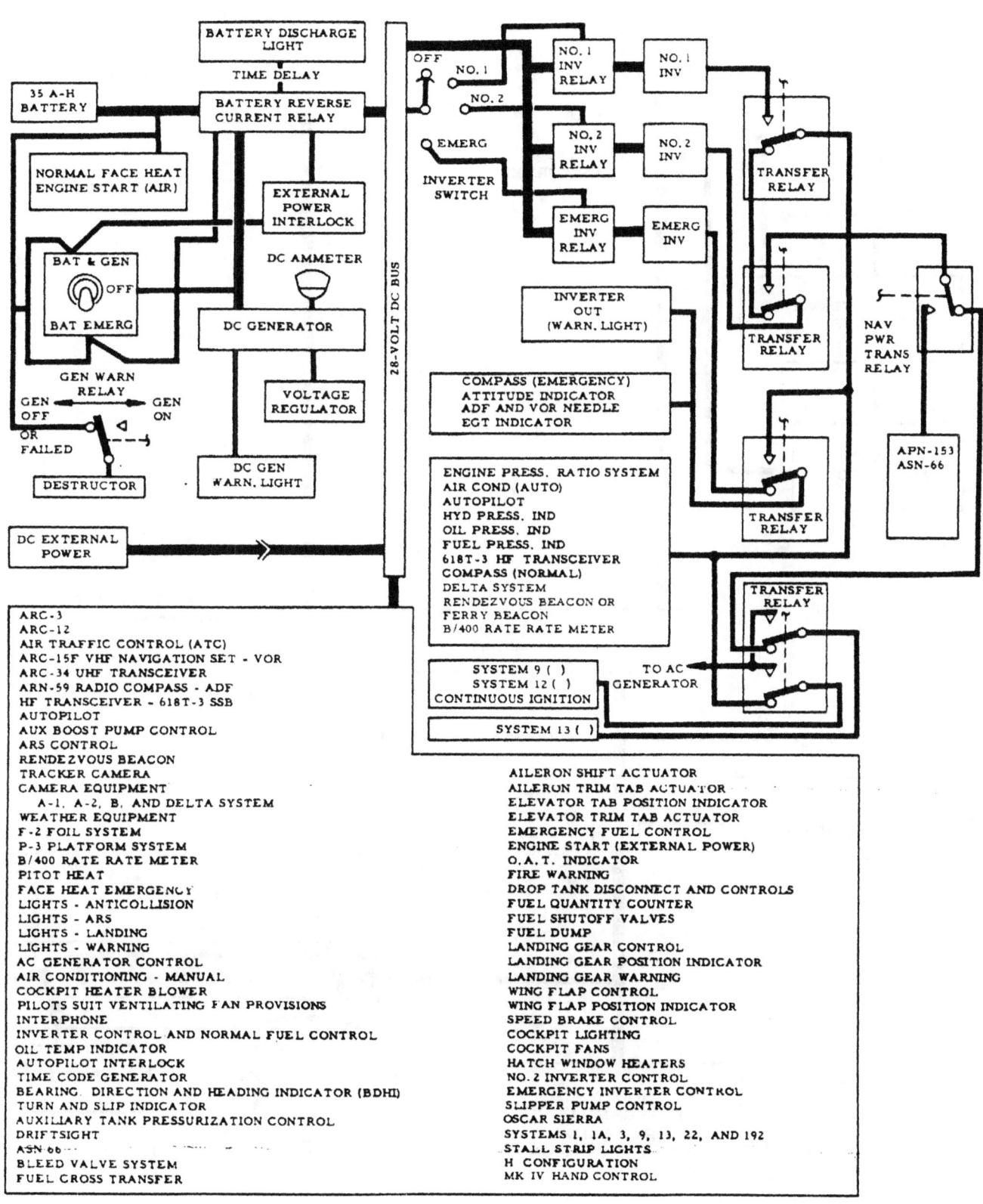

Figure 1-10

AC GENERATOR POWER DISTRIBUTION

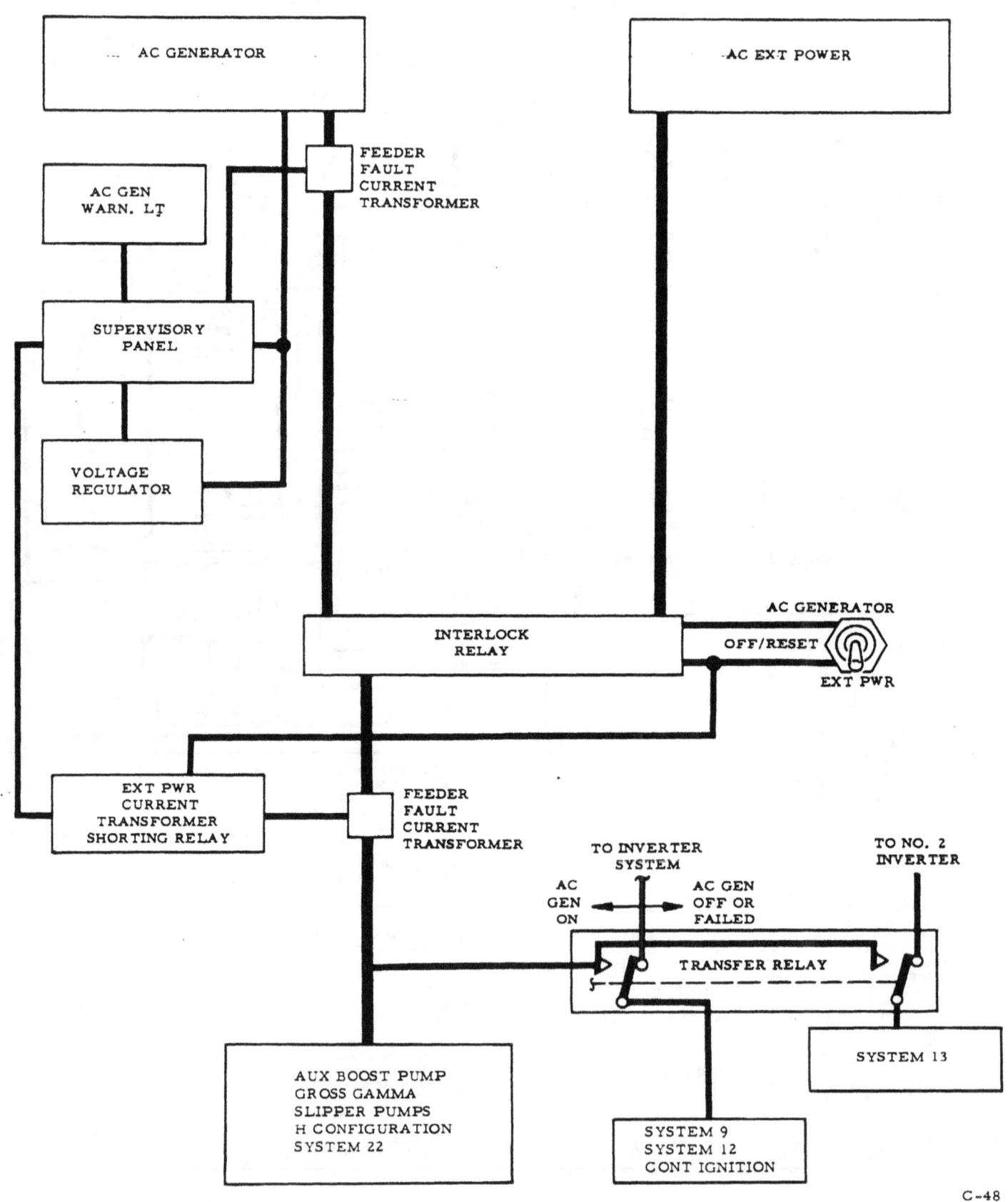

Figure 1-11

ANNUNCIATOR PANEL

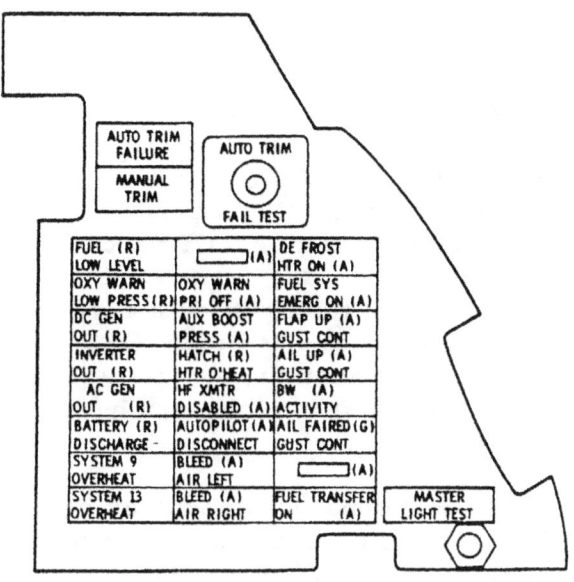

Figure 1-12

GENERATOR OUT WARNING LIGHT

This red warning light, in the annunciator panel, glows whenever the generator is inoperative and the DC bus is energized. The battery switch must be in the BAT EMERG or BAT & GEN position, or the external power must be connected.

BATTERY DISCHARGE WARNING LIGHT

The red warning light will illuminate when the battery has been discharging at a rate of 5 amperes or more continuously for at least 15 seconds. The light indicates that the generator has insufficient output or is switched off the line. It is therefore possible for the battery discharge warning light to glow and the generator out warning light not to glow. With this condition, the action to be taken is the same as that for a DC generator failure.

NOTE

Illumination of the battery discharge light may occur during low engine RPM operation. If this occurs, ground personnel should be requested to check the DC voltage regulator adjustment.

AC ELECTRICAL POWER SYSTEM

AC electrical power for certain electronic intelligence systems, the auxiliary boost pump, and normal power for continuous ignition, is provided by the engine-driven AC generator. The generator is rated at 10 KVA, 208/120 V, 320-450 cycle AC.

A voltage regulator and supervisory panel are mounted in the wheel well. Feeder fault current transformers are mounted at the AC bus and as an integral part of the AC generator. The supervisory panel provides protection against overvoltage, undervoltage, underfrequency, and feeder fault. The AC generator system provides a source of 28-volt DC power which is used in the system control circuits and therefore does not require ships 28-V DC for normal operation. However, ships 28-V DC is also supplied through a 5-ampere breaker in the equipment bay as backup in case of failure of the DC power generating circuits in the AC generator system. A three-position switch, (AC GENERATOR, OFF/RESET, EXT PWR) is on the lower left side of the instrument panel. An AC GEN OUT light (red) is on the annunciator panel. The light will come on whenever the generator is tripped off the line. The light will go out whenever the switch is placed in the OFF/RESET position.

SECTION I   AF (C)-1-1

The AC generator will trip off whenever it is operated below 320 CPS (approximately 80% engine RPM) and automatically reset whenever the frequency is above 320 CPS. A generator trip due to undervoltage, overvoltage, or feeder fault requires that the AC generator switch be placed in the OFF/RESET position and then returned to AC GENERATOR position in order to reset the generator.

NOTE

In order to prevent repeated momentary feeding into a faulted system, only one reset of the AC generator system should be attempted.

INVERTER SWITCH

The AC inverter power is controlled by a four-position, (OFF, 1, 2, EMERG) inverter selector switch on the lower left instrument subpanel below the battery-generator switch.

INVERTER OUT WARNING LIGHT

This red warning light glows whenever the inverter selected is inoperative or when the inverter switch is turned to OFF.

AC GENERATOR SWITCH

AC generator power is controlled by a three-position (AC GENERATOR, OFF/RESET, EXT PWR) switch on the lower left instrument panel. The center position is OFF/RESET. The switch is moved up to the AC GENERATOR position; and down to the EXT PWR position.

AC GENERATOR OUT WARNING LIGHT

This red warning light is OUT during normal operation of the AC generator. The light comes ON to indicate a failure of the AC generator due to overvoltage, undervoltage, feeder fault, or underfrequency. In the event of an AC generator failure for any other reason, the AC generator switch should be placed in the OFF-RESET position and then returned to AC GENERATOR in an attempt to reset the system. If this does not extinguish the light (with engine RPM above 80%), place switch in OFF-RESET position. When the switch is in the OFF-RESET position the warning light is OFF.

During low RPM conditions, such as before takeoff and during descent, the AC generator OUT light will be ON if the switch is in the AC GENERATOR position. This is normal, and no corrective action is required.

HYDRAULIC SYSTEM

The hydraulic system as shown in figure 1-13 operates the landing gear, speed brakes, wing flaps, fuel boost pump drive motor, and on the Model U-2F, the latch and receptacle mechanism of the ARS. There is a hydraulic quick disconnect receptacle for use with a hydraulic ground cart when operating the hydraulic system on the ground.

The aircraft system is a constant 3,000 psi pressure type, incorporating an accumulator and self-regulating engine-driven pump. The air charged accumulator stores pressure for peak demands and thus reduces fluctuation in pump loading. The system relief valve serves as a safety device to bypass oil back to the tank and prevent excessive system pressure.

HYDRAULIC SYSTEM

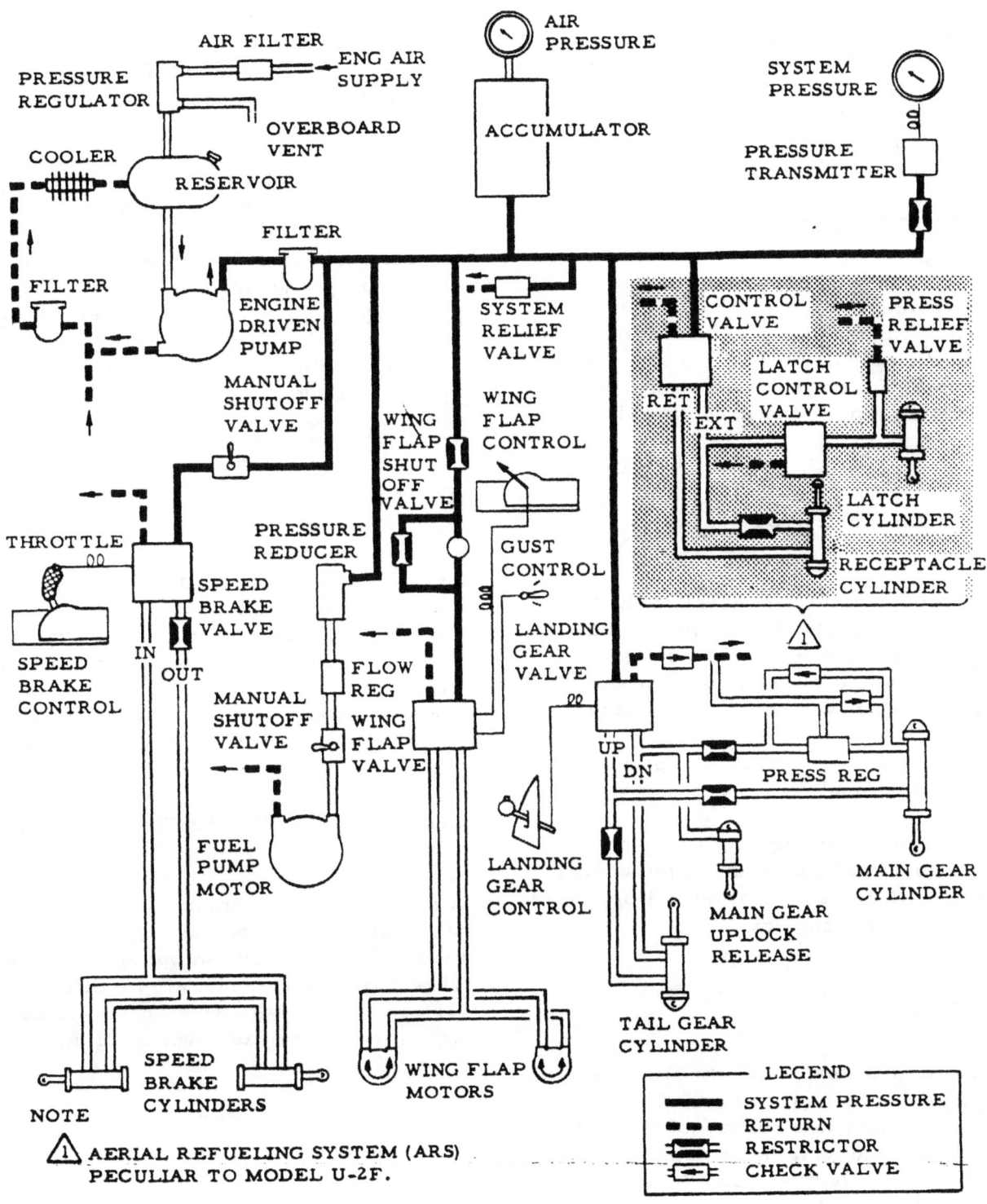

Figure 1-13

The hydraulic oil is cooled as it circulates through a heat exchanger in the engine oil ram air cooling scoop on the lower right side of the fuselage.

Engine compressor bleed air is used to pressurize the hydraulic fluid reservoir to reduce foaming and increase pump efficiency at high altitude.

There is no emergency hydraulic system.

HYDRAULIC PRESSURE GAGE

This instrument, on the right side of the instrument panel, indicates hydraulic system pressure. The normal pressure range is 2850 to 3150 psi. This is a remote indicating system and is powered by either normal inverter.

FLIGHT CONTROL SYSTEM

The flight controls are conventional, consisting of rudder pedals and a wheel mounted on a control column. All surfaces are directly connected to the cockpit controls by cables. No power boost is provided.

There are no surface locks except as provided by the ground crew.

Electric trim is provided in pitch and roll. There is no directional trim in the cockpit. However, with the autopilot operating, there is effective yaw damping.

GUST CONTROL

The most unique feature of the flight control system is the gust control. This device makes it possible to simultaneously shift both ailerons and wing flaps to an up position. This is called the "Gust" position. Ailerons are shifted 10 degrees and flaps are shifted 4 degrees. This has the dual purpose of reducing both wing and tail structural loads. The gust control is used to shift the surfaces up when flying in turbulent air or when flying at higher speeds in smooth air. The aileron motion is effected by an electric shifting mechanism located under the cockpit floor. The wing flaps are moved by their usual hydraulic motors.

The gust control switch is on the left console aft of the throttle. Two amber lights are provided on the annunciator panel to indicate that the ailerons and flaps have shifted up. The lights glow when the aileron actuator and flaps have reached the gust position. A green light on the annunciator panel indicates when the ailerons are faired. In addition, the wing flap position indicator shows the flaps are shifted up.

CONTROL WHEEL

The control wheel has several switches installed. On the right-hand grip is the elevator trim tab control switch and the autopilot disengage button. On the left-hand grip is the trim tab power switch and a radio button. Manual trim tab actuator power goes through the trim power switch.

CONTROL WHEEL POSITION INDICATOR

A small pointer is mounted on the top center of the control wheel. A degree scale is mounted on the column opposite the pointer. This provides an accurate method of monitering left or right-wheel displacement which may occur as a result of uneven fuel feeding.

AILERON CONTROL SYSTEM

The ailerons are controlled through rotation of the wheel. Surface travel is 16 degrees up and 14 degrees down. With actuation of the gust control, these travels are reduced to 6-1/2 degrees up and 5 degrees down from the shifted neutral position. An amber light illuminates when the ailerons have shifted to the gust position, and a green light illuminates when they are in the faired position. The electrically actuated trim tab is on the left aileron. The trim tab switch is on the left console. A ground adjustable bend tab is installed on the right aileron.

ELEVATOR CONTROL SYSTEM

The elevator system is conventional with column control. Surface travel is 30 degrees up and 20 degrees down. The electrically actuated pitch trim tabs are at the inboard end of each elevator.

Two control switches are on the wheel. The left-hand switch controls power to the manual and automatic elevator trim system. The right-hand switch positions the tab. In addition to electric trim operation, the tabs have automatic servo operation to reduce longitudinal control forces.

ELEVATOR TRIM TAB POSITION INDICATOR

A position indicator below the gear handle gives the pilot a visual reference for trim tab setting. The indicator is powered by 28 V DC and the circuit is controlled through the elevator trim tab power switch on the left-hand grip of the control wheel.

Elevator trim tab position is indicated even though the trim power switch is off.

RUDDER CONTROL SYSTEM

The rudder control system is conventional with pedal controls. Surface travel is 30 degrees left and right. A ground adjustable bend tab is installed on the rudder.

The rudder pedals may be adjusted fore and aft to accomodate different leg lengths. The adjustment lock lever on the outboard side of each pedal can be released by outboard toe pressure. There is no rudder centering device in this system.

COLLAPSIBLE RUDDER PEDALS

Collapsible rudder pedals are installed to reduce pilot fatigue by permitting free extension of the pilot's legs. In the collapsed position the upper portion of each pedal is rotated to a horizontal position, permitting the pilot to extend and stretch his legs.

Operation:

1. To collapse rudder pedals, push rudder pedal pusher forward with toe and steady opposite rudder pedal at the same time with opposite foot. Repeat on opposite side.

2. To return rudder pedals to normal position, lift collapsed rudder pedal with toe or swab stick while steadying opposite pedal with other foot.

Changed 1 September 1968

WING FLAPS

The wing flaps are actuated by two hydraulic motors which are interconnected by a flexible shaft. If one motor should fail, the other motor will operate both flaps, but at a slower speed.

A second function of the interconnecting shaft is to maintain synchronization of left and right flap positions. If the left and right flaps become unsynchronized for any reason, asymmetry is limited to a maximum of 5 degrees by an automatic switch device. This safety feature precludes the possibility of large differences in flap position due to malfunction or failure. After the cutout switch has been actuated, it is no longer possible to move the flaps until reset on the ground.

There is no emergency method of operating the wing flaps. However, flaps may be lowered with normal engine windmilling in the event of an engine failure.

WING FLAPS CONTROL

An electrical solenoid valve is used to control the direction of flow of hydraulic fluid to the reversible hydraulic drive motors. This valve is controlled by the flap switch on the left console outboard of the throttle. The switch has three positions: OFF, UP, and momentary DOWN.

NOTE

The gust control switch overrides the wing flap switch. Thus the wing flaps are inoperable with the gust control in the gust position. Also, if the flaps are down when the gust control switch is moved to the gust position, the flaps will retract to the gust position.

WING FLAP POSITION INDICATOR

The wing flap position indicator is on the left lower instrument subpanel. It has a range from minus 4 degrees to plus 35 degrees, which coincides with the maximum flap travel. This is an electric instrument operating from DC power.

RETRACTABLE STALL STRIPS

Manually operated retractable stall strips are installed on the leading edges of the wing about midway between the wing root and the wing tip. They provide better stall characteristics when extended at low altitude.

The stall strips protrude approximately one-half inch when extended and are flush with the wing leading edge when retracted. They are extended by pulling the white STALL STRIP T-handle on the outboard side of the right-hand kneehole and moving outboard into a detent. They are retracted by pulling the T-handle out of the detent and allowing the handle to return slowly by resisting the spring return force of the system.

RETRACTABLE STALL STRIP POSITION INDICATOR LIGHTS

There are two amber lights (left and right stall strip) on the right-hand instrument panel which illuminate whenever the stall strips are retracted and the landing gear is down and locked. The lights go out when the stall strips are extended. When the landing gear is not down and locked, the lights will not illuminate regardless of the position of the stall strips.

SPEED BRAKES

The speed brakes are a drag device used to decrease speed. The speed brakes extend to a maximum of 50 degrees. The direction of travel may be reversed at any time.

SPEED BRAKE CONTROL

The speed brake control is a three position center-off slide switch on the top of the throttle lever. The hydraulic fluid flow to the actuating cylinders is solenoid valve controlled. To OPEN the speed brakes, the switch is pulled back; to CLOSE, the switch is pushed forward. The speed brakes may be stopped in any position on either the extension or retraction cycle by returning the switch to its OFF position. However, air loads may slowly close the speed brakes from any partially open position if the switch is positioned OFF. This will not occur with the switch in the OPEN position. In the event of engine failure, normal engine windmilling will develop sufficient hydraulic pressure to actuate the speed brakes.

LANDING GEAR

The airplane is equipped with a hydraulically actuated bicycle landing gear which is controlled by a lever on the left forward side panel. Both main gear and tail gear have dual wheels and retract forward. Lateral balance is provided by left and right droppable auxiliary gear (pogos) located outboard on the wing. An electrically operated indicator and warning system is provided for the main and tail gear only. An emergency drop system is provided to extend the main and tail gear in case of failure of the normal system.

EMERGENCY EXTENSION SYSTEM

This is a free fall system. Both the main and tail gear are mechanically released by a cable system, which is connected to the emergency gear release handle on the lower center subpanel. Gravity and slipstream force the gear to the down position, the gear is mechanically locked in place by a spring-loaded downlock mechanism.

LANDING GEAR CONTROL LEVER

The landing gear is controlled by a lever on the left forward side panel. This lever actuates switches that operate an electric solenoid valve which directs the hydraulic fluid to the landing gear cylinders. The lever is held down by a solenoid lock controlled by a scissors switch on the main landing gear. The lock is automatically released for inflight retraction. For emergency gear retraction with the airplane weight on the gear or in case of solenoid lock malfunction in flight, the lock can be manually released by pressing the red button immediately above the landing gear lever.

To extend the gear, the pilot must release a spring-loaded latch by thumb pressure on the small switch protruding from the handle.

LANDING GEAR WARNING SYSTEM

The gear unsafe warning red light is in the translucent end of the gear control lever. It glows whenever the landing gear is not locked in the position selected by the control lever.

The gear unsafe warning horn sounds whenever the throttle is retarded below the position for 75% to 80% RPM if the landing gear is not down and locked; also the unsafe warning light will glow. A warning horn cutout button directly to the right of the gear handle may be used to silence the horn. The throttle switch will be automatically reset each time the throttle is advanced past 80% RPM.

LANDING GEAR POSITION INDICATOR

The landing gear position indicator on the upper left side of the instrument panel shows the position of both landing gear. When the DC electrical bus is not energized, diagonal stripes appear in each window. When the bus is energized, a wheel appears in each window for each gear that is down and locked, and the word UP appears for each gear that is up and locked. If the gear is at any intermediate position, diagonal stripes appear.

POGOS

Pogos are installed about half-way out on each wing and can swivel 360 degrees. The pogos provide lateral support during ground handling and takeoff. If the pogos are not extracted prior to takeoff they should fall out of their sockets as the wings begin to lift. Pogos can be retained for flight by leaving safety pins installed.

No cockpit indication is provided.

STEERING SYSTEM

The airplane is steerable by means of the tailwheel which has an angular range of 6 degrees each side of center. The pilot controls the steering through the rudder pedals which are directly connected to the tail gear by means of a cable system. When the tail gear is retracted, this system is inactive.

WHEEL BRAKES

A pair of spot and disc type brakes are installed on each main wheel. The brake system is an independent manually operated hydraulic system as shown on figure 1-14. A small reservoir just forward of the cockpit feeds a single conventional master cylinder. The brakes are operated by conventional toe pedals. Since there is no differential braking, either or both pedals will operate the single master cylinder. The pressure from the master cylinder is carried to all brakes by means of a pressure manifold. A pressure relief valve limits the maximum brake pressure.

DRAG CHUTE

A 16-foot drag chute may be installed in the compartment just above the tailpipe nozzle. Operation of the release mechanism is manually controlled by a cable from the cockpit. The chute is installed in the compartment with the attaching hook unlocked. This allows the chute to leave the airplane in case of accidental opening of the doors during takeoff or flight. The chute is operated by a tee handle on the side of the console below the throttle. The tee handle is pulled, rotated 90° clockwise, and placed on the "step" in order to deploy the chute. The handle is automatically locked on the step to prevent it from snapping back into place and releasing the chute. A lever adjacent to the handle is operated to release the handle for return to the stowed position. Rotating the handle another 90° clockwise and pulling it out further unlatches the hook and allows the chute to jettison.

BRAKE DIAGRAM

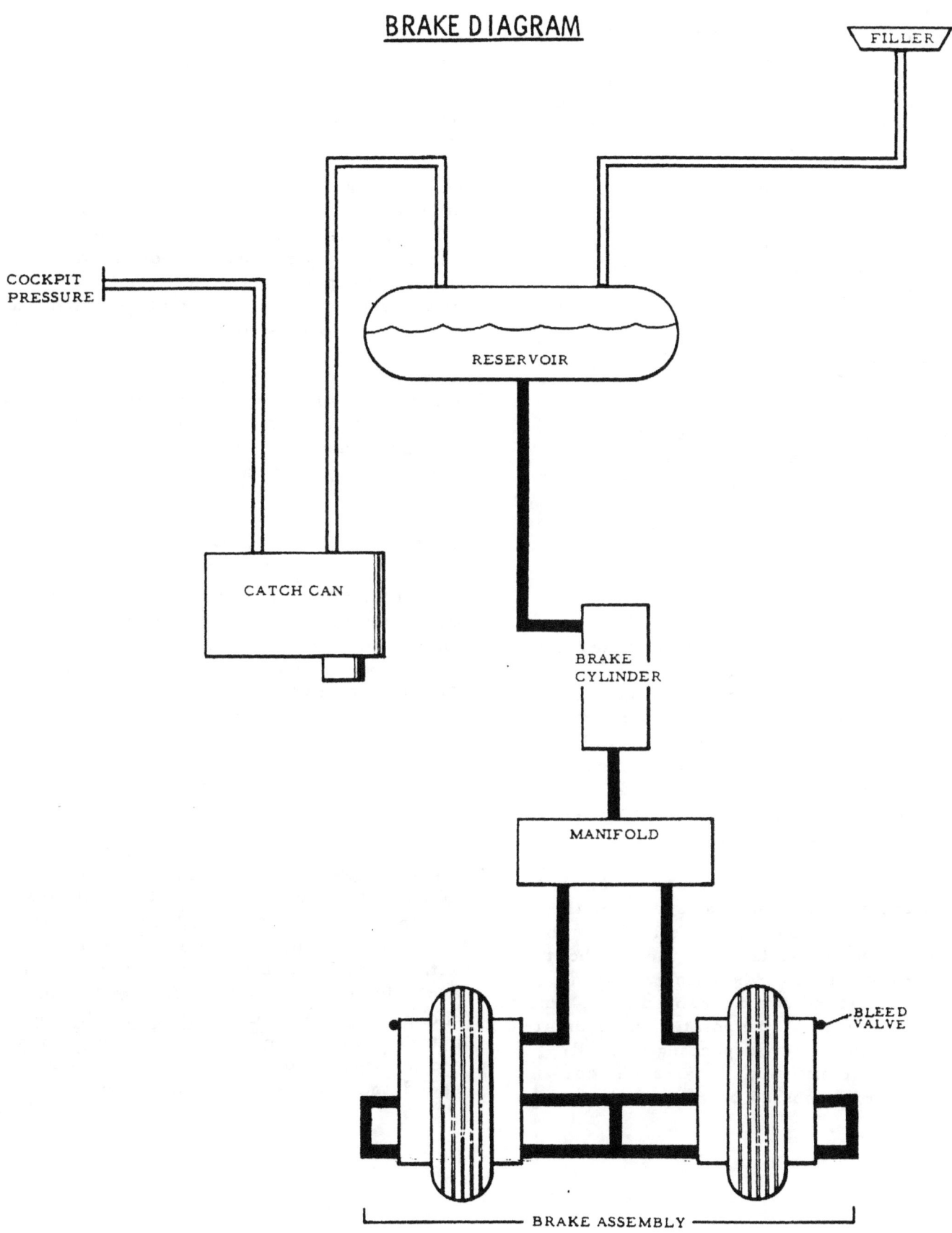

Figure 1-14

If the drag chute is deployed in normal flight, it will not stay intact but will soon disintegrate due to engine jet stream effects. On takeoff or go-around the chute should be jettisoned since it may not always disintegrate immediately.

NOTE

A drag chute is not installed if defensive systems are installed in the drag chute compartment.

INSTRUMENTS

Those instruments which are not properly a part of a complete system are covered below. (See figures 1-15, 1-16, and 1-17 for panel arrangement.)

AIRSPEED INDICATOR

A conventional 40-to 400-knot airspeed indicator is on the upper left side of the instrument panel. Two pointers simultaneously indicate actual airspeed and maximum allowable airspeed. The outer dial is graduated in 5-knot increments. A rotating drum, visible through a window in the dial, is graduated from 0 to 100 in 2-knot increments for more accurate indication.

The maximum allowable airspeed pointer shows maximum structural speed for the airspeed corresponding to the limit Mach number, whichever is less. Below about a 40,000-foot altitude, the pointer will indicate the limit indicated airspeed of 240 knots. Above 40,000 feet, the pointer will indicate lower airspeeds corresponding to limit Mach number of 0.80. An airspeed correction card is posted in the cockpit.

ALTIMETERS

Two altimeters are installed in the cockpit. The airplane altitude is shown by the altimeter on the left side of the instrument panel. It is a conventional sensitive type with a range from zero to 80,000 feet. The 10,000 foot pointer has a special notched disc and a pointer extension for better readability. A striped section warning indicator appears through the notched disc at altitudes below 16,000 feet. An altimeter correction card is posted in the cockpit.

The second altimeter is on the lower right instrument subpanel. By means of a manual selector valve above the right side console, the instrument will read either cabin altitude or equipment bay altitude.

VERTICAL SPEED INDICATOR

The vertical speed indicator is conventional and shows the rate at which the airplane is climbing or descending, based on the rate of change of atmospheric pressure. The dial is graduated from zero to 6000 feet per minute on two adjoining scales.

PITOT STATIC SYSTEM

Two separate pitot static systems are installed. Each system contains one pitot tube and two static ports. The pitot tubes are located on the bottom fuselage just aft of the nose. The static ports are flush with the skin, each system having one on each side of the fuselage, forward of the cockpit. One system is used only by the autopilot Mach sensor; the other provides a pitot static source for altimeter, airspeed and vertical speed indicators.

AF (C)-1-1   SECTION I

COCKPIT ARRANGEMENT

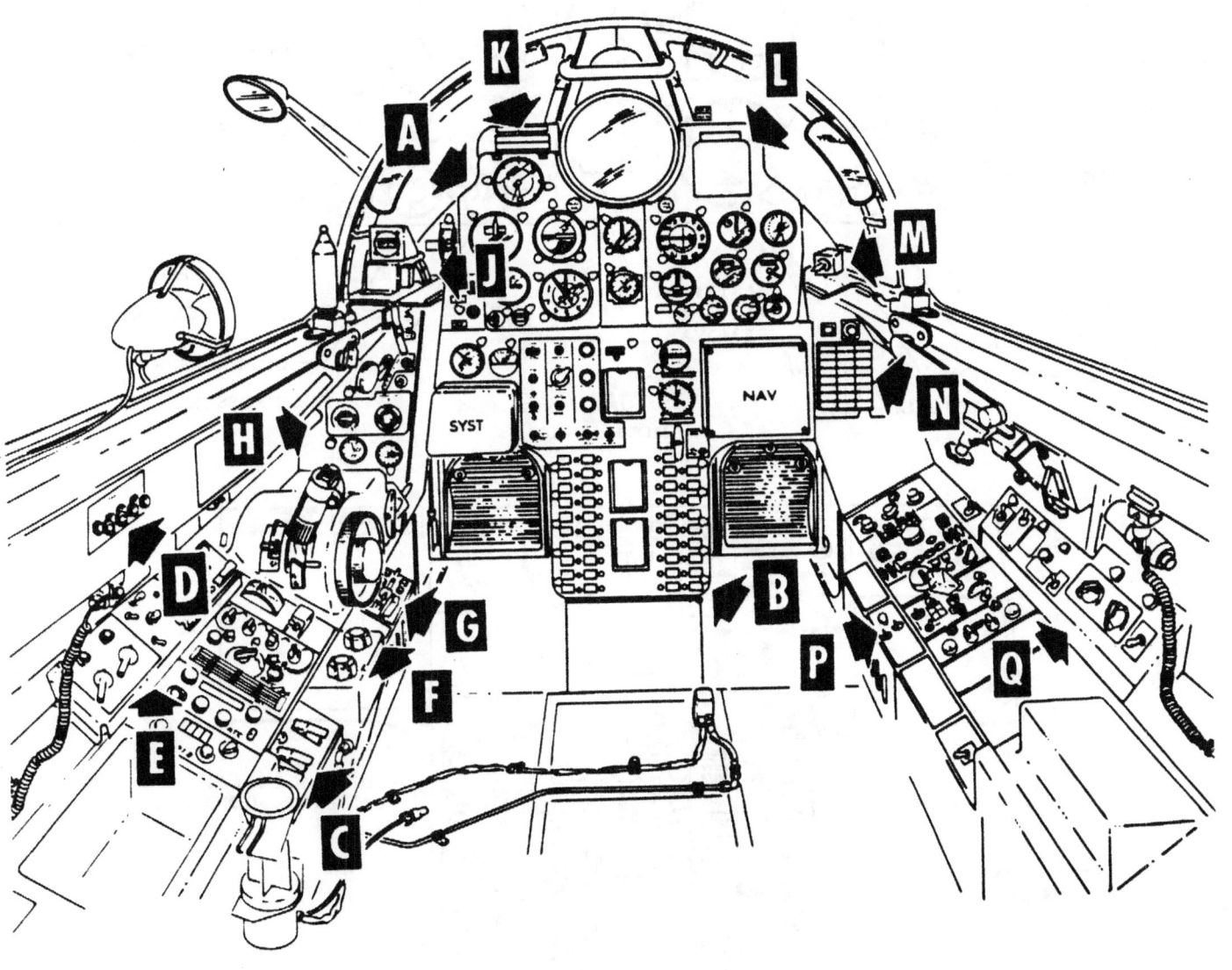

Figure 1-15 (Sheet 1)

SECTION I AF (C)-1-1

COCKPIT ARRANGEMENT

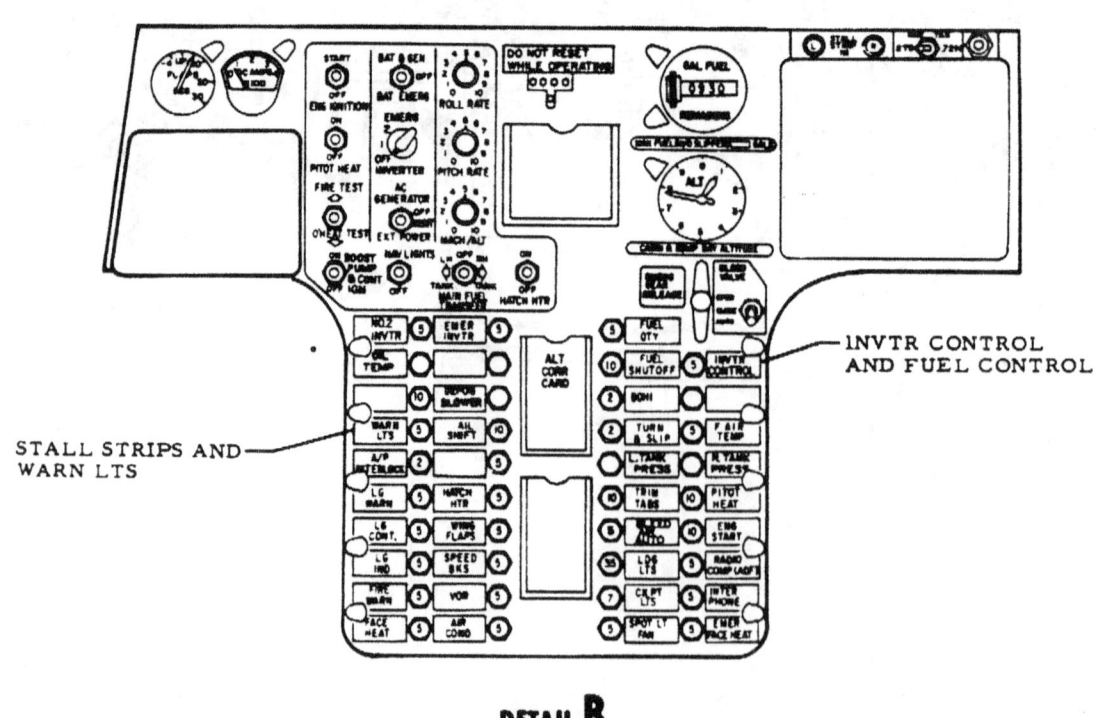

DETAIL A
MODEL U-2C

DETAIL B
MODEL U-2C

Figure 1-15 (Sheet 2)

Changed 1 September 1968

COCKPIT ARRANGEMENT

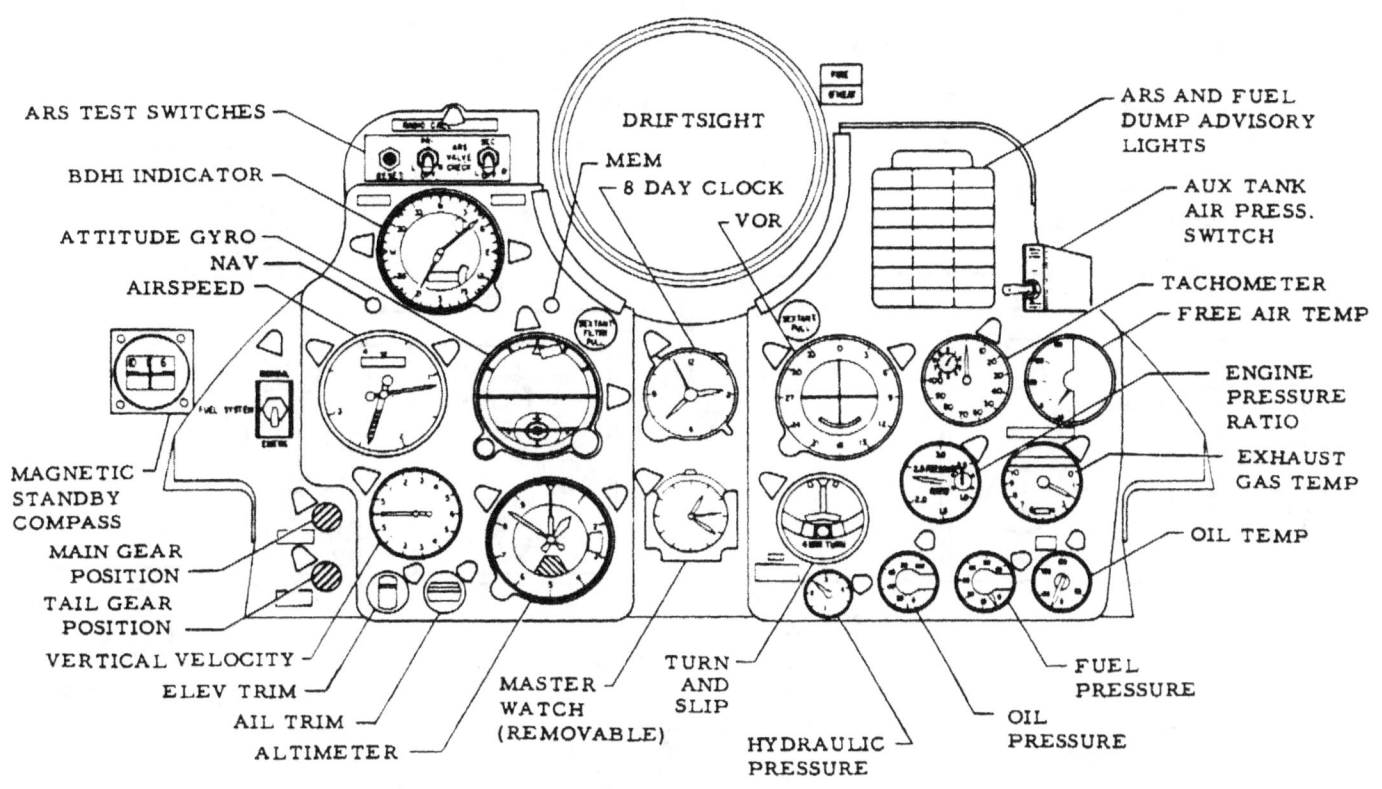

DETAIL A
MODEL U-2F

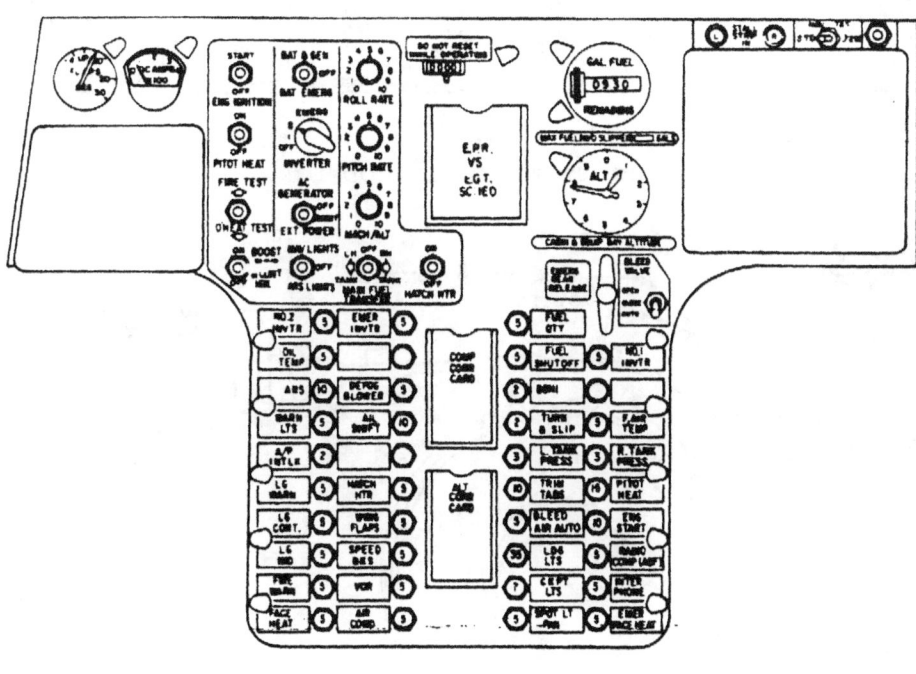

DETAIL B
MODEL U-2F

Figure 1-15 (Sheet 3)

COCKPIT ARRANGEMENT

DETAIL C

FUEL DUMP SWITCHES

DETAIL D

FUEL DUMP
CIRCUIT BREAKERS

DETAIL E

LEFT SIDE SWITCH PANEL MODEL U-2C

DETAIL E

LEFT SIDE SWITCH PANEL MODEL U-2F

DETAIL F

THROTTLE QUADRANT
AND ADJACENT SWITCHES

C-51

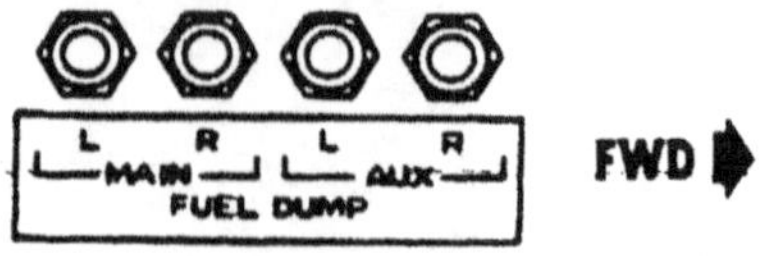

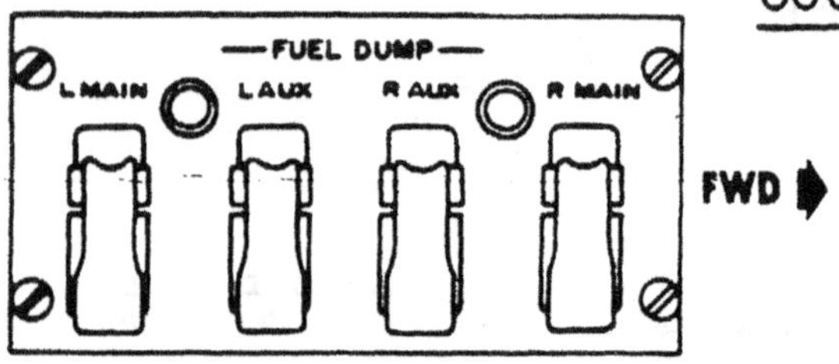

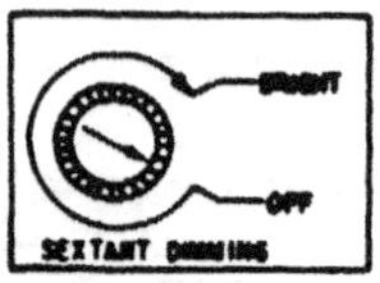

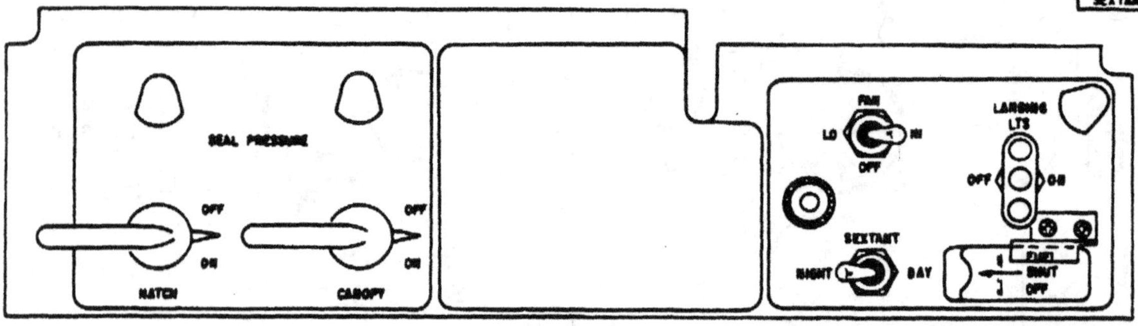

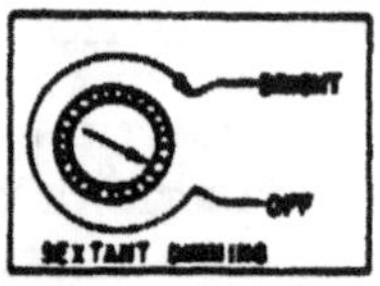

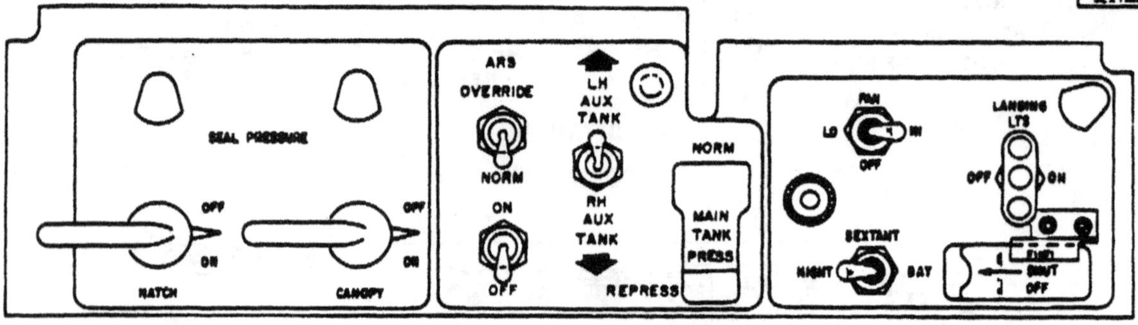

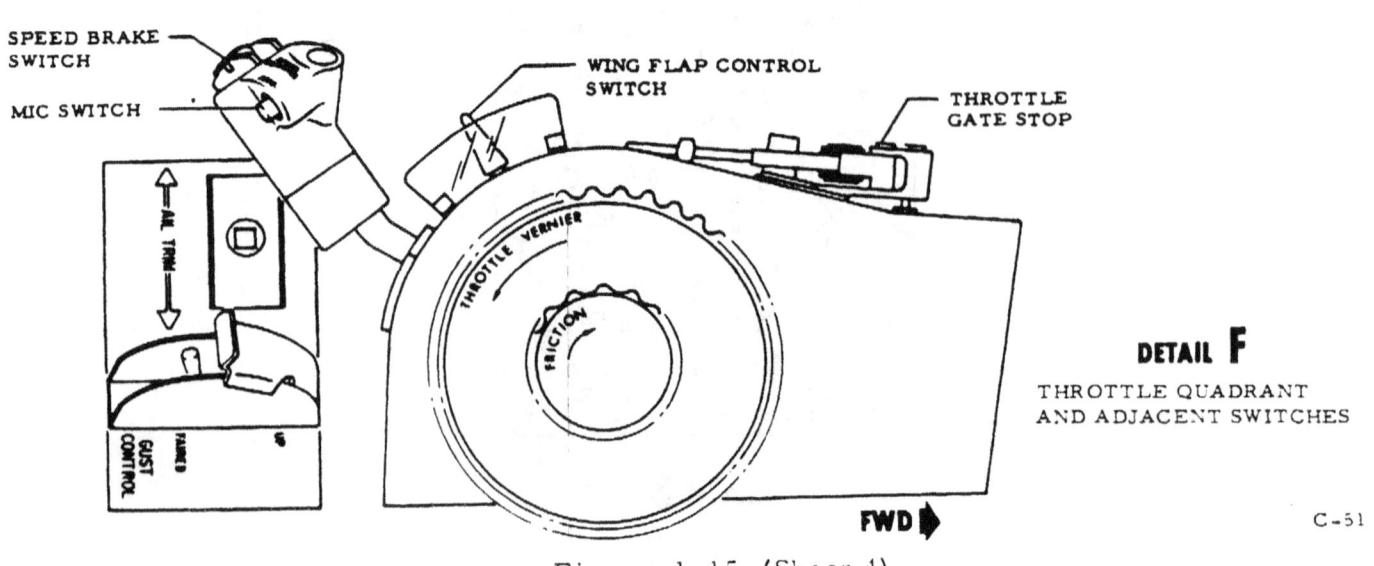

Figure 1-15 (Sheet 4)

COCKPIT ARRANGEMENT

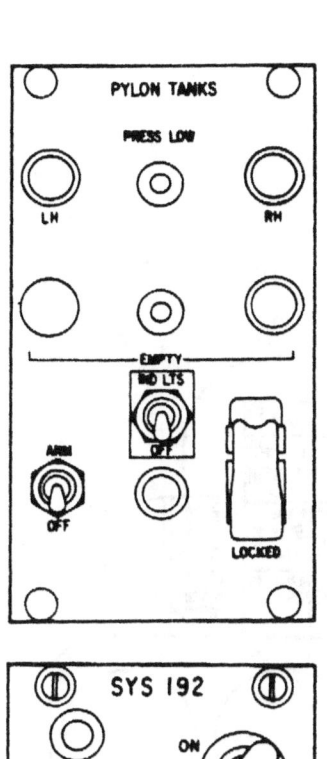

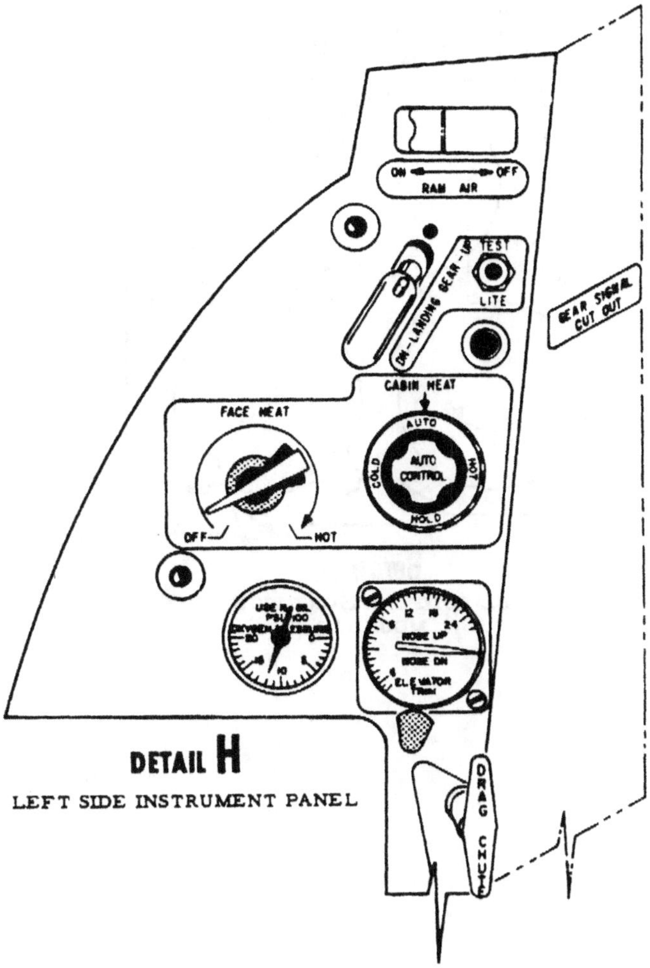

DETAIL H
LEFT SIDE INSTRUMENT PANEL

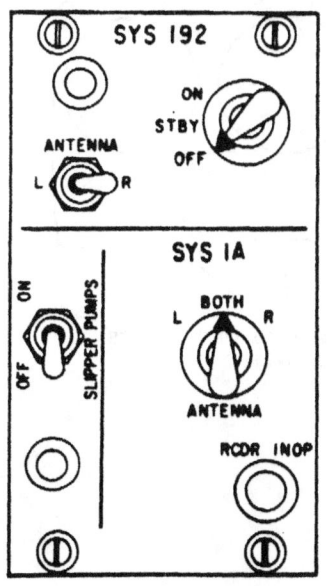

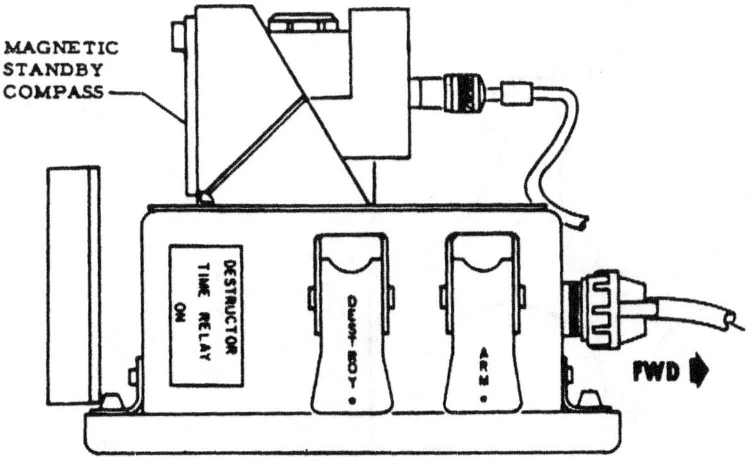

DETAIL G
DROP TANK, SYSTEM 1A AND 192, AND SLIPPER PUMP CONTROL PANELS

DETAIL J
DESTRUCTOR CONTROL

Figure 1-15 (Sheet 5)

SECTION I AF (C)-1-1

COCKPIT ARRANGEMENT

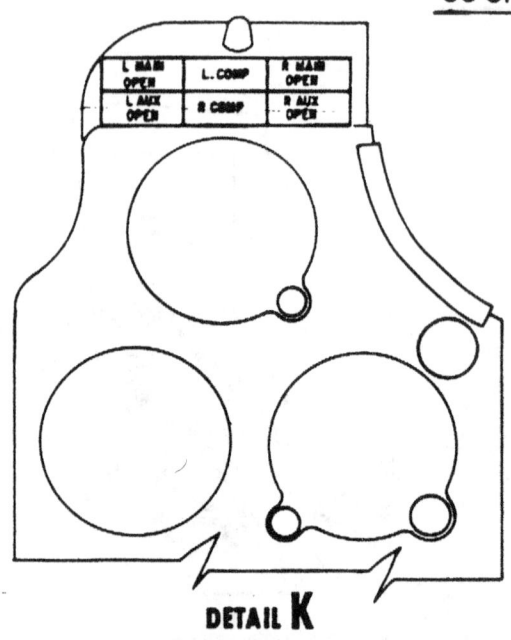

DETAIL K
FUEL DUMP ADVISORY LIGHTS
MODEL U2-C

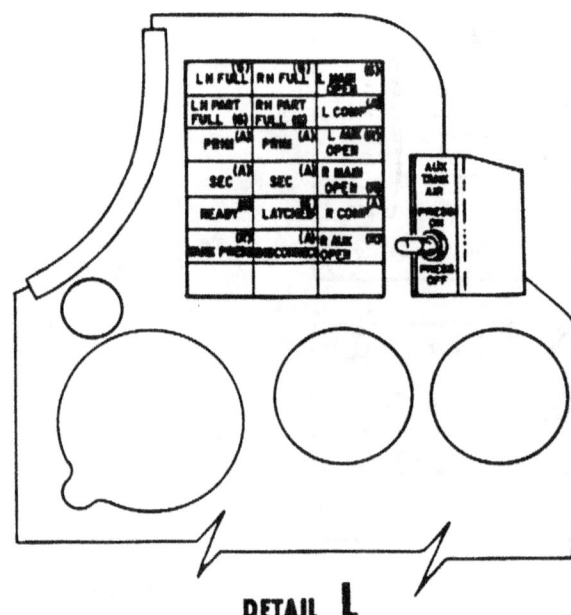

DETAIL L
ARS AND FUEL DUMP ADVISORY LIGHTS
MODEL U-2F

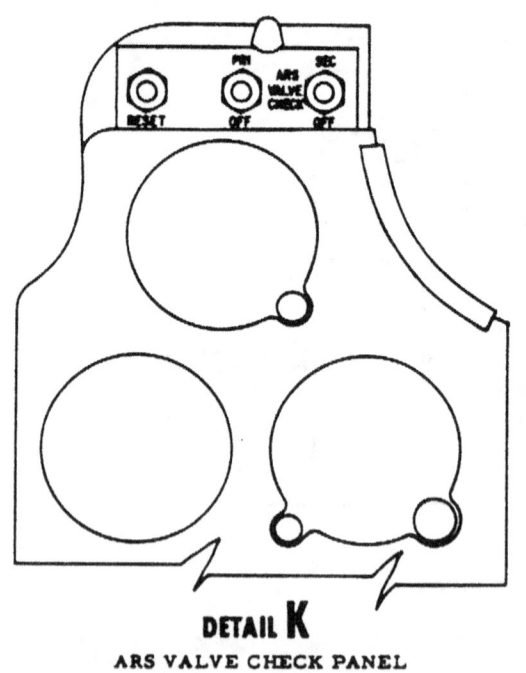

DETAIL K
ARS VALVE CHECK PANEL
MODEL U-2F

Figure 1-15 (Sheet 6)

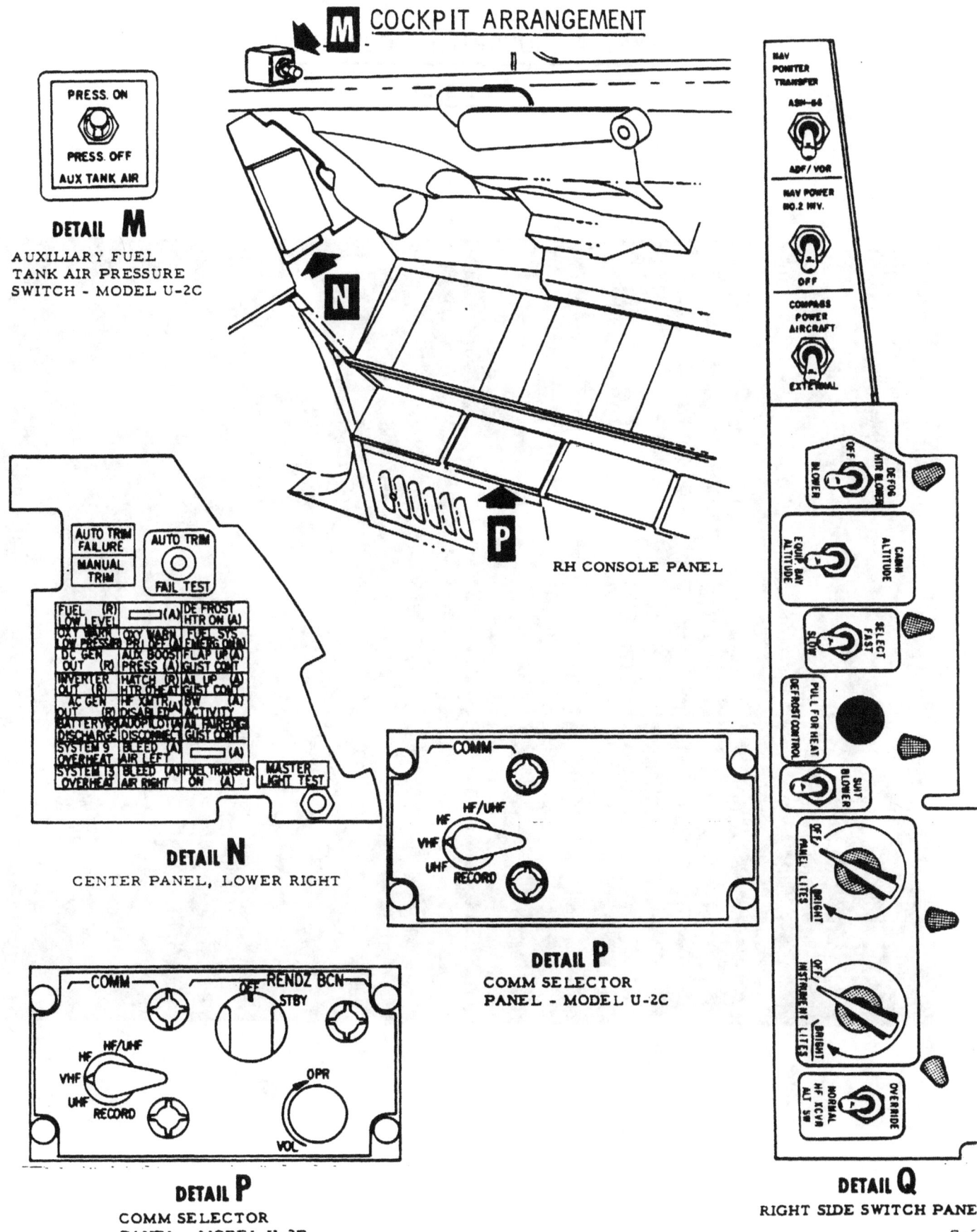

Figure 1-15 (Sheet 7)

SECTION I AF (C)-1-1

TYPICAL COCKPIT - MODEL U-2C
LEFT SIDE

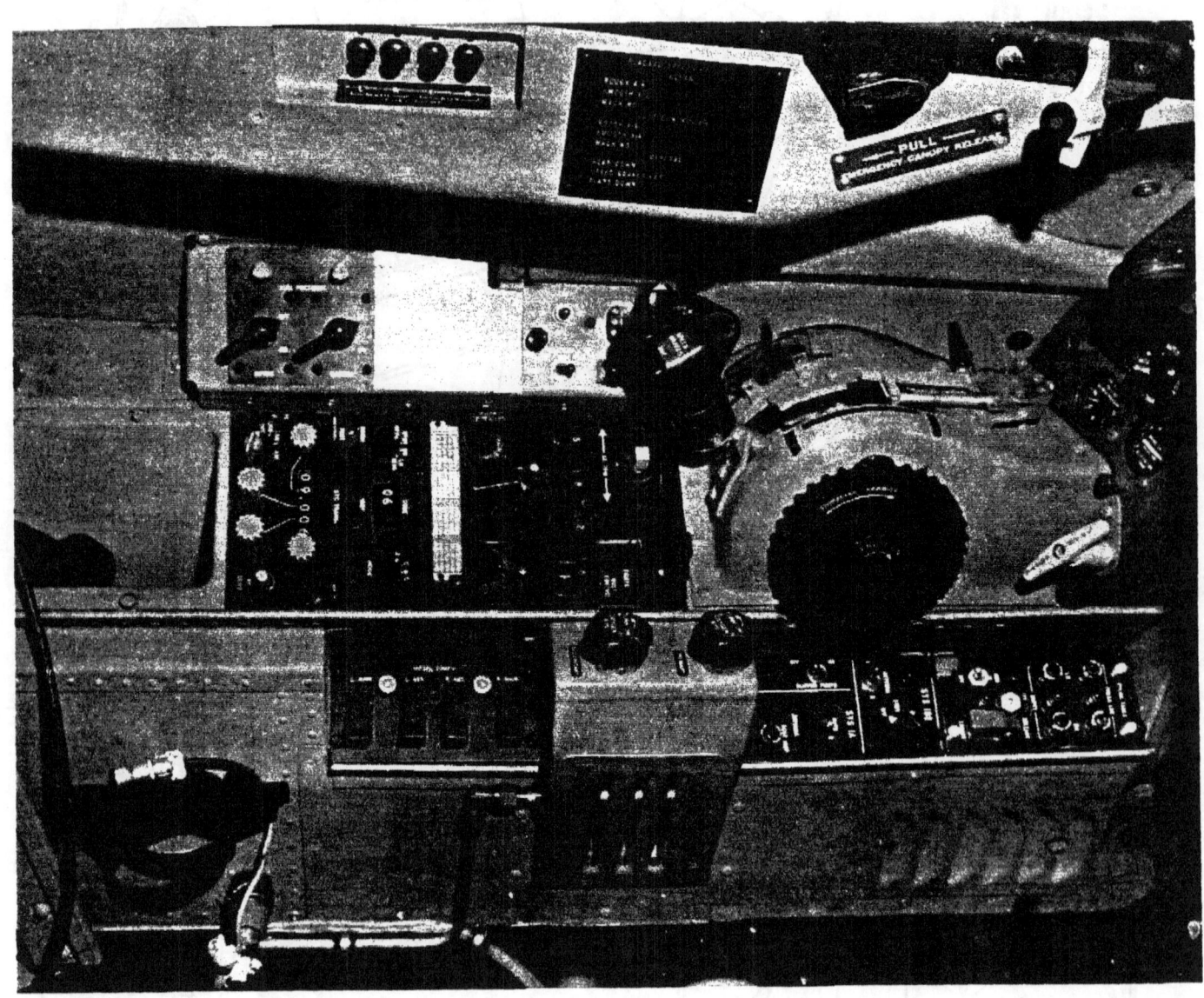

Figure 1-16 (Sheet 1)

AF (C)-1-1 SECTION I

TYPICAL COCKPIT - MODEL U-2C

CENTER INSTRUMENT PANEL

Figure 1-16 (Sheet 2)

1-39

SECTION I AF (C)-1-1

TYPICAL COCKPIT - MODEL U-2C

RIGHT SIDE

Figure 1-16 (Sheet 3)

TYPICAL COCKPIT - MODEL U-2F

LEFT SIDE

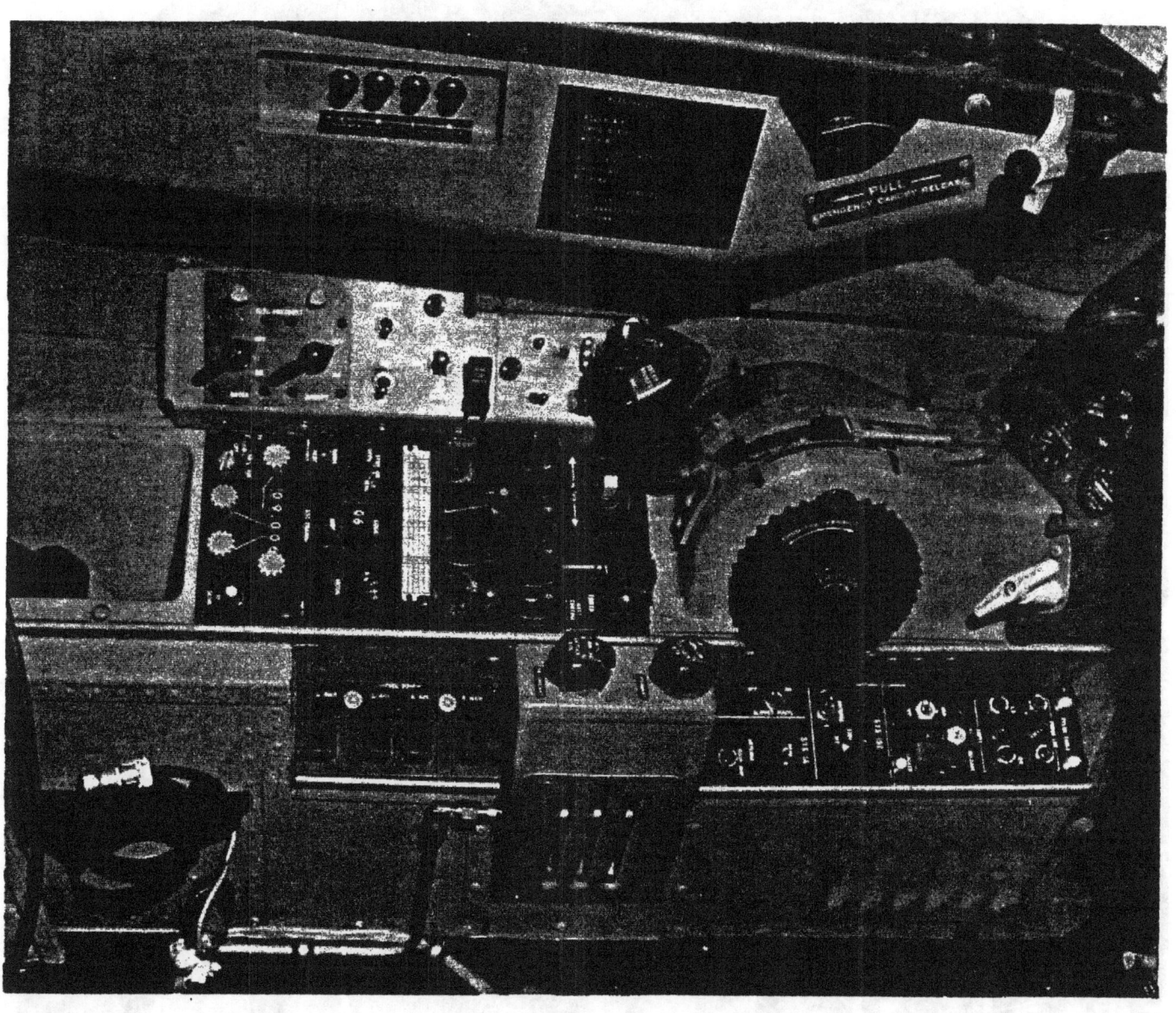

Figure 1-17 (Sheet 1)

TYPICAL COCKPIT - MODEL U-2F

CENTER INSTRUMENT PANEL

Figure 1-17 (Sheet 2)

AF (C)-1-1　　　　　　　　　　　　SECTION I

TYPICAL COCKPIT - MODEL U-2F

RIGHT SIDE

Figure 1-17 (Sheet 3)

OUTSIDE AIR TEMPERATURE SYSTEM

The outside air temperature indicator is on the upper right instrument panel. It is connected to a standard temperature element in a specially constructed probe on the left side of the fuselage nose. Use figure A1-6 to obtain true OAT from observed temperature.

TURN AND SLIP INDICATOR

The turn and slip indicator is a conventional DC powered instrument which shows airplane rate of turn and slip or skid. This instrument gives a 360-degree turn in four minutes at one needle width. It is on the right side of the instrument panel.

TYPE J-8 ATTITUDE INDICATOR

The attitude indicator is located on the left center of the instrument panel. The pitch attitude of the aircraft is indicated in climb or dive by displacement of the horizon bar with respect to the miniature adjustable airplane. The minature airplane is above the horizon bar in a climb and below the horizon bar in a dive. The roll attitude of the aircraft is shown by the relation of the bank index to the degree markings on the instrument face.

A Pull to Cage knob is on the lower right corner of the instrument case. To cage the gyro, the knob should be pulled smoothly and gently to avoid damaging the instrument. The gyro may be caged before takeoff to expedite normal erection. It may also be caged in flight if the aircraft is straight and level, by visual reference to a true horizon.

A knob is provided on the lower left corner of the instrument to adjust the miniature airplane for zero pitch indication. The adjustment can be made for any pitch attitude from 5 degrees dive to 10 degrees climb.

The J-8 indicator is operated on main or emergency inverter power. If power is off, a warning flag appears on the face of the instrument.

NOTE

The J-8 attitude indicator is subject to bank errors during and following prolonged turns. This installation is also unreliable for detecting banked attitudes of less than 3 degrees.

COURSE INDICATOR (VOR)

This instrument is part of the ARC-15F VHF navigation set. Refer to Section IV A for description and operation.

BEARING-DISTANCE-HEADING INDICATOR

This instrument combines the indications from the attitude heading reference system (compass), VHF navigation equipment (VOR), radio direction finder (ADF), and the dead reckoning navigation system. It is on the left side of the instrument panel and requires both AC and DC power. Alternating current requirements are supplied by either of the normal inverters or from the emergency inverter.

Heading information is supplied to the rotating face of the instrument by the attitude heading reference system (AHRS). An index at the top of the instrument shows the magnetic heading as the card rotates. A flux valve in the left wing tip furnishes magnetic heading information to the AHRS amplifier and gyro platform which in turn feeds the information to the face of the indicator. In case of a malfunction of the gyro platform, the gyro can be bypassed by selecting COMP on the mode selector; the flux valve heading information is then amplified only (not stabilized) by the AHRS and fed to the card on the indicator. (The gyro may also be used as a "free gyro" for navigation in high latitudes.)

The AHRS controls are on the right-hand console. A complete explanation of the AHRS will be found in Section IV A.

There is a preselect heading feature in the indicator which works in conjunction with the autopilot. A "bug" controlled by a knob in the lower right-hand corner of the instrument, indicates the preselected heading. The autopilot will turn the aircraft to the preselected heading whenever the HDG mode (on the autopilot control panel) is selected. The "bug" can be used as an adjustable lubber line when not in HDG mode.

The instrument has two pointers labeled 1 and 2. The pointers can display either radio navigation system information (ADF/VOR) or dead reckoning navigation steering information (ASN-66) depending upon the position of the ADF/VOR-ASN-66 pointer transfer switch:

In the ADF/VOR position, the NO.1 pointer displays ADF information and the NO.2 pointer displays VOR information. When the airplane is heading directly toward the station, the needles will point to the index mark at the top of the instrument. The actual magnetic bearing of any station may be read as the needle point on the rotating card.

In the ASN-66 position, the NO.1 and NO.2 pointers display dead reckoning navigation steering information as described in Section IV.

The instrument has a red OFF flag which appears whenever the heading information is doubtful or incorrect due to the gyro platform not being errected properly, or due to a loss of power to the flag. If heading information is questioned select COMP mode on compass control panel.

The instrument also has nautical miles distance-to-go counters which function when the navigation system is operating.

MAGNETIC STANDBY COMPASS

The standby compass is on the left sill. This is a standard compass for checking the AHRS and for emergency heading reference.

CLOCKS

An eight day clock is on the center instrument panel. A bracket is also provided for a master watch.

FIRE WARNING SYSTEM

A dual warning light above the instrument panel and right of the driftsight will indicate a fire or overheat condition. Either condition is indicated by a steady light. Thermoswitches in the forward engine compartment provide fire warning and switches in the aft fuselage section provide overheat warning. The two circuits are not interconnected. The thermoswitches in both circuits are set to close at the same temperature. Therefore, either group could be affected by either fire or overheat. A circuit test switch on the lower left switch panel tests the circuits up to but not including the thermoswitches. The switch is moved up to the FIRE TEST position and down to the O'HEAT TEST position.

SECTION I   AF (C)-1-1

SPEED WARNING SYSTEM

The speed warning system warns the pilot of excessive speeds for a given gust control configuration by activating the gear warning system. It was necessitated by the rapid accelerating characteristics of the aircraft and the serious structural consequence of excessive speed. The indicated airspeed is monitored by a sensitive pressure switch in the nose section. If the aircraft is flying with the gust control faired, the warning occurs at 220 ($\pm$ 10) knots. If the aircraft is flying with the gust control in gust, the warning occurs at 250 ($\pm$ 10) knots.

CANOPY

The airplane has a one-piece, jettisonable canopy hinged on the left-hand side. The canopy is raised and lowered manually and has a locking handle on the right-hand side. There are three latches in the right-hand cockpit sill for locking the canopy. The canopy is locked by pulling aft on the handle on the right side of the cockpit.

EXTERNAL RELEASE HANDLE

The external handle is located on the right side of the fuselage near the canopy leading edge. It is normally flush and pops out for use. It can be used for either locking or unlocking the canopy from the outside.

MANUAL CANOPY RELEASE

A manual release handle is stowed in the cockpit on the right-hand aft side. To manually release the canopy, insert this handle in the fitting on the left side of the cockpit. Forward movement of the handle unlocks the left-hand canopy latches. For manual release, both left and right handles are simultaneously pushed forward. When released, the canopy is restrained from moving sideways by a tubular structure in the aft end.

EMERGENCY CANOPY JETTISON HANDLE

An emergency canopy jettison handle, painted yellow, is located on the left side of the cockpit near the forward edge of the canopy. It is protected from inadvertent release by a cover while on the ground. To jettison the canopy ONLY, depress the handle release button and pull the handle aft. This fires the torque rod thrusters and releases the canopy mechanically on both sides. The aft portion of the canopy rises and is snubbed by a catch plate at the top aft center to provide pitch and lateral control of the canopy. The forward thrusters fire and rotate the forward portion of the canopy up and aft around the catch plate to clear the cockpit. The entire sequence takes approximately 1/4-second.

PRESSURE SEAL

A tubular rubber seal is provided to seal the gap between the canopy frame and the cockpit sill and windshield. Engine compressor bleed air, suitably regulated, is normally used to inflate the seal. It is controlled by a manual valve on the left side console.

A nitrogen gas supply is installed to automatically increase canopy seal pressure under conditions where the engine compressor bleed source is not at a high enough pressure to seal properly. Refer to Section IV for description and operation.

EJECTION SEAT

The ejection seat is of simple lightweight design using a "low-g" catapult to minimize the possibility of ejection injury. The seat has no adjustment provisions. Pilot position in the seat is adjusted by use of wood blocks. A shoulder harness lock-and-release lever is on the left-hand side. The ejection seat is equipped with an automatic release seat belt. There are no arm rests on the seat and ejection sequence is started by pulling up on a D-ring located at the front of the seat between the pilot's legs. See figure 1-18.

Pulling the D-ring actuates an M3 initiator which starts the jettison/ejection cycle. The first initiator provides gas pressure to start the canopy sequence and to disconnect the elevator control and stow the wheel forward to clear the pilot's knees and feet. It also actuates a T-34 initiator. The T-34 initiator has a one-second delay before simultaneously locking the shoulder harness, actuating the catapult and another T-34 initiator also with a 1-second time delay to release the seat belt. The catapult imparts a maximum of 15-g acceleration to a 350 pound seat-man. This is sufficient to clear the vertical fin by 7 feet at 260 knots. The additional one-second delay in the seat belt release system allows the seat to be well clear of the airplane prior to opening of the belt.

One safety pin is provided in the M3 initiator on the right-hand side of the seat pan. This must be removed for the seat to fire.

AUXILIARY EQUIPMENT

Refer to Sections 4 A and 4 B for description and operation of the following auxiliary equipment:

Air Conditioning and Pressurization System

Windshield and Canopy Defrosting System

Oxygen System

Personal Equipment

Communications and Associated Electronic Equipment

Lighting Equipment

Driftsight System

Sextant System

Hatch Window Heater System

Photographic Equipment

Electronic Intelligence Systems

F-2 Foil System

P-3 Platform System

B/400 Rate Rate Meter

Automatic Observer

Weather Survey Equipment

Destructor

Compass System

Autopilot System

Miscellaneous

Aerial Refueling System (ARS) Ⓕ

Dead Reckoning Navigation System

CANOPY JETTISON AND SEAT EJECTION SYSTEM (SHEET 1)

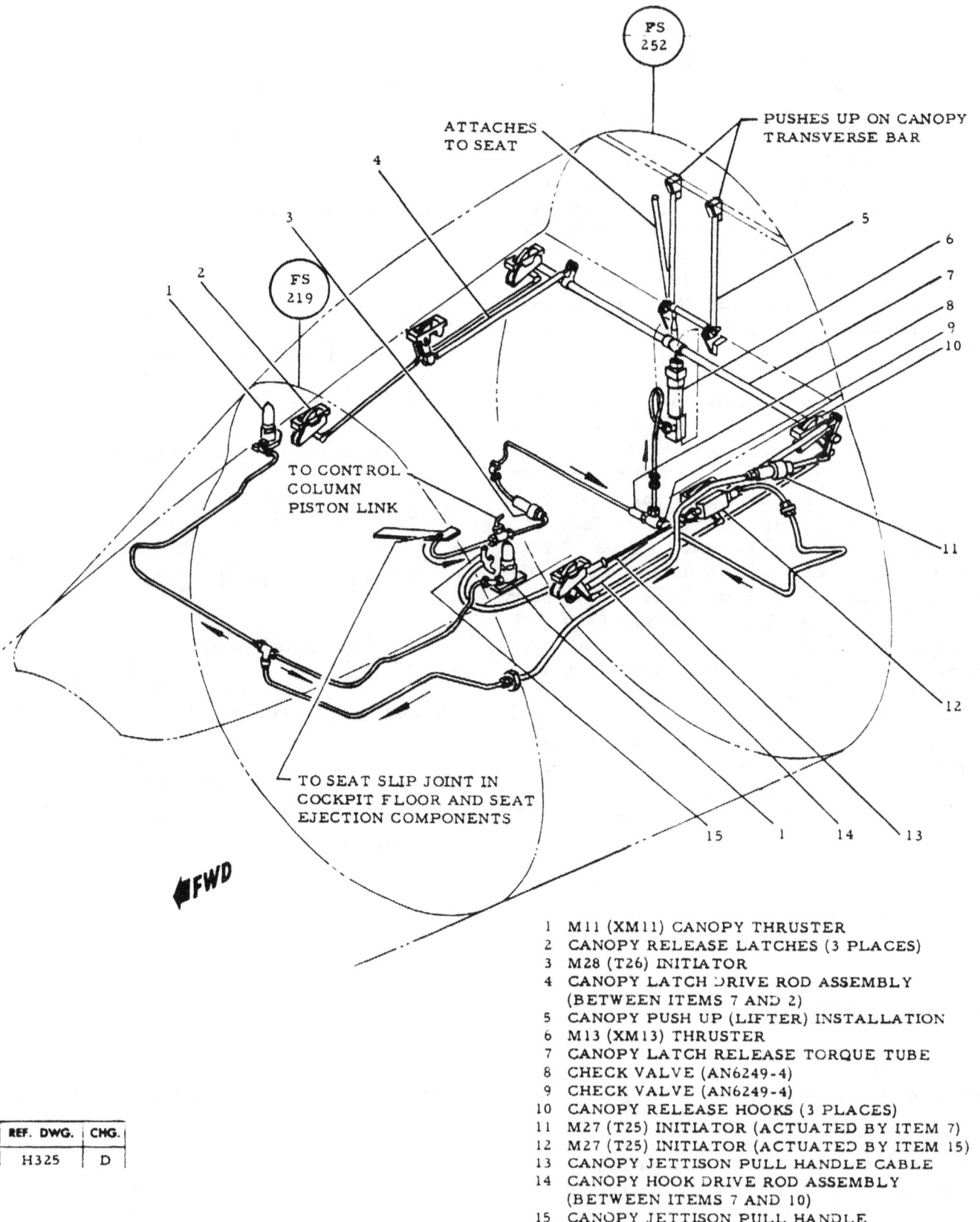

1. M11 (XM11) CANOPY THRUSTER
2. CANOPY RELEASE LATCHES (3 PLACES)
3. M28 (T26) INITIATOR
4. CANOPY LATCH DRIVE ROD ASSEMBLY (BETWEEN ITEMS 7 AND 2)
5. CANOPY PUSH UP (LIFTER) INSTALLATION
6. M13 (XM13) THRUSTER
7. CANOPY LATCH RELEASE TORQUE TUBE
8. CHECK VALVE (AN6249-4)
9. CHECK VALVE (AN6249-4)
10. CANOPY RELEASE HOOKS (3 PLACES)
11. M27 (T25) INITIATOR (ACTUATED BY ITEM 7)
12. M27 (T25) INITIATOR (ACTUATED BY ITEM 15)
13. CANOPY JETTISON PULL HANDLE CABLE
14. CANOPY HOOK DRIVE ROD ASSEMBLY (BETWEEN ITEMS 7 AND 10)
15. CANOPY JETTISON PULL HANDLE

REF. DWG.	CHG.
H325	D

Figure 1-18 (Sheet 1)

CANOPY JETTISON AND SEAT EJECTION SYSTEM (SHEET 2)

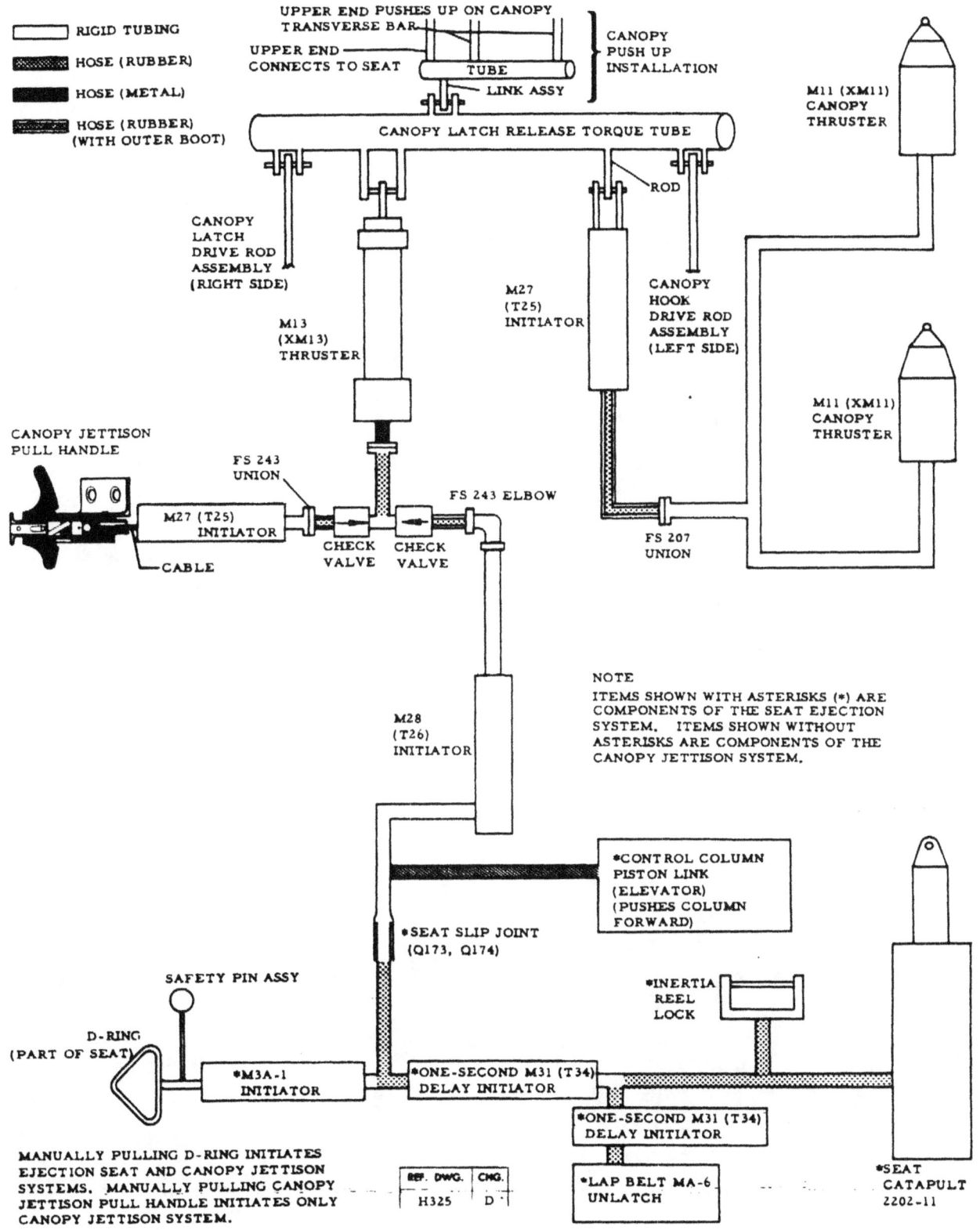

Figure 1-18 (Sheet 2)

AF (C)-1-1

NORMAL PROCEDURES SECTION 2

TABLE OF CONTENTS

	PAGE		PAGE
STATUS OF THE AIRPLANE	2-1	CRUISE CLIMB	2-15
BEFORE ENTRANCE INSPECTION	2-2	SYSTEMS OPERATION	2-20
PRELIMINARY COCKPIT CHECK	2-3	LEVEL OFF	2-22
PRIOR TO BOARDING AIRCRAFT	2-3	DESCENT	2-22
PILOT EQUIPMENT CHECK	2-4	DESCENT CHECK	2-23
HIGH	2-4	5,000-FOOT CHECK (DESCENDING)	2-25
LOW	2-5		
PILOTS COCKPIT CHECK	2-5	BEFORE LANDING	2-25
START ENGINE	2-8	LANDING	2-28
AFTER START	2-9	AFTER LANDING CHECK	2-33
BEFORE TAXIING	2-9	TAXIING	2-33
TAXIING	2-10	AFTER PARKING	2-34
BEFORE TAKEOFF	2-10	FUEL DUMPING	
NORMAL TAKEOFF	2-11	MODEL Ⓒ	2-34
AFTER TAKEOFF	2-13	MODEL Ⓕ	2-35
CLIMB	2-14	CONDENSED CHECKLIST (Deleted) Issued in Card Form Only	
CLIMB CHECK	2-14		

STATUS OF THE AIRPLANE

FLIGHT RESTRICTIONS

Refer to Section V for all operating limitations.

FLIGHT PLAN

Accomplish the following:

1. The necessary fuel, airspeed and power settings required to complete a proposed mission should be determined by using the operating data from the Appendix. Proper charts should be prepared in order to keep a running check during flight of the fuel and oxygen remaining, distance covered and other important variables.

2. Ascertain which type of tactical equipment is installed for the mission and check any special operating instructions or techniques.

3. Check that the required communications channels and radio equipment will be available for the proposed mission.

WEIGHT AND BALANCE

Refer to Handbook of Weight and Balance T.O. 1-1B-40 for weight and balance data. Also refer to Section V for additional information.

WARNING

Do not attempt takeoff without a tactical load or ballast equivalent installed in the equipment bay. Verify that the aircraft loading falls within established limits.

Changed 6 November 1967

2-1

ENTRANCE

Entrance to the airplane is gained from the right side of the fuselage. Use of an external ladder is necessary. The canopy can be unlocked from the outside of the aircraft by utilizing the exterior canopy locking handle.

BEFORE ENTRANCE INSPECTION

PILOT EQUIPMENT INSTALLATION

1. If the flight is to be made to altitudes in excess of 45,000 feet, a partial pressure suit must be worn. The airplane is equipped to accomodate a partial pressure suit and helmet. The pilots oxygen and electrical disconnects are special types and are matched to the seat pack and facepiece. If the maximum flight altitude will not exceed 45,000 feet, a suitable flying suit may be worn instead of the partial pressure garment. When the partial pressure garment is not worn, the normal seat pack with suitable adapters will allow use of either a P-4 or HGU-2/P flying helmet and the MBU-5/P or A-13 oxygen mask as described in Section IV.

NOTE

Uncomfortable breathing oxygen pressures and garbled radio transmissions will be experienced with use of the MBU-5/P or A-13 oxygen mask unless a pressure reducing valve is used between the mask and seat pack.

2. The lower cavity of the ejection seat is ordinarily filled with the special seat pack which is an integral part of the partial pressure suit. Some pilots find their comfort on long missions materially increased by placing wooden blocks underneath the front of the seat pack. This changes the angle of the seat pack to provide support to the upper legs.

3. The ejection seat does not afford a height adjustment to accomodate individual pilots; therefore, precut plywood sheets or some suitable substitute should be placed under the seat pack to provide for correct seating height of the pilot flying the aircraft.

WARNING

The ejection seat height adjustment to accomodate individual pilots must be made with noncompressible material such as precut plywood sheets. The chance of vertebral injury is increased considerably when the pilot sits on a thick, compressible mass, such as soft cushions placed on top of or under the seat pack. When such compressible items are used, the seat will not exert a direct force on the pilot until it has moved 2 or 3 inches. After this amount of travel, the seat has gathered such momentum that excessive impact is produced when the seat initially lifts the pilot. The chance of injury is also increased during a crash landing.

4. The seat pack should be installed in the airplane by a personal equipment specialist.

5. A back type parachute must be worn.

6. When flights are conducted in certain remote areas it is sometimes advisable to provide the pilot with additional survival equipment not contained in the seat pack. If this equipment is installed, check to ascertain that all required items are present in the aircraft and properly secured.

PREFLIGHT

The preflight inspection by the flight crew is based on the fact that maintenance personnel have completed the maintenance preflight contained in the Handbook of Inspection Requirements and the airplane has been cleared for flight by authorized maintenance personnel prior to takeoff. Preliminary exterior inspection and Preliminary Cockpit Check are normally performed by a qualified assistant prior to the pilot's arrival at the aircraft. This procedure is dictated by the limitations placed upon the pilot's movements in the partial pressure suit. The exterior inspection of the aircraft is an acceptance check for flight with emphasis on the items which affect safety. Attention will be directed toward checking hatch security, engine intakes, wings, flaps, tail section, fuel caps, leaks, loose panels, and airplane general condition.

PRELIMINARY COCKPIT CHECK

1. Form 781 - Checked for aircraft status and servicing.

2. Seat ejection system - Check.

    a. Seat ejection D-ring secured in full down position by D-ring retainer spring.

NOTE

If any discrepancy is noted during inspection of this system, do not enter the airplane until the system is checked by a maintenance technician.

   b. Seat ejection safety pin installed.

3. Canopy jettison T-handle safety cover installed.

4. Appropriate FLIP - Checked and current.

5. Cockpit fan - Condition checked.

6. Aircraft call sign and data cards installed and current.

    The following data cards will be displayed in the cockpit:

    Climb Schedule - Altimeter, Airspeed, EPR, and Heading Indicator correction cards.

7. Oxygen quantity - 1800 psi minimum.

8. Fuel totalizer - Set.

9. Master watch - Installed (if applicable).

10. Equipment bay altimeter - Set 29.92.

11. Driftsight - Checked for operation.

12. Sextant - Checked for operation of all mechanical components.
    (Check lighting if required on mission.)

13. Sextant switch and dim rheostat - Checked.

14. Cockpit lights and landing lights - Checked (if applicable).

15. Canopy manual release handle - Stowed properly.

PRIOR TO BOARDING AIRCRAFT

When the pilot arrives at the aircraft, qualified personnel will be available to assist him in performing checklists up to engine start.

Changed 1 September 1968

SECTION II AF (C)-1-1

NOTE

In hot or extremely cold weather, every effort should be made to keep the pilot at a comfortably cool temperature prior to starting engine.

1. Form 781 - Checked by pilot and Part 2 signed.

2. Shoulder straps and safety belt - Fully extended to facilitate pilot hookup.

3. Relief bottle - Installed.

4. Seat blocks and back cushion - Installed as required.

5. Seat pack - Checked and installed.

6. Seat pack oxygen system - Flow rates to mask and pressure readings to mask and suit within tolerance.

7. Seat pack quick disconnect and connections - Recheck properly connected to aircraft systems.

8. Quick disconnect - Check for locking of manual release plunger knob. Advise crew chief if separation occurs.

9. Check aircraft oxygen supply - 1800 psi minimum.

10. Check oxygen supply in seat pack - 1800 psi.

11. Parachute - Fitted to pilot.

12. Underarm preserver - Flotation kits pulled through parachute sling assembly and positioned outside the parachute harness. (Applicable to overwater flight only.)

13. Partial pressure suit capstan and breathing bladder hoses - Extended and free of parachute harness. (Applicable for high flight only.)

PILOT EQUIPMENT CHECK

A qualified personal equipment technician will be available to assist in the hookup and checking of these items.

PILOT EQUIPMENT CHECK - HIGH FLIGHT

1. Seat pack strap, left - Attach to parachute, adjust.

2. Seat pack strap, right - Attach to parachute, adjust.

3. Capstan hose - Connect to seat pack suit hose and install safety clip.

4. Bladder hose - Connect to seat pack mask hose and install safety clip.

5. Facepiece hose - Connect to seat pack mask hose T-block and install safety clip.

6. Electrical leads - Connect Viking plug to seat pack lead and lock.

7. Shoulder harness - Straighten and connect to lap belt.

8. F-1 release key - Connect to lap belt.

9. Lap belt - Connect and lock, adjust.

10. Secondary oxygen valve - Turn full on (leave valve loose).

11. Facepiece - Attached to helmet and locked.

12. Oxygen red warning light - Check OXY WARN LOW PRESS out.

13. Primary oxygen valve - Turn full on (leave valve loose).

14. Oxygen warning lights - Check OXY WARN PRI OFF (amber) and OXY WARN LOW PRESS (red) off.

15. Radio check - Request pilot to check radio.

16. Radio bypass cord - Under left half of lap belt.

17. Face heat - Turn rheostat half on.

18. Face heat EMER FACE HEAT circuit breaker - Pulled. Feel facepiece for heat.

19. Press-to-test - Check for oxygen leaks and for suit inflation in 7 seconds or less.

20. Low altitude lanyard - Connect to parachute. (D-ring or T-handle.)

21. Oxygen supply - Recheck aircraft and emergency bottle pressure.

22. Quick disconnect - Recheck for fully locked position.

23. Suit ventilation system - CONNECTED and ON (if applicable).

PILOT EQUIPMENT CHECK - LOW FLIGHT

1. Seat pack strap, left - Attach to parachute, adjust.

2. Seat pack strap, right - Attach to parachute, adjust.

3. Shoulder harness - Straighten and connect to lap belt.

4. F-1 release key - Connect to lap belt.

5. Lap belt - Connected and locked, adjust.

6. Low altitude lanyard - Attach to parachute. (D-ring or T-handle.)

7. Capstan hose - Capped under lap belt.

8. Low altitude pressure reducer - Attach to plate on parachute and lock.

9. Seat kit mask hose - Connect to low altitude adapter.

10. Oxygen mask hose - Connect to low altitude adapter.

11. Radio adapter cord - Connect to seat kit plug and lock.

12. Helmet radio cord - Connect to adapter cord.

13. Radio bypass cord - Under left half of lap belt.

14. Oxygen flow - Pilot check primary and secondary - ON.

15. Radio - Request pilot to check radio.

16. Oxygen - Recheck aircraft and emergency supply.

17. Quick disconnect - Recheck for fully locked position.

PILOTS COCKPIT CHECK

It is important that the cockpit check be carefully completed in sequence with no items left until later. Once the engine is started, the airplane is ready for takeoff and close attention is required to flying the aircraft during the early climb. The pilot's action is shown in capital letters.

1. Flight plan - ON REQUEST.

2. Attitude indicator - CAGED.

3. Emergency inverter - CHECKED. (Check inverter light out, attitude indicator OFF flag not showing.)

4. Main inverters - CHECKED.
   (Check inverter light out, attitude indicator OFF not showing.)

5. Inverter switch - No. 1 SELECTED.

6. Compass power switch - EXTERNAL.

7. NAV POWER switch - NO. 2 INV (if applicable).

8. NAV POINTER transfer switch - as required.

9. ASN-66 function switch - TEST (if applicable).

10. Doppler function switch - TEST (if applicable).

11. Autopilot power switch - ON.

12. Navigation aids - ON.

13. Fuel dump circuit breakers - IN.

14. Fuel dump switches - CLOSED (Covers down, lights OFF).

15. Seal valves - OFF.

16. Fuel transfer switch(es) - OFF.

17. Main tank pressure switch - REPRESS. (F)

18. Landing lights - OFF.

19. Emergency fuel shutoff - COVER DOWN.

20. Gust control - FAIRED.

21. Oxygen valves - CHECKED.
    Check both oxygen valves on and finger tight.

22. Wing flap switch - UP.

23. Friction lock - SET.

24. Throttle - OFF. (Gate CLOSED.)

25. Speed brake switch - IN.

26. Slipper pump switch - OFF.

27. Drag chute handle - IN.

28. Oxygen pressure - CHECKED.
    (Minimum for takeoff - 1700 psi.)

29. Face heat - SET.
    Rheostat control is normally set from 9 to 12 o'clock.

30. Cabin heat selector - AUTO.

31. Cabin temperature - SET.
    Rheostat control is normally set from 9 to 12 o'clock for takeoff.

32. Landing gear handle - CHECKED.
    Check indicators for down and locked position. Check gear handle warning light and horn for operation.

33. Ram air switch - COVER DOWN.

34. Emergency fuel switch - NORMAL.

35. Master watch and aircraft clock - CHECKED.
    Check operation of aircraft clock, master watch set for proper time.

36. Pitot heat - OFF.

37. Fire and overheat warning circuit - CHECKED.

38. Boost/Ignition switch - OFF.

39. Battery and generator switch - OFF.

40. AC generator - AS REQUIRED.
    If AC generator power is required in flight, turn switch ON at this time.

41. NAV light switch - OFF.

42. Hatch heater switch - OFF.

43. Cabin and equipment bay altimeter - SET - 29.92.

44. Emergency face heat circuit breaker - PULLED.

45. All other circuit breakers - SET.

46. Fuel totalizer - SET.

47. Bleed valve - CLOSE.

48. AUX TANK PRESS switch - OFF if there is no fuel in auxiliary tanks. ON if there is fuel in auxiliary tanks.

49. Panel and Master Caution lights - CHECKED. (Master Caution light out prior to check.)
Check for illumination of lights in annunciator panel, master caution switch, fuel dump indicator panel and appropriate system lights.

50. Stall strip handle - IN and LIGHTS ON.

51. Compass system - CHECKED.
Compass control panel:

    a. DG-SLA-COMP mode switch in SLA position.

    b. SYNC IND - Check needle centered.

    c. LAT - Set for proper latitude and hemisphere when operation in DG mode is planned.

    d. Check BDHI for proper heading against standby compass.

    e. BDHI OFF flag not showing.

52. Trim power switch - ON.

53. Autopilot - CHECKED.

    a. Move control yoke aft of neutral, wait 6 seconds and check elevator trim index centered. Engage autopilot. Engagement should be smooth.

    b. Rotate pitch command knob fore and aft, and note that yoke follows.

    c. Rotate turn knob right and left and note that wheel follows.

    d. Overpower with nose up yoke - elevator trim tab should slowly move in nose down direction.

    e. Overpower with nose down yoke - elevator trim tab should slowly move in the nose up direction. Press AUTO TRIM FAIL TEST switch while overpowering. Trim tab should stop moving and AUTO TRIM FAILURE light should illuminate. Release overpower.

    f. Press AUTO TRIM FAILURE light switch to select MANUAL trim.

    g. Auto trim light should go out.

    h. Check manual trim operation with yoke trim switch.

    i. Select auto trim.

    j. Engage Mach, move Mach trim fore and aft and note that yoke follows. After initial movement, the yoke may continue to creep.

    k. Disengage autopilot using yoke disengage switch. Autopilot disconnect light and Master Caution light should glow and a dual frequency tone should be heard in the headset.

54. Elevator trim tab indicator - 2 DEGREES NOSE DOWN.

55. Equipment master switch - OFF.

56. Equipment mode selector switch - OFF.

57. Heater-blower defroster - OFF.

58. Altimeter selector switch - EQUIP BAY ALT.

59. Defroster - OFF.

60. Instrument and panel lights - SET. Rheostats should be turned OFF for VFR day flights.
For night or instrument operation, set as desired.

SECTION II   AF (C)-1-1

61. HF XCVR ALT switch - NORMAL.

62. Auxiliary lights - SET.

63. Rudder pedals - SET.
    Adjust rudder pedals while crew chief holds rudder neutral.

64. Aileron trim - SET.

65. Flashlight - CHECKED (if required).

66. Mission special equipment - CHECKED.
    Preset as briefed or as required from appropriate equipment checklist. Press-to-test lights of applicable special equipment.

67. Oxygen quick disconnect - CHECKED.

68. Suit connections - CHECKED.

69. Low altitude escape lanyard - CHECKED.

70. F-1 release - CHECKED.

71. Facepiece latch - CHECKED.

72. Emergency face heat cord - CHECKED.
    The assistant should insure that both quick disconnects are properly connected.

73. Green apple - CHECKED.
    Check for evidence of having been inadvertently pulled.

74. External rear view mirror - CHECKED.

75. ADF and VOR - CHECKED.
    Check all functions of ADF and VOR.

76. Canopy handle - DOWN.

77. Canopy - CLOSED AND LOCKED.

78. Canopy jettison handle cover - REMOVED.

START ENGINE

The Start Engine and After Start checklists will be performed by the pilot without the use of an assistant. The canopy will be closed and locked prior to engine start because of the possibility of engine damage if loose items should be drawn into the air intake.

1. Brakes - HOLD.
   Hold brakes until ready to taxi.

2. Starting unit - SIGNAL START.
   Give windup signal (circular motion of index finger to start auxiliary starting unit.)

3. Starting air - SIGNAL ON.
   After starting unit reaches full RPM, raise hand with thumb up to signal crew chief that you are ready for starting air.

NOTE

During start observe emergency fuel system light. Light should go out between 3 to 10% RPM. If light is not out by 10% RPM, discontinue start and ground check system.

4. Ignition Switch - START AND HOLD (at minimum of 12% RPM).

5. Throttle - IDLE.
   Place throttle in idle when engine RPM reaches 18%. The EGT indication will be very slow.

CAUTION

Although the desired start RPM is 18% pneumatic starter limitations must be observed. The throttle may be moved to idle at a minimum of 16% RPM if necessary.

If normal start is not indicated by a rise in EGT and/or oil pressure within 30 seconds after the throttle is placed in idle, release ignition switch, return throttle to OFF, and signal starting air OFF. Allow excess fuel to drain from engine prior to attempting next start.

If EGT exceeds 400° C during start release ignition switch, throttle OFF, and allow starting air source to rotate engine for 30 seconds.

6. APU and starter air - SIGNAL DISCONNECT.
   At 35% RPM, release ignition switch. Signal with thumb out to disconnect APU and starter air.

7. Generator-Battery Switch - BAT & GEN.
   Place in BAT & GEN position after auxiliary electrical power is disconnected.

AFTER START

1. Idle RPM - CHECKED.
   Idle RPM will be approximately 45%.

2. Engine instruments - CHECKED.
   Check that all engine instruments and hydraulic pressure gage indicate within normal operating range.

3. Canopy and equipment bay seal valves - ON.

4. Pressurization - CHECKED.
   Check for decrease in cabin/equipment bay altimeter of at least 300 feet.

5. Flight instruments - CHECKED.

   a. Altimeter set to current station altimeter setting and indicating within 75 feet of station altitude.

   b. Vertical speed indicator indicating zero.

   c. Airspeed indicator indicating approximately 30 knots.

   d. BDHI compass indication corresponds approximately with the magnetic compass. Compass synchronization needle centered. No off flag.

   e. Turn and slip indicator centered.

   f. Attitude indicator erected and miniature aircraft set level with 90-degree indices on the sides of the case.

6. Flight controls - CHECKED.
   AILERON FAIRED light on.
   Check ailerons and elevators for freedom of movement and full range of travel. Rudder movement and range of travel will be restricted with airplane stationary due to tail wheel steering.

7. Generator-Battery - CHECKED.

   a. Select BAT-EMERG position and check battery for electrical output by noting proper operation of inverter and AC instruments.

   b. Select BAT & GEN position and check ammeter for normal indication of generator electrical output. (50 to 100 amperes.)

8. Wing flaps - UP.
   Check that wing flap position indicator reads zero.

9. Fan - ON.
   Normally, the HI position will be used

10. ATC - STBY.

11. HF Radio - ON (if required).

12. NAV light switch - ON.

13. Master Caution light - OUT.

14. Autopilot Power switch - ON.

15. Ejection seat safety pin - REMOVED. (Inform mobile.)

BEFORE TAXIING

This check will be performed by a qualified assistant.

1. Nephographic window covers - REMOVED.

2. Landing gear pins - REMOVED.

3. Engine access doors - CLOSED.

SECTION II   AF (C)-1-1

4. Navigation lights - CHECKED.

5. Wheel chocks - REMOVED.

6. Route of taxi - Check clear and signal pilot clearance for taxi.

TAXIING

A normal brake check should be made during initial taxi roll.

The aircraft can be taxied normally if sufficient space is available. The minimum turning radius with pogos installed is approximately 300 feet in zero wind. Winds up to 30 knots from any direction can be compensated for with slight increase in turning radius. Because of the large turning radius and large wing span, care must be exercised to ensure that turns are started with sufficient space available.

If the situation requires, the aircraft will be lined up for takeoff by the ground crew.

Ground and taxi time should be held to a minimum to reduce the amount of fuel used before flight.

Taxiing at light weights over uneven surfaces or at too fast a speed will cause the pogos to fall out of their sockets if the safety pins have been removed. With heavy weights, taxi at reduced speed and exercise caution to avoid possible rough areas. Also extreme caution must be exercised when taxiing on icy surfaces.

BEFORE TAKEOFF

Special effort should be made to ensure that a comprehensive pretakeoff check is performed. Due to the fast acceleration, short takeoff roll, and steep climb angle, insufficient time is available after takeoff to accomplish checks which might have been neglected prior to beginning the takeoff run.

This portion of the checklist will be performed with the aircraft at a complete stop. It may be accomplished on the runway or, all items up to and including No. 9 just prior to taking the runway.

1. Canopy - CLOSED AND LOCKED. Recheck canopy closed, locked and with seal valves ON.

2. ATC - As required (Mode and Code set).

3. Trim - SET.

4. Pitot heat switch - AS REQUIRED.

5. Boost/Ignition switch - ON.

6. Hatch heater switch - ON. (If applicable.)

7. Bleed valves - AUTO. (Bleed valves open lights - ON.)

8. Equipment master switch - ON. (If applicable.)

9. Special equipment - As briefed.

10. Pogo locking pins - REMOVED. Crew chief will remove pins and display to mobile and ascertain that locking plunger is flush on underside of wing.

Changed 1 September 1968

11. Shoulder harness - LOCKED.
    Move locking handle forward to lock shoulder harness.

12. Takeoff clearance - OBTAINED.

13. Engine instruments - CHECKED.
    Recheck all engine and flight instruments for proper indications.

NORMAL TAKEOFF

Line speed and refusal speed computation and utilization are not normally feasible with this aircraft due to the rapid acceleration, extremely short takeoff run and preoccupation of the pilot with directional control. However, the tactical situation might require the use of takeoff data for very short runways.

Advance the throttle to 80% RPM. Release the brakes.

The brakes must not be used to hold more than 80% RPM or landing gear failure may result. Engine RPM increase may lag behind throttle advancement due to slow engine acceleration. Care must be used to avoid overshooting desired RPM.

Continue throttle advance to the gate. This will give approximately 93%-94% RPM. The throttle gate stop is important because the automatic sensing devices have been removed from the fuel control and the stop protects against overspeeding and overtemping the engine.

When using the gate stop, it is not necessary to closely monitor engine RPM and EGT during takeoff.

Full throttle takeoff should be avoided since the engine is fuel flow limited at sea level when operating at maximum power. The throttle will have to be retarded to control the EGT to 630°C and/or the RPM to 100% at about the time the airplane becomes airborne. Full attention must be paid to the control of the aircraft.

Hold control column neutral during initial takeoff roll, maintaining directional control with tail wheel steering and rudder. At approximately 50 knots the wings will start to rise and allow the pogos to fall off. When this happens use care to keep the wings level.

If one wing is allowed to rise prematurely, directional control difficulties will be aggravated. Mobile control will inform the pilot when pogos are clear of the aircraft. This procedure will enable the pilot to better concentrate on the remainder of the takeoff.

At approximately 70 knots the tail will start to rise. At this point a slight increase in back pressure may be required to maintain this pitch attitude. The desired attitude is a near two-point attitude which will allow the aircraft to become airborne when takeoff airspeed is reached.

Takeoffs will normally be made with gate power. The technique is the same for both light and heavy fuel loads. With a light load, the acceleration to takeoff speed will be faster, and a more concerted effort must be made to keep the aircraft in the near two-point attitude.

It is possible to develop a porpoise on takeoff. The usual cause is the airplane becoming airborne without the pilot realizing it and the control column being pushed forward, or rapid movements being made with the control column just prior to leaving the ground. If a porpoise should develop on takeoff, bring the control column back easily to keep the airplane in the air. Trying to fight the porpoise by pumping the column back and forth will only aggravate the condition.

If the yoke is allowed to move forward of the neutral position, the tail will rise rapidly to an abnormally high position. If this occurs, the aircraft will be in the first stage of a porpoise. Therefore, porpoise recovery procedures must be used to prevent the aircraft from bouncing back onto the runway.

Because of the rapid acceleration of the U-2, the importance of maintaining the proper pitch attitude for takeoff cannot be overemphasized.

The recommended takeoff speeds allow 15% margin over stall speeds and are as follows:

FUEL (Gal)	TAKEOFF SPEED (Knots)
400 - 600	95 - 98
600 - 800	98 - 100
800 - 1100	100 - 105
1100 - 1345	105 - 110
1345 - 1545	110 - 115

On heavy weight takeoffs do not assume the very steep climb angle to which you may be accustomed from flying at light weights, since an accelerated stall may be encountered under these conditions.

NOTE

With light fuel loads, lining up on the crown of the runway may result in one pogo falling out prior to start of the takeoff run. In this event, takeoff can be made with one pogo by holding that wing down until ample aileron control is assured.

Although pogos are normally installed for all takeoffs, an alternate procedure may be employed using a hand launch in lieu of pogos. This procedure should be limited to fully checked-out pilots. Proper fuel balance is essential for this procedure.

CROSSWIND TAKEOFF

In addition to the procedures given for normal takeoff, the following steps are included for crosswind takeoffs.

1. Raise tail slightly to preclude the possibility of becoming airborne before full control is gained. With a 20-knot crosswind component, full rudder will be required to maintain a straight line prior to leaving the ground.

2. Hold the upwind wing down slightly.

3. Counteract drift when aircraft becomes airborne by skidding with rudder and holding wing low into wind. This will enable you to hold heading in case gusty air causes unintentional contact with runway.

NOTE

Only a slight bank is required to compensate for a yaw, so there is no danger of a wing tip striking the ground while correcting for drift with rudder.

NIGHT TAKEOFF

Night takeoff procedures are the same as those employed for daylight takeoff.

AFTER TAKEOFF

CLIMB SPEED

The airplane accelerates rapidly. Close attention is required after takeoff to hold the speed to the proper climb speed of 160 knots IAS.

WARNING

With full power, the aircraft will quickly accelerate past the gust control FAIRED placard speed of 220 knots if attention is diverted after takeoff. The airspeed limitations must be observed or structural failure may result. The speed warning system will sound the warning horn at approximately 220 knots IAS.

LATERAL TRIM

After stabilizing at climb speed for a few minutes, it may be necessary to adjust aileron trim tab to relieve any wing heavy condition. This should only be done with full fuel loads. With partial fuel loads, if it is suspected that the fuel load may not be evenly balanced, the aileron tab should not be used. In this event, the fuel cross transfer system should be used to alleviate the wing heaviness.

Once the aircraft is in trim, note position of the wheel indicator. During the remainder of the flight, use fuel cross transfer, if necessary, to trim the aircraft.

TURBULENT AIR

If turbulence is encountered, or anticipated, move gust control to GUST position and maintain climb speed. Otherwise the gust control should be left in FAIRED position for the climb.

Upon reaching smooth air conditions, return gust control to FAIRED position.

In shifting from FAIRED to GUST, an abrupt nose-up trim change occurs. This may be compensated with moderate elevator force until the elevator trim tab action catches up. In shifting from GUST to FAIRED, a nosedown trim change occurs.

Actuation of the gust control from GUST to FAIRED should be done at an indicated airspeed of 150 knots or less.

Always monitor the wing flap position indicator when shifting from GUST to FAIRED. If flaps do not stop at FAIRED, return switch to the GUST position.

CLIMB

ENGINE OPERATION

Maximum allowable engine power of 630°C EGT can be used for climb if desired up to 40,000 feet, but in order to avoid engine roughness, banging, or flameout between 40,000 and 60,000 feet it will necessary to reduce power to 485°C maximum EGT. If 630°C EGT is used up to 40,000 feet, almost constant manipulation of the throttle will be required.

Reducing the engine power to 540°C for takeoff and initial climb will simplify throttle management and will provide the pilot additional freedom for the conduct of the climb.

When power is increased at 60,000 feet, slowly advance the throttle to an EGT of 500°C, or limit maximum EPR and hold this throttle position. The EGT will then come up in a short time due to increase in altitude. Throttle adjustments will be required to maintain EGT and engine pressure ratio within prescribed limits, and to avoid airframe buffet in climb and turns. (See Section V for engine operation limits.)

When the OAT is extremely cold; i.e., -50°C or colder, it is advisable to wait until an altitude of 62,000 feet is reached before advancing throttle from the 485° setting. This will minimize the possibility of a flameout.

AIRCRAFT ATTITUDE

The aircraft climb attitude with maximum power is very steep up to 20,000 to 25,000 feet. Close attention is required to hold the proper climb speed of 160 knots from ground level to 50,000 feet. Above this altitude, the climb speed gradually decreases to 92 knots at 74,000 feet.

CLIMB CHECKS

INITIAL CLIMB CHECK

1. Landing lights - OFF.
2. Landing gear - UP, INDICATORS CHECKED.
3. Climb speed - 160 K.

NOTE

If the fuel low level light comes on during or shortly after takeoff, it may be due to the fact that at high power, and low altitude, the engine can use fuel out of the sump tank faster than it can be replenished from the wing tanks. Note the totalizer reading, throttle back, stay in the area of the field. If the fuel low level light does not go out after 20 more gallons have been used, execute emergency procedure in accordance with Section III.

4. Engine and flight instruments - CHECKED.

    Check all engine and flight instruments and annunciator panel for proper indications.

5. Bleed valve lights - OUT.
6. Aileron wheel position - NOTE.
7. Defroster - ON.
8. ATC - CHECKED.

NOTE

As soon after takeoff as flight conditions permit, positive operation of the ATC should be established with an Air Traffic Control facility if the flight route will require an operative ATC.

9. Low altitude escape lanyard - UNHOOK AND STOW at 10,000 feet MSL.

WARNING

This check is of utmost importance on high altitude missions, since ejection at high altitude with the lanyard connected would result in excessive parachute opening shock and prolonged exposure.

10. Special equipment - ON as briefed.

18,000-FOOT CHECK

1. Altimeter - SET AT 29.92.
2. Pressurization - CHECKED. (Approximately 10 M cabin.)

40,000-FOOT CHECK

1. Oxygen quantity - CHECKED.
2. Oxygen system - PRESS-TO-TEST.
3. Pressurization - CHECKED (Cabin altitude approximately 20,000 feet).
4. Main pressure switch - NORMAL (if fuel is carried in auxiliary tanks). Ⓕ
5. Autopilot - CHECKED.
   At pilot's discretion, engage autopilot. Check autopilot trim and flight operation during climb between 40,000 and 45,000 feet.

WARNING

Prior to engaging the autopilot, the airplane should be trimmed near hands off. Do not engage autopilot if either trim index shows full deflection.

Prior to engaging or disengaging autopilot, the autopilot trim indexes should be observed for near neutral position. Minor oscillations may be present but should present no problem. If trim indexes are not near neutral, transients should be anticipated.

WARNING

Subcomponent failures can occur in the autopilot system which can render any one axis or all three axes inoperative without benefit of the MASTER CAUTION light, the AUTOPILOT DISCONNECT warning light, or the dual frequency audio tone in the headset. Do not rely exclusively on these indications to determine that the autopilot is engaged and controlling properly.

55,000-FOOT CHECK

1. Gust control - FAIRED.
   If initial climb was made in GUST.
2. NAV light switch - OFF.
3. Bleed valves - CLOSE WHEN BLEED VALVE LIGHTS ILLUMINATE. (For maximum range reduced power operation) observe minimum EPR limits for bleed valves closed.
4. Autopilot - ON.
5. Mach Hold - ON (at 60,000 feet).
6. Equipment mode selector - STANDBY. (If applicable.)
7. Special equipment - ON AS BRIEFED.
8. Pressurization - CHECKED.
9. Slipper pump switch - ON as required.

WARNING

Do not operate slipper pump until fuel load is below 800 gallons in U-2C or 300 gallons in U-2F.

CRUISE CLIMB

The cruise climb will start at an altitude of 67,000 to 69,000 feet, except for the maximum range mission which starts at 57,000 to 59,000 feet. The cruise climb comprises the major portion of the total flight time and is flown in a manner to obtain maximum performance.

Changed 1 September 1968

AIRCRAFT CONTROL

Normally, the autopilot with mach hold is used to control the aircraft above 60,000 feet. If turbulent conditions are encountered engagement of the Mach hold may be delayed. The desired cruise climb airspeed schedule is not quite a constant Mach number; therefore, the Mach sensor will not control the airspeed exactly.

WARNING

Flight instruments, trim indicator, and autopilot trim indices should be frequently scanned to insure that the desired flight attitude and airspeed is being maintained.

Use the Mach trim wheel as necessary to maintain the proper indicated airspeed.

It is very important to fly the recommended climb speed schedule. At higher altitudes when in a maximum power cruise climb the aircraft must be flown between very narrow speed limits just under the Mach buffet limit. See figures 6-2, 6-3, and 6-4. At speeds 1 to 3 knots faster than the standard climb speed schedule, typical Mach effects of Mach buffet or nose tucking may be encountered. If any of these symptoms are encountered, and if the IAS is at or above the standard speed schedule, reduce speed no more than 2 knots below climb speed schedule and continue climbing. Stall buffet or stall will be encountered at speeds 5 to 8 knots slower than the standard climb speed schedule when flying in a maximum power, maximum altitude cruise climb. If buffeting is encountered and IAS is below the standard climb speed schedule, increase the speed to climb speed schedule.

WARNING

During maximum power cruise climb, it is dangerous to allow the aircraft to approach speeds of 5 knots or more below the standard climb speed schedule, because a stall can result. Recovery from a stall at altitude may be extremely difficult due to lack of aileron response when the wing is stalled.

High speed Mach buffet and low speed stall buffet feel similar. Aileron response is good in Mach buffet and very poor in the low speed stall buffet. The airspeed indicator must be relied on to aid in determining whether buffeting is due to high or low speed. If there is any question regarding the accuracy of the airspeed indicator, or the ability to discern whether buffet is Mach or stall buffet, always test by increasing airspeed. Never test by reducing airspeed and risk stalling. The aircraft can be flown safely into buffet 4 knots (as shown on figure 6-3) in order to establish positively that buffet is high speed Mach buffet and obtain a check on the airspeed indicator. The severity of buffet may increase to moderate or "heavy buffet" when making this check. Also, a tucking or pitchdown tendency may develop, requiring a back stick force of 10 to 20 pounds to prevent the speed from building up.

Increasing the speed more than 4 knots into the Mach buffet region will increase the stick forces (tuck) and may result in wing rolloff. Do not penetrate the Mach buffet more than the amount stated, as severe compressibility effects will be encountered.

WARNING

If on autopilot when making the above checks, hands should be on the wheel in the event of possible clutch slippage. If the autopilot is disconnected, stick forces must be maintained to avoid rapid pitchdown.

If the aircraft cannot be flown without Mach buffet, reducing power slightly will provide relief. The aircraft will descend several hundred feet when power is reduced. Power should be restored as soon as the aircraft can be flown without buffet to obtain maximum altitude. If the aircraft cannot be flown 2 knots below standard speed schedule without high speed buffet, the airspeed/altimeter systems should be thoroughly checked after the flight.

NOTE

Minor random buffet disturbances due to either transient autopilot operation of the Mach sensor or clear air turbulence should not be a cause for reducing 2 knots below schedule airspeed.

At high altitude the "g" capability is limited because the aircraft is operating at high lift coefficients. Under some conditions buffet will be encountered at 1.1 g. In a turn at bank angles in excess of the standard 12° "bug" turn it is important to maintain airspeed and tolerate Mach buffet to avoid stalling.

When operated at maximum allowable engine power, the airplane flies at a combination of altitude and wing lift coefficient that affects the flying characteristics. It can be flown manually, but not for extended periods of time since it requires the pilot's full attention. Therefore, a properly functioning autopilot is necessary to satisfactorily accomplish the mission.

Fueled slipper tanks also have an adverse effect on the flight characteristics during cruise, particularly at maximum power and heavy weight. With slipper tanks it is absolutely essential that the autopilot be operating at peak efficiency for satisfactory mission performance.

In the event the flight must be continued without autopilot, it may become necessary to reduce power and continue cruising at a lower altitude where the airplane is easier to fly.

The best cruise performance is obtained by flying the speed schedule as precisely as possible. If the speed is too fast, the airplane will not climb at the proper rate. This can increase fuel consumption and decrease range. The optimum procedure is to climb, gradually bleeding airspeed so as to pass through each altitude with the airspeed specified for that altitude. This can be accomplished by interpolating the speed schedule for altitude increments of 500 to 1000 feet. If the autopilot does not control the airspeed with reasonable accuracy, the expected performance will not be realized. This should be reported for corrective action.

ENGINE OPERATION

The engine may be operated at maximum permissible power throughout the cruise climb. The primary instrument used for power setting is the exhaust gas temperature gage except as limited by maximum allowable engine pressure ratio. Engine speed may also be used for cross checking. See Section V for engine operation limitations.

At times, the EGT indication can be slow in responding to small cruise power adjustments and engine RPM may be more responsive to the throttle movement. If this occurs, the power can be adjusted in small RPM increments, pausing between each adjustment to allow time for the EGT to respond and stabilize. An RPM change of 1/4% can be used to approximate a 5° EGT change.

Changed 1 September 1968

SECTION II AF (C)-1-1

It is possible to vary the engine power from the minimum EPR/EGT up to maximum permissible EPR/EGT. On many types of missions this will be required. However, when accelerating the engine from lower powers, the rise in engine speed may not follow the throttle movement. The throttle should be moved forward very slowly and the engine allowed time to catch up, otherwise a flameout may occur.

ALTITUDE CONTROL

The P-13 engine will normally never reach an altitude where idle RPM is encountered. However, maximum altitude is limited to the point where maximum EGT is reached and power cannot be further reduced without dropping below minimum EPR requirement.

When this condition occurs, prompt action must be taken to prevent overtemperature operation of the engine. Initially, the condition may be corrected merely by allowing the aircraft speed to increase while maintaining constant altitude. When the speed has increased to the needle, aerodynamic drag should be added. The initial drag is created by using the variable speed brakes or extending the landing gear. As the aircraft descends, engine power is increased to maintain engine operation at maximum limitations. Eventually, the aircraft will cease descent and again begin a cruise climb. When the aircraft again reaches the altitude of limiting condition, the remaining drag device will be used.

AIRSPEED CONTROL

The airspeed schedule must be followed closely to obtain maximum range and altitude. One knot indicated airspeed at 70,000 feet is equivalent to four knots true airspeed. At this altitude the climb schedule is approximately 1 knot IAS below the buffet boundary of the aircraft. Close attention must be given to pitch and altitude control.

A precise airspeed schedule is important on the U-2C and U-2F. It has been carefully selected and tested as the best compromise. At a faster speed the airplane will encounter wing buffet. At a slower speed the airplane is approaching a stalled condition. Careful attention should be given, therefore, to correction for airspeed and altimeter instrument errors. Position error has been taken into account in arriving at the recommended IAS schedule. Every attempt should be made to follow the climb speeds. They are as follows:

MAXIMUM ALTITUDE CRUISE

PRESS. ALT (Corrected for Instrument Error Only)	CLIMB SPEED IAS - Knots (Corrected for Instrument Error Only)	STD DAY TAS - Knots
50	160	391
55	145	398
60	130	403
62	124	403
64	118	402
66	112	401
67	107	394
68	105	397
69	103	397
70	99	392
71	97	392
72	95	394
73	93	394
74	92	397
75	91	402
76	89	402

MACH 0.72 MAXIMUM RANGE CRUISE

ALTITUDE	IAS
55	152
56	148
57	144.5
58	141
59	138
60	135
61	131.5
62	128
63	125
64	122
65	119
66	116
67	113
68	111
69	108
70	105

The airspeed schedule is not a constant Mach number and it is necessary to reset the autopilot Mach trim quite often, particularly during the early cruise climb. If care is not taken, it is possible to inadvertently enter a Mach buffet region by engaging autopilot Mach hold on speed schedule at one altitude, for example 66,000 feet (Mach 0.698) and neglecting to trim Mach down (to stay on speed schedule) as altitude is gained. If this happens, the autopilot will maintain the 0.698 Mach which will result in Mach buffet at 67,000 feet (see figure 6-1).

Wing buffet should not occur at these climb speeds under steady one-G conditions. Variations in airplanes and/or airspeed indicator errors may cause encounter with the buffet boundary. If buffet is encountered, reduce speed 2 knots in accordance with instructions under AIRCRAFT CONTROL and continue the climb.

This airspeed reduction should be made by slowly decreasing Mach with the Mach trim wheel until the desired airspeed is obtained. This reduction may also be obtained by disengaging Mach hold and slowly increasing pitch angle until desired airspeed is set, then reengaging Mach hold. In either case, corrections must be made slowly, to avoid increasing G-forces and subsequently increasing buffet.

If after reducing airspeed 2 knots below standard climb schedule, and in steady one-G flight, buffet is still present, a slight reduction in power and subsequent descent below optimum altitude will increase the boundary between climb schedule and buffet. The recovery to landing base should be made at maximum-range cruise altitude and airspeed if operational conditions permit.

NOTE

Buffet might occur, even though on climb schedule, under slight G-accelerations such as are produced when pulling the nose up to hold speed. In this event, speed should not be reduced but the corrections should be minimized by proper autopilot adjustment if possible.

On missions where maximum power is not used, buffet is not so critical and airspeed need not be held so precisely. The speed can be allowed to increase a few knots towards the end of the flight when the weight has been reduced. The engine is sensitive to airspeed at high altitude. Low speed may induce compressor stall and consequent flameout.

AIRCRAFT TRIM

The autopilot incorporates an automatic pitch trim feature which eliminates the need for manual pitch trim corrections when using the autopilot. The auto trim disconnects in case of a runaway tab.

NOTE

If the autopilot is overpowered by manual force in the pitch axis, the auto trim function will automatically feed in elevator tab, opposing the manual force. Therefore, the pilot should not override the autopilot in pitch longer than is necessary, otherwise the stick force may be unduly high when the autopilot is eventually disengaged.

If the aircraft becomes out of trim laterally, it is probably caused by uneven fuel feeding from the wing tanks. This situation should always be corrected by use of the fuel cross transfer system. With the autopilot engaged, wheel position or the roll trim indicator may be observed to determine when enough fuel has been transferred. The aileron trim tab should not be used to compensate for uneven wing tank fuel feeding.

NOTE

If an uneven fuel feeding condition is allowed to persist, the autopilot clutch will eventually slip and manual assistance will be required to maintain a wings-level attitude until the condition is corrected by use of the fuel transfer switch. Normally, a wing heavy condition with the autopilot engaged will be noted by the indication of the gradual rotation of the wheel from the neutral position.

CRUISE CHARTS

During the course of long missions, which may last many hours, it is advisable to keep a running check of fuel and oxygen consumption. This is necessary in order to disclose excessive consumption as early as possible so that corrective action may be taken. The simplest method is by use of a graph which shows the anticipated usage as a function of time. Examples are shown in figure 2-1. During the mission, periodically read and plot the oxygen pressure and fuel totalizer reading.

PILOT COMFORT

During the long cruise climb portion of a normal mission, pilot comfort assumes considerable importance. Pilot comfort can be substantially increased if a proper balance is achieved between cabin temperature and facepiece heat. Best results are obtained utilizing the following procedures:

1. Keep facepiece heat as low as possible without inducing fogging.

2. Use the defroster fan as required.

NOTE

At high altitude, airflow through the air conditioning system is less. The defroster control should be full open.

3. Keep cabin temperature control at warmest setting, consistent with pilot comfort, to aid in keeping feet warm and eliminating windshield and canopy frosting.

As an aid to pilot comfort, and also as a check on the oxygen system, the seat pack Press-to-Test button may be actuated at pilots discretion.

SYSTEMS OPERATION

Refer to Section VII for information regarding systems operations.

FLIGHT CHARACTERISTICS

Refer to Section VI for information regarding flight characteristics.

TYPICAL CRUISE CHARTS
FUEL AND OXYGEN VS TIME

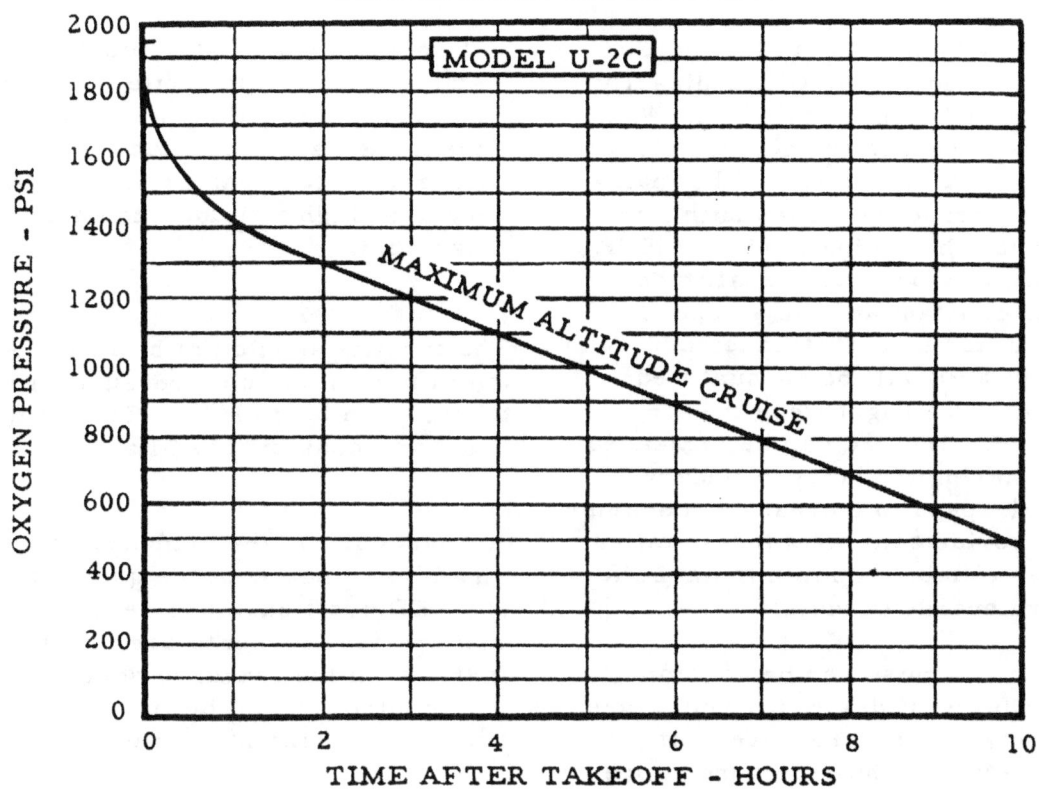

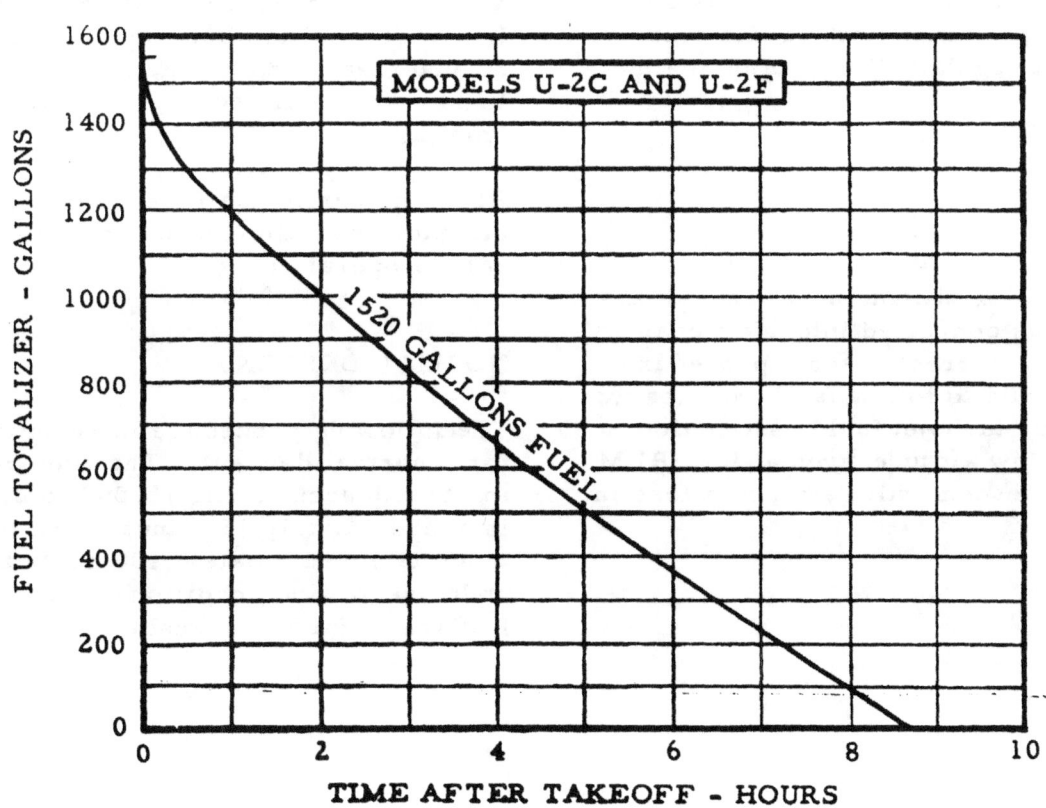

Figure 2-1

SECTION II AF (C)-1-1

LEVEL OFF

The mission may be planned to level off at a preselected altitude. This will ordinarily be done to gain an increase in range. When the cruise climb reaches the desired altitude, the schedule airspeed should be maintained by power reduction. Eventually, the power may reach the engine minimum EPR setting. When this occurs, the altitude should be held constant and the airspeed allowed to increase as fuel is consumed. Since minimum EPR will be encountered toward the end of the flight when the airplane is light, the airspeed can be increased without encountering the buffet boundary. Allowing the airspeed to increase at constant altitude will provide somewhat better miles per gallon than climbing on minimum EPR at the standard speed schedule.

For ferry flights, cruise can be accomplished at 45,000 feet if desired. This gives several advantages: no pressure suit or prebreathing required; the zero wind range is still appreciable if proper powers and speeds are used; possible tail winds are available; approximately the same true speed as on a regular climb can be maintained.

DESCENT

The high cruising altitude near the end of a mission can be converted into additional range by using correct descent procedure. The general rule to follow is to avoid early descent. Since turbojet fuel consumption is excessive at low altitude, even at idle RPM, a premature descent will result in a loss in range.

NOTE

Descents should not be made on the standard speed schedule with the gust control in the GUST position. Due to the higher angle of attack in this configuration, it is easier to stall the airplane at high altitude and cause engine flameout.

The throttle should not be retarded below the minimum engine pressure ratio values tabulated in Section V. This is to preclude any possibility of flameout on descent.

All descents on the standard speed schedule should be made with the gust control in the FAIRED position. The recommended normal and fast descents are made at speeds faster than the standard schedule and the gust control should be in the GUST position to minimize Mach number effects.

For defrosting, run auxiliary blower during descent. If necessary, use the HTR-BLOWER position but use sparingly. Below 20,000 feet, keep engine speed up to 60% RPM or more if required to help with defrosting.

At the higher indicated airspeeds, use elevator trim tab cautiously due to its fast rate of operation.

NORMAL DESCENT

The following procedure is recommended for a normal descent. The time required for this descent from 75,000 to 20,000 feet is approximately 15 minutes and the distance is approximately 100 nautical miles under zero wind conditions. See Appendix I, Part 7, for more exact information.

2-22 Changed 6 November 1967

Make descent just below speed for start of buffet. This is approximately 10 to 15 knots below the warning needle, varying with weight and altitude. When 200 knots IAS is reached, continue descent at this speed unless turbulence is present.

MAXIMUM RANGE DESCENT

In order to obtain maximum range, remain at cruise altitude until approximately 75 nautical miles from the landing base. Then start a Descent for a straight-in approach. This will bring you to pattern altitude at the field. In using this procedure, only a small increase in range is realized; however, premature descent may cause a loss in range due to encountering headwinds.

EMERGENCY FAST DESCENT

Refer to Section III for information on this subject.

CONTROLLABILITY

During descent, if the speed is allowed to get above the Mach warning needle, buffeting and tuck will be noted and can be corrected for by slowing down. If in a turn at the time, decrease angle of bank. See Section VI for complete information on flight characteristics.

Never extend the wing flaps above 45,000 feet since dangerous pitching moment effects are produced.

DESCENT CHECK

Accomplish descent in accordance with the following procedures to ensure safe aircraft operation and to preclude damage to special equipment:

1. Oxygen - PRESS-TO-TEST.
   Check oxygen and pressure suit by use of PRESS-TO-TEST before retarding throttle for descent.

2. Bleed valves - OPEN.
   Bleed valve lights - ON.

3. Defroster - ON.

4. Heater-blower defroster - ON.

5. Pitot heat - ON (if required).

6. Slipper pump switch - ON (for 5 minutes if necessary to strip tanks completely). AC generator power is required.

7. Landing gear - EXTENDED.

8. Speed brakes - EXTENDED.

9. AUX TANK AIR switch - PRESS OFF. ©

NOTE

For Model U-2F the AUX TANK AIR switch must remain in the PRESS ON position for descent if auxiliary tanks contained fuel at takeoff.

SECTION II   AF (C)-1-1

10. Throttle - Retard to minimum EPR.

WARNING

With throttle at or near idle cabin pressure may slowly rise to ambient pressure. Example: Prolonged holding at 20,000 feet.

11. Airspeed schedule - MAINTAIN.
    Maintain appropriate speed schedule for the type of descent being made.

12. Equipment mode selector - OFF.

13. Special equipment - SET.
    Set as briefed or as indicated by special equipment checklist.

14. NAV light switch - ON (at 55,000 feet).

15. AC generator - OFF/RESET (when light comes on or when not needed).

16. Gust control - GUST AT 45,000 FEET.
    (If not done prior.)

17. Stall strips - EXTEND -
    L & R amber lights OUT (with landing gear down and locked).

CAUTION

If aircraft has been in heavy moisture, ice may have formed in the stall strip wells. If this occurs, it may be necessary to delay extension of the stall strips until the aircraft descends below the freezing level. If ice is suspected, do not exert abnormal pull force on the stall strip T-handle to preclude damage to the stall strip extension mechanism.

18. Gust control - FAIRED (at 20,000 feet).
    Reduce speed to approximately 150 knots or slower if air is turbulent and place the gust control in the FAIRED position.

CAUTION

Always monitor the wing flap position indicator when shifting from GUST to FAIRED. If the flaps do not stop at the faired position, return switch to GUST.

19. Lateral trim - CHECKED.
    As speed is reduced, check the lateral trim by noting the control wheel position. (Should be the same as it was on the after takeoff trim check.) Correct for any wing heaviness by transferring fuel from the heavy wing before landing.

20. Facepiece - REMOVED AND STOWED.
    Remove and stow in bag below cabin altitude of 10,000 feet.

21. Oxygen valves - OFF.

22. ATC - CHECKED.

23. Altimeter - RESET TO STATION ALTIMETER SETTING AT 18,000 FEET.

24. Low altitude escape lanyard - HOOKED.

25. Slipper pump switch - OFF. (If applicable.)

5,000-FOOT CHECK (DESCENDING)

1. Fuel counter - CHECKED, SPEED COMPUTED.
   Check the fuel remaining and determine the threshold speed for landing approach. The correct threshold speed can be obtained from Part 8 of the Appendix as a function of fuel remaining.

When gusty wind conditions exist, add one half of the wind gust factor with reported gusts in excess of five knots. The gust correction factor is applicable only to the final approach airspeed.

2. Throttle gate - CLOSED.

BEFORE LANDING

Although the airplane is not difficult to land, it does have certain characteristics that differ from other jet aircraft. It is essential that the pilot have a thorough knowledge of these characteristics in order to be able to accomplish landings with a high degree of precision and safety under all runway and weather conditions.

With wing flaps up, the total drag is so low that the airplane will maintain level flight at pattern altitude with engine speed only a few percent above idle. Even with landing gear and partial wing flaps extended, the glide-angle is shallow, necessitating a flat approach. However, with full flaps a power approach can be made with as much as 85% engine RPM. Upon reducing power to idle, touchdown will occur with a minimum float distance when using the proper threshold speed.

NOTE

There is no loss of rudder or aileron effectiveness due to using full flaps. However, there is slightly more buffeting at times and slight directional hunting in rough air. This is not considered objectionable.

A typical landing pattern, using the 360° overhead approach, is shown in figure 2-2. Flap position for landing may be determined by the pilot with regard to mission requirements. Using full flaps will result in a more conventional approach, with shorter float distance and a shorter ground roll.

WARNING

Do not exceed 30° of bank during any part of the traffic pattern due to the decreased stall margin as bank is increased. Aircraft control is also difficult under conditions of moderate turbulence.

INITIAL APPROACH

Make initial approach with sufficient engine RPM to maintain threshold speed plus 30 knots at an altitude of 1,500 feet above terrain.

Changed 1 September 1968

NORMAL LANDING PATTERN

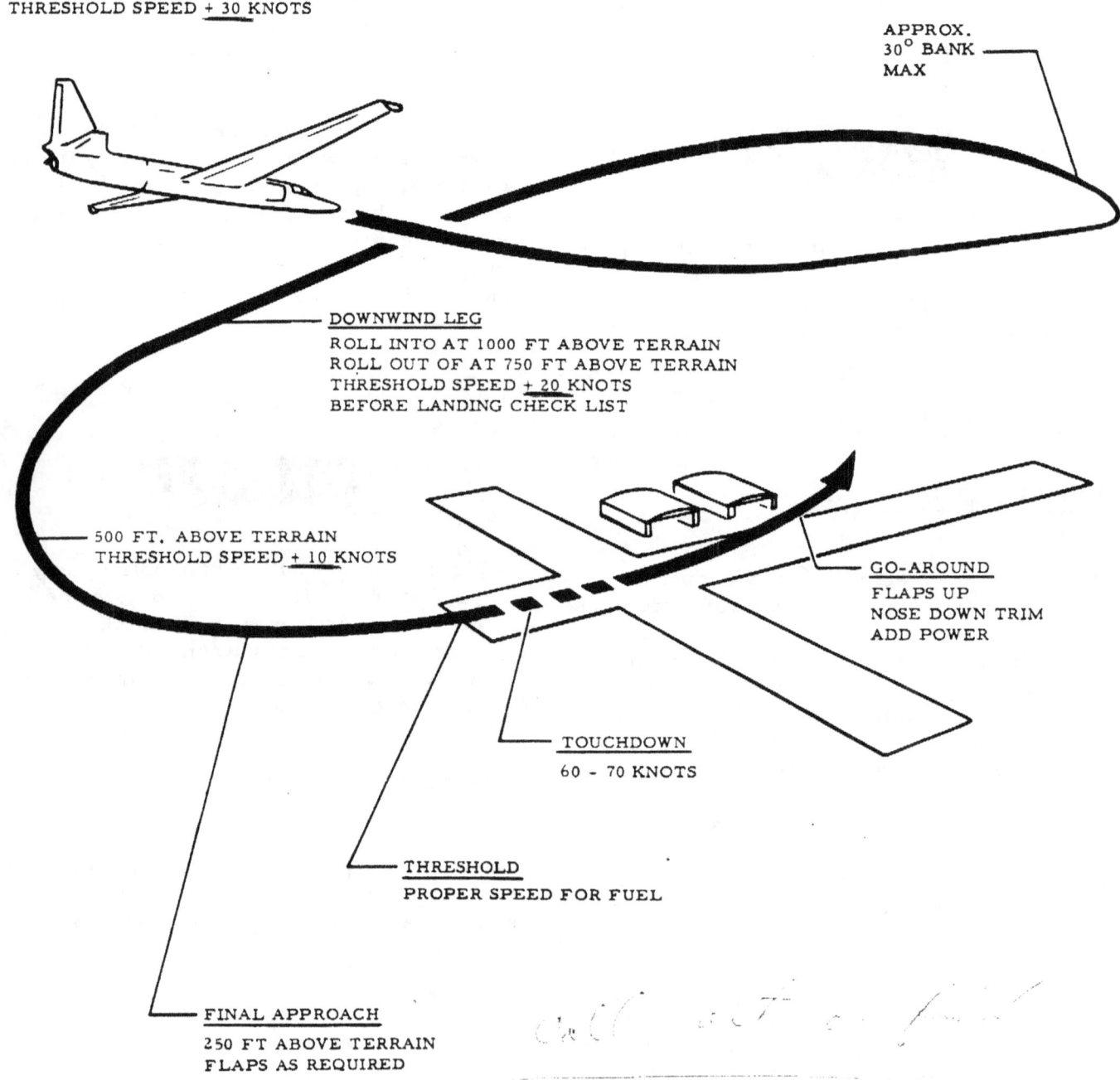

Figure 2-2

180° TURN TO DOWNWIND LEG

At a point near the end of the runway, the exact location of which will be determined by traffic density and headwind conditions, begin a slow descending turn of approximately 30° bank to enter the downwind leg. Slightly reduce engine RPM or extend partial wing flaps.

DOWNWIND LEG

Roll out on downwind leg approximately 1000 feet above terrain with enough power to maintain threshold speed plus 20 knots in a slight descent. The Before Landing checklist shall be completed as soon as possible after turning on downwind leg.

BEFORE LANDING CHECK

1. Landing gear - CHECKED DOWN AND LOCKED.
   Recheck landing gear down and locked using the landing gear position indicators, landing gear unsafe warning horn, and landing gear unsafe warning light.

2. Brakes - CHECKED.
   Press brake pedal to check system pressure.

3. Gust control - FAIRED.

4. Control wheel - RECHECK.
   Determine the fuel load is evenly balanced by checking the control wheel position.

5. Speed brakes - EXTENDED.

6. Wing flaps - SET.
   Opposite approach-end of runway extend wing flaps to desired position for landing.

Assure that flap switch does not return to up position when released.

WARNING

The wing flap extension placard speeds must be carefully observed since operation at too high a speed may result in structural failure of the horizontal stabilizer.

7. Threshold speed - RECHECKED.
   Check fuel quantity and recheck computed threshold speed.

8. Stall strips - RECHECK EXTENDED.

9. Landing lights - AS REQUIRED.

10. Bleed valves - RECHECK - OPEN.

SECTION II AF (C)-1-1

180°-TURN TO FINAL APPROACH

Roll into the 180°-turn to final approach at approximately 750 feet above terrain and decrease speed to threshold speed plus 10 knots. The more flaps used, the more power required to maintain proper speed, making it easier to judge the touchdown point when power is reduced. The altitude midway around this turn should be approximately 500 feet above the terrain.

FINAL APPROACH

The rollout onto final approach should be completed at approximately 250 feet and at threshold speed plus 10 knots. This speed will be gradually "bled off" so as to attain threshold speed at threshold point (this point is defined as the beginning of that portion of the runway useable for landings) where the throttle is placed in idle. The altitude at the end of the runway should not exceed 10 feet. It is very important that the airplane be brought to the threshold with the proper speed and altitude to ensure a good landing.

LANDING

TOUCHDOWN

Touchdown should be made on the main and tail gear together. In order to do this, the airplane must be flown down to a height of one foot above the runway. As the speed bleeds off, the tail will come down until the airplane settles onto the ground if the main gear has been held off long enough. When using full flaps, most of the excess speed can be bled off during the final 10 feet of descent and the airplane will reach the ground level just a few knots above the touchdown speed.

WARNING

Extreme care must be used to correct for the least bit of drift just prior to touchdown. If the main and tail gear are aligned with the direction of landing at touchdown, directional control problems on the runway are simplified.

The main gear is well forward of the center of gravity position and if allowed to touch down first, a skip will normally occur. This usually is a result of too much speed and not holding the aircraft off long enough. To correct from a skip, hold whatever elevator control you have already applied and when the aircraft begins to settle towards the runway again, resume the application of normal back pressure. When a bounce occurs on landing, it usually is the result of allowing the aircraft to abruptly contact the runway main gear first. A bounce can result in a dangerous situation and the best corrective action is to immediately start applying power and initiate a go-around. If impossible, give a slight forward pop with the elevator control to return the aircraft closer to the runway and to avoid encountering a stall condition. This action should be immediately followed by a continuation of a landing flareout so as to prevent a second contact with the runway main gear first.

CAUTION

If a bounce should degenerate into a porpoise, a go-around should be initiated immediately since a porpoise can cause aircraft structural damage.

If the flareout is made too high above the runway or in gusty air, a wing may drop and the airplane will yaw. This requires an immediate correction with aileron and rudder. If immediate contact is made with the ground, the correction must be removed very quickly. Holding a steering correction in too long will cause a ground loop.

NORMAL LANDING RUN

After touching down, the wings should be held level with aileron only, and directional control maintained with the rudder. Do not attempt to raise a wing by steering into it while on the high speed portion of the roll. There is danger of oversteering and causing a ground loop. The important factor is to keep the airplane going straight. Directional control is improved if flaps are retracted after touchdown; however, the landing roll can be reduced by leaving the wing flaps full down. If landing has been made with less than full flap, avoid extending flaps immediately after touchdown to preclude the possibility of becoming airborne again.

After touchdown, the control column should be brought back smoothly and held in order to keep the tail wheel in firm contact with the runway for optimum directional control during the landing roll.

NOTE

After touchdown, exercise caution in aft movements of the control column. By making fast or abrupt movements to the rear, it is possible to pull the aircraft back into the air in a nose high attitude.

Heavy braking should be avoided to prevent tire and brake damage. Ordinarily the brakes are moderately applied when the airplane has decelerated to approximately 50 knots; then when it is decided to stop, at approximately 30 knots, they are applied moderately hard if wings are level, and lightly if one wing is on the ground.

Care must be exercised in applying the brakes when one wing is low. Under this condition, the main gear tire on the high wing side is lightly loaded and as brake pressure is applied this wheel is easily skidded, causing a flat spot or a blowout.

Ailerons should be used to hold the wings level as long as possible. This requires light force but quick action. As aircraft slows down, full aileron may be required. One wing will drop to the runway when speed diminishes and the aircraft should be stopped soon thereafter; however, the aircraft can still be turned in either direction. Normally, a wing tip dragging on a hard surface will not be harmful to the aircraft at light fuel loads.

Assure that the wing tips are clear of obstructions before the skid contacts the ground.

CROSSWIND LANDING

A crosswind landing can be accomplished with wind velocity and direction that results in a runway crosswind component of 15 knots or less. Directional control is improved if flaps are retracted after touchdown.

Do not attempt a normal crosswind landing if more than three quarters of the rudder travel is applied to correct for wind drift just prior to touchdown.

NOTE

Do not land using crab techniques.

Landing With Crosswind Component Up to 15 Knots

Runway alignment should be maintained by slipping, utilizing a combination downwind rudder and enough bank into the wind to maintain heading and prevent aircraft drift. The bank angle required is slight for 3/4 rudder application.

Once the aircraft has touched down, rudder control only should be sufficient to maintain directional alignment with the runway. If difficulty is encountered, the following technique should be used.

Landing With Crosswind Components in Excess of 15 Knots

Successful landings, with full rudder application to correct for drift, can be accomplished with runway crosswind component of 20 knots because tail wheel steering will help when on the ground. However, landings under these conditions should only be considered in actual emergencies when no suitable runway more nearly aligned with the wind is available.

After landing with crosswind components in excess of 15 knots immediately put the downwind wing on the ground. The additional drag of the downwind wing skid coupled with available rudder control and steering will permit a safe after-landing roll.

HEAVY WEIGHT LANDING

Reasonable care in landing will not result in any structural difficulty when landing with less than 550 gallons of fuel. However, if it is not possible to burn out or dump excess fuel and a landing is necessary at a heavier weight, every effort must be made to make as smooth a landing as possible. The rate of sink at touchdown must be reduced to a minimum in order to avoid structural damage to the landing gear.

CAUTION

Be careful not to cause a bounce and then allow the airplane to stall and drop in. Stalling speed will be correspondingly higher due to the heavier weight. The stall warning buffet will be less noticeable.

NIGHT LANDING

The technique for accomplishing night landings is essentially the same as for daylight landings. A skip or bounce landing can cause more difficulty at night because it is harder to ascertain the height of the skip and the extent of the necessary correction.

NOTE

The sextant knob should be out for night landings, to preclude the possibility of the landing lights shining through the driftsight head and into the pilot's eyes.

LANDING WITH DRAG CHUTE DEPLOYMENT

The use of the drag chute is not necessary for a normal landing, but it will materially reduce float distance and landing roll. This will permit the use of shorter runways and/or steeper approaches as the tactical or emergency situation may dictate. The landing approach should be planned the same as for a no-chute landing so that if for some reason the chute doesn't deploy, the landing can be completed.

Desired procedure is to deploy the drag chute between 2 and 5 feet above the approach end of the runway at the threshold speed for a given weight and configuration as shown in the Appendix. If a go-around is necessary, jettison the drag chute.

When deployed at moderate gross weights, the drag chute will feel light and there will be ample time to flare. When deployed at light gross weights, the drag is very noticeable and will require an immediate flare because touchdown will occur much sooner. When using the chute on a heavier than normal landing, wait until the airplane is on the ground before deployment.

If the drag chute is deployed at slightly high speed, it is very easy to balloon too high in correcting for drag due to chute. If the drag chute is deployed at slightly slow speed and particularly light weights, it will be difficult to keep from hitting main wheel first. However, the chute will stop any porpoise that starts.

It is not necessary to jettison the chute before taxiing. However, when desired, it can be jettisoned at approximately 20 knots or after stopping by momentarily increasing throttle. The chute must be blossomed to jettison properly.

CAUTION

When deployed at too high an altitude and too slow a speed, the drag chute will cause excessive rates of sink from which it may be impossible to flare.

Do not deploy the drag chute in a crosswind with a component greater than 10 knots.

The drag chute may not come out of the container when deployed with the engine shut off or after touchdown due to insufficient pilot chute drag.

MINIMUM RUN LANDING

The minimum run landing is accomplished by proper use of the following procedures:

1. Extend full wing flaps on approach and leave down.

2. Consider threshold to be 500 feet prior to end of runway. Establish proper threshold speeds and altitude at this point.

3. Deploy drag chute over end of runway and leave on until stopped.

4. Use brakes earlier and harder than normal.

5. Shut off engine as early as possible. This will normally mean immediately after touchdown; however, the engine may be shut down at the threshold if lack of available runway should dictate this procedure. Engine idle thrust is a large factor in length of float distance and ground roll.

CAUTION

If the decision is made to shut down the engine at threshold, extreme care is required during the touchdown since engine power is not available for go-around, and the drag chute may not come out of the container if deployed with engine shut off.

GO-AROUND

Due to the large amount of excess thrust, extra caution must be exercised to maintain complete control of the airplane. The power should be added slowly and only as much power as is required. The 93% to 94% RPM stop should be observed except in case of dire necessity. The application of power causes a moderate nose up trim change. This must be compensated for by application of nose down trim and moderate control push forces. It is not necessary to retract the wing flaps to reduce airplane drag, but they are usually retracted in order to eliminate the need to observe the flap placard speed. Retraction of the wing flaps also causes a light nose up trim change. It is not necessary to retract the landing gear or speed brakes.

Before Touchdown

If the decision is made to go around while still in flight, the following procedures are used:

1. Apply power as rapidly as desired. Resist the trim change.

2. Wing flaps may be retracted immediately, since the airplane will be accelerating and altitude loss due to flap retraction will be negligible.

After Touchdown

If the touchdown has been made, use the following procedures:

1. If time permits, start wing flaps up.

2. Start application of nose down trim.

3. Advance power rapidly to 85% RPM.

4. As aircraft speed increases and elevator becomes effective, increase power as desired.

5. Do not allow the nose to rise abruptly upon breaking ground.

Do not exceed the wing flap placard speed with flaps extended.

TOUCH-AND-GO-LANDINGS

Touch-and-go landings with fuel loads of 400 gallons or less are authorized in this aircraft and can be routinely accomplished without difficulty. Takeoff acceleration is rapid and runway length is not a critical factor during touch-and-go operations. After normal touchdown, the aircraft should be allowed to decelerate to 50 knots, at which time the flaps should be retracted and the elevator trim reset toward the neutral position. Due to the rate of change in position of the elevator trim tab, approximately 8 to 13 seconds will be required to obtain neutral trim position. Usually it is sufficient to apply nose down trim during the time required for the flaps to retract. When the takeoff is initiated, engine power should be advanced to at least 85% RPM and the control column placed forward of the neutral position to raise the tail approximately 1 foot above the runway. As takeoff speed is reached, back-pressure should be applied gently to the control column to fly the aircraft from the runway. Until experience is gained in the aircraft, considerable attention is required to prevent excess speed. There is a common tendency to gain excess altitude in the closed traffic pattern. When making a closed pattern, use the Before Landing check after turning downwind.

CAUTION

A combination of excessive power application and use of full forward control column movement will result in the aircraft becoming airborne and flying back onto the runway before sufficient elevator back-pressure can be obtained to keep the aircraft airborne.

TAXIING WITHOUT POGOS AFTER LANDING

The aircraft can be easily taxied without pogos installed, even with a crosswing component of 20 knots. It is possible to turn the aircraft and come to a stop downwind under these conditions. If it is desired to taxi the airplane off the active runway after landing, the following procedure can be used:

1. Slow aircraft to 30 to 40 knots.
2. Jettison drag chute.
3. Add slight amount of power to maintain speed.

To aid in turning 90 degrees with minimum radius, hold the wing low on the side of the desired turn for at least 1000 feet prior to turning. This keeps the desired wing tip on the ground during the turn since the fuel will have run to the low wing. If this is not done, centrifugal force in the turn will throw the outside wing down and increase the turn radius. The minimum turning radius is approximately 150 feet.

Take full advantage of intersections by getting to opposite edges and cutting inside corner of intersection. Take care not to hit runway and taxi lights with the wing.

AFTER LANDING CHECK

This checklist should be completed as soon as possible after the landing is completed and the aircraft brought to a stop:

1. Ejection seat initiator safety pin - INSTALLED.
2. Seal valves - OFF.
3. ATC - OFF.
4. HF radio - OFF.
5. Landing lights - OFF.
6. Pitot heater - OFF (if applicable).
7. Auxiliary Boost/Continuous ignition - OFF.
8. Hatch heaters - OFF.
9. Bleed valves - CLOSE.
10. Stall strips - RETRACT.
11. Heater-blower defroster switch - OFF.
12. Equipment master switch - OFF.

TAXIING

If it is desired to taxi the aircraft with pogos after landing, follow the same procedures as outlined on page 2-10.

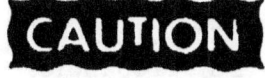

Do not taxi with canopy open.

AFTER PARKING

This check should be completed as soon as possible after aircraft is parked:

1. Flaps - DOWN.

2. Cockpit defroster fan - OFF.

3. Inverter - OFF.

4. NAV lights - OFF.

5. Navigation equipment - OFF.

6. Canopy defroster - OFF.

7. Throttle - OFF.
   If just prior to shut down the engine has been operated for more than 2 minutes at less than 70% RPM, operate at 70% for at least 30 seconds. Shut down the engine upon signal from crew chief or when desired by pilot.

8. Command radios - OFF.

9. Battery - Generator - OFF.

10. Canopy jettison handle cover - INSTALLED.

11. Canopy - UNLOCKED AND OPEN.

12. Personal equipment - UNHOOKED.
    Insure that all personal equipment is unhooked prior to attempting to get out of the cockpit.

13. Form 781 - COMPLETED.
    Write up all discrepancies in detail.

FUEL DUMPING

If it is necessary to dump fuel prior to landing, accomplish the following procedure:

1. Level flight, 150 to 180 knots, flaps faired. Note wheel position.

2. AUX TANK AIR switch - PRESS ON.

3. Position both L and R auxiliary tank dump switches to DUMP, L AUX OPEN, and R AUX OPEN lights should illuminate. Make visual check.

If the wheel position moves appreciably while dumping, the valve in the light wing should be closed. Check for dumping from heavy wing by wheel returning to the original position. If wheel remains in displaced position, dumping should be discontinued and fuel balanced with the transfer switch.

4. When L COMP and R COMP lights come on, close respective dump valve.

5. AUX TANK AIR switch - PRESS OFF.

6. Position both L and R main tank dump switches to DUMP, L MAIN OPEN, and R MAIN OPEN lights should illuminate. Make visual check.

CAUTION

If the wheel position moves appreciably while dumping, the valve in the light wing should be closed. Check for dumping from heavy wing by wheel returning to the original position. If wheel remains in displaced position, dumping should be discontinued and fuel balanced with the transfer switch.

7. When L COMP and R COMP lights come on, close respective dump valve.

8. Upon completion of final tank, pull fuel quantity circuit breaker, set counter to 300 and reset circuit breaker.

If minimum dump time is essential, all dump switches may be opened simultaneously but when the COMP lights illuminate, they must be closed selectively to ensure that both main and auxiliary tanks are completed.

If dumping is accomplished in other than level flight attitude, or if the valves are left open after the completion lights illuminate, it is possible to dump below the 130 gallon level in the main tanks (auxiliary tanks on U-2F).

FUEL DUMPING (F)

Use above procedure except dump main tanks first and leave AUX TANK AIR switch ON.

AF (C)-1-1

EMERGENCY PROCEDURES SECTION 3

TABLE OF CONTENTS

	PAGE
INTRODUCTION	3-3
ENGINE FAILURE	3-3
ABORT BEFORE LEAVING GROUND	3-4
GROUND EGRESS	3-4
COMPLETE POWER FAILURE (IMMEDIATELY AFTER TAKEOFF)	3-4A
PARTIAL POWER LOSS AFTER TAKEOFF	3-4A
ENGINE FAILURE ABOVE 45,000 FEET	3-5
ENGINE VIBRATION	3-7
AIRSTART	3-7
NORMAL FUEL SYSTEM	3-8
EMERGENCY FUEL SYSTEM	3-8
LOW ALTITUDE AIRSTART	3-10
ENGINE FUEL SYSTEM MALFUNCTIONS	3-10
ENGINE OVERSPEED AT HIGH ALTITUDE	3-10
ENGINE OVERSPEED AT LOW ALTITUDE	3-10
FLAMEOUT LANDING PROCEDURE	3-11
FLAMEOUT LANDING PATTERN	3-13
FLAPS EXTENDED PATTERN	3-13
FLAPS RETRACTED PATTERN	3-15
LANDING EMERGENCIES	3-17
LANDING WITH WING FLAPS OF LESS THAN 15°	3-17
LANDING ON UNPREPARED SURFACE	3-19
DITCHING PROCEDURES	3-20
EMERGENCY DESCENT	3-20
HIGH MACH RECOVERY	3-21
FIRE	3-21
GROUND FIRE	3-21
FIRE OR OVERHEAT LIGHT	
DURING TAKEOFF ROLL	3-22
WHILE AIRBORNE	3-22
COCKPIT FOG OR SMOKE ELIMINATION	3-22
COCKPIT SMOKE	3-22
EMERGENCY CANOPY REMOVAL	3-23
EJECTION	3-23
PRIOR TO EJECTION	3-24
EJECTION PROCEDURES (With Ejection Seat)	3-24
AFTER EJECTION	3-24
BAILOUT PROCEDURES (Ejection Seat Failure)	3-25
OXYGEN SYSTEM MALFUNCTION	3-26
OXYGEN DIFFICULTY	3-27
HYDRAULIC SYSTEM EMERGENCY OPERATION	3-28
LANDING GEAR EMERGENCY EXTENSION	3-29
LANDING GEAR CONTROL LEVER LATCH MALFUNCTION	3-29

Changed 1 September 1968

SECTION III AF (C)-1-1

	PAGE
AIRCRAFT FUEL SYSTEM MALFUNCTION	3-29
FUEL LOW LEVEL INDICATION	
DURING CLIMB	3-30
IN LEVEL FLIGHT AT ALTITUDE	3-30
DURING DESCENT	3-31
DROP TANK SYSTEM MALFUNCTION	3-31
DROP TANK FUEL DOES NOT FEED	3-31
DROP TANK(S) DOES NOT RELEASE	3-32
ELECTRICAL SYSTEM MALFUNCTION	3-32
HIGH AMMETER READING	3-32
DC GENERATOR FAILURE	3-32
COMPLETE ELECTRICAL FAILURE	3-33
NO. 1 AND/OR NO. 2 INVERTER FAILURE (Main Inverter)	3-34
NO. 1, NO. 2, AND EMERGENCY INVERTER FAILURE	3-34
ENGINE OIL SYSTEM MALFUNCTION	3-35
TRIM TAB MALFUNCTION	3-35
RUNAWAY TRIM ON MANUAL FLIGHT	3-35
RUNAWAY TRIM ON AUTOPILOT	3-36
GUST CONTROL MALFUNCTIONS	3-36
WING FLAPS MALFUNCTION	3-37
RETRACTABLE STALL STRIP MALFUNCTION	3-37
FUEL DUMP SYSTEM MALFUNCTION	3-37
AIR CONDITIONING MALFUNCTION	3-37
CABIN COOLER FAILURE	3-38
COOLER BYPASS VALVE FAILURE	3-38
PRESSURIZATION SYSTEM EMERGENCY OPERATION	3-38
FACEPIECE HEAT FAILURE	3-39
COMMUNICATIONS FAILURE	3-39
POGO RELEASE FAILURE	3-39
EPR SYSTEM MALFUNCTION	3-40
CONDENSED CHECKLIST (Deleted) Issued in Card Form Only	

INTRODUCTION

This section includes procedures to be followed to correct an emergency condition. The procedures, if followed, will ensure safety of the pilot and airplane until a safe landing is made or other appropriate action is accomplished. Multiple emergencies, adverse weather, and other peculiar conditions may require modifications of these procedures. Therefore, it is essential that the pilot determine the correct course of action by use of common sense and sound judgement.

Procedures appearing in bold face capital letters are considered critical. Procedures appearing in small letters are considered noncritical. Each is defined as follows:

CRITICAL: Those steps of procedures which must be performed immediately without reference to written checklists. These critical steps should be committed to memory.

NONCRITICAL: All other steps of procedures wherein time may be available to consult a checklist before attempting to alleviate an emergency condition.

To assist the pilot when an emergency occurs, basic rules are established which apply to most emergencies occuring while airborne. They should be remembered by the pilot. The rules follow:

1. Maintain aircraft control.
2. Analyze the situation.
3. Take proper action.

ENGINE FAILURE

The majority of engine malfunctions in this airplane are flameouts at altitudes of approximately 40,000 to 60,000 feet. Tests show that climbing at reduced power will prevent flameout in this altitude region. Other flameouts are caused by engine compressor stall and are usually accompanied by a loud "bang-bang" noise. This banging does not harm the engine. Compressor stall may be caused by improper throttle management, allowing the airspeed to drop too low, failure to observe EPR limits and other causes. If flameout is due to the above conditions, a restart can usually be accomplished. If a flameout is caused by mechanical or material failure within the engine, a restart may be possible, depending on the seriousness of the failure, and flight continued at lower power settings and altitudes.

Varying degrees of low compressor instability or flameouts may occur at engine pressure ratios below the minimum EPR versus altitude schedule for bleed valves closed shown on figure 5-1. If the bleed valve lights fail to illuminate when the minimum EPR for bleed valves closed operation is reached, the valves should be manually opened to permit further throttle reduction for descent and avoid flameout.

Flameout as a result of low compressor stall is without warning and too sudden to be avoided by correcting action from the pilot. Normal restart can be accomplished after a flameout due to these conditions.

The flight characteristics with a dead engine are normal, and rapid trim revisions are not necessary. The glide ratio is approximately 20 to 1.

ABORT BEFORE LEAVING GROUND

In the event of a complete power failure during the takeoff ground roll, an abort must be initiated promptly. All available drag devices must be utilized due to the low drag of the aircraft. Accomplish as much of the following as possible:

1. THROTTLE - OFF.

2. DEPLOY DRAG CHUTE (if installed).

3. APPLY BRAKES.

    Use brakes as much as possible without sliding the tires.

4. Extend drag devices (flaps and speed brakes).

5. Bat & Gen switch - off.

6. Evacuate aircraft.

GROUND EGRESS

The following procedure will ensure the quickest and safest evacuation of the aircraft.

With Partial Pressure Suit

1. Seat belt and shoulder harness - disconnect.

2. Seat pack quick disconnect - disconnect.

3. Facepiece hose - disconnect.

4. Chest bladder hose - disconnect.

5. Capstan hose - disconnect.

6. Electrical leads - disconnect.

7. Release both seat pack clips from parachute clip retainers.

8. Cockpit seal pressure - off.

9. Canopy - open or jettison.

10. Exit aircraft from safest side.

Without Partial Pressure Suit

1. Seat belt and shoulder harness - disconnect.

2. Seat pack quick disconnect - disconnect.

3. Breathing hose - disconnect from pressure reducer.

4. Release both seat pack clips from parachute clip retainers.

5. Cockpit seal pressure - off.

6. Canopy - open or jettison.

7. Exit aircraft from safest side.

COMPLETE POWER FAILURE IMMEDIATELY AFTER TAKEOFF

If a complete power failure should occur after leaving the ground, the procedures for effecting a safe landing will depend upon the position of the aircraft relative to the runway. The airspeed, altitude, and length of remaining runway will directly affect the pilot's decisions. Should engine failure occur when the aircraft is several feet in the air, land straight ahead if possible. If there is sufficient altitude to maintain directional control, it may be possible to land the aircraft on another runway or in an adjacent field. A wheels down landing without power is made the same way as a normal landing. For a wheels up landing, insure that the aircraft touches at a slightly higher speed than for a normal landing, so that the aircraft contacts the ground in a level attitude. If a power failure occurs after leaving the ground, accomplish as much of the following as possible:

NOTE

The following minimum safe speeds (10 knots above stall) are recommended for the accompanying takeoff fuel loads.

FUEL LOADS (GALS)	MINIMUM SAFE FLYING SPEED (Knots IAS)
1520	109
1320	106
1020	102
895	100
695	97

1. THROTTLE - OFF.
2. LANDING GEAR - DOWN.
   (If time and conditions permit.)
3. Stall strips - extend.
4. Speed brakes - extended.
5. Wing flaps - as required.
6. BAT & GEN SWITCH - OFF.
   (Before making contact with the ground.
7. Drag chute - as required. (If installed.
8. Canopy open or jettison after stopping.

PARTIAL POWER LOSS AFTER TAKEOFF

The pilot should check that the throttle is full open. If power loss is still evident, he must make the decision to continue or abort takeoff. If power loss occurs during the takeoff roll, proceed as described under Complete Power Failure Before Leaving the Ground. If the power loss occurs after takeoff and the decision is to continue flight, retract the landing gear as soon as possible. The engine fuel control should be switched to emergency since this may correct the difficulty. Flight can be sustained with engine speed as low as 70% RPM. Attempt to make a flameout pattern from the low key point and land as soon as possible. Lower the landing gear when the runway is assured.

NOTE

Do not reduce power until absolutely necessary. You may not be able to regain the previous level of power.

1. THROTTLE - FULL OPEN.
2. SELECT EMERGENCY FUEL SYSTEM
3. FUEL DUMP SWITCHES - DUMP.
4. Stall strips - extended.
5. Land as soon as practicable.

ENGINE FAILURE DURING FLIGHT

In case of an engine failure during flight there may be only a minimum of time and altitude available to determine a course of action. The following factors should be considered in deciding whether to make a flameout landing, an airstart attempt, or an ejection:

a. An airstart can be readily accomplished providing the cause for the failure has been corrected and the engine RPM is at least 10% to 12%.

b. If engine RPM has been allowed to drop 7% or lower (due to low airspeed), at least 3000 feet of altitude will be consumed in increasing the engine speed to 10% to 12% for an airstart.

c. A safe ejection is doubtful at very low altitude. See Ejection.

d. A manual bailout decision normally should be made above 5000 feet.

e. A flameout landing pattern normally should not be attempted below 1000 feet.

ENGINE FAILURE ABOVE 45,000 FEET

If the engine failure occurred above 45,000 feet, the protection of your pressure suit will be necessary. In anticipation of suit inflation, it is advisable to tighten the helmet tiedown cables. Place your hand on the seat pack Press-to-Test button and if your pressure suit does not inflate at a cabin pressure of 42,000 feet, press the button until inflation is obtained. If there is still no action, pull the emergency oxygen supply green apple.

NOTE

The cabin pressure decay rate will vary depending on the effectiveness of cabin sealing. It may be several minutes after engine shutdown before the pressure suit is automatically actuated.

Refer to figure 3-2 for dead engine glide distance and speed. If desired, the rate of descent can be increased by extending the gear and speed brakes. If necessary, a fast descent can be made as described under Emergency Descent.

WARNING

During a fast descent, care must be exercised that the airspeed limitations are not exceeded or structural failure may result. The gust control must be actuated to GUST at 45,000 feet or the speed warning system, if installed, will sound the horn as the speed builds up to approximately 220 knots IAS. The speed warning system will be inoperative if the battery generator switch is off.

Canopy/windshield fog and icing will occur in descent and may become quite severe under conditions of cold temperature and high humidity. The windshield heater/blower should be operated as long as the DC generator stays on the line. When the DC generator drops off the line, only the blower portion will operate and has a 5-ampere current draw.

Changed 1 September 1968

If an immediate landing has to be made, the canopy should be jettisoned as a last resort when there is no forward visibility. Extending the time in the air by using 115 K for maximum glide distance will of course give more time for the windshield to clear.

The battery is a 35-ampere-hour battery but will last an estimated 20 to 30 minutes depending on the essential items being operated, such as:

a. UHF radio on - 15.0 amperes.

b. UHF radio transmitting - 20.0 amperes.

c. Windshield blower - 5.0 amperes.

d. Emergency inverter - 9.0 amperes.

   1. Emergency compass.

   2. Attitude indicator.

   3. ADF/VOR needle.

e. Pitot heat - 13.0 amperes.

f. Interphone (required for radio) - 1.0 ampere.

g. Face heat - 1.0 ampere.

h. Lights.

   1. Anticollision   7.0 amperes.

   2. Landing   18.0 amperes.

   3. Cockpit   5.0 amperes.

i. Cockpit fan - .60 amperes.

j. Engine start - 10.0 amperes.

k. Turn & Slip Indicator - .2 amperes.

l. Wing flap control - .7 amperes.

m. IFF - 2.5 amperes.

If the engine RPM is high enough (28 to 35%) after flameout, the DC generator may stay on the line. However, if the battery discharge light comes on the generator may not be contributing any power to the bus and in fact may be drawing power in the form of reverse current from the battery. It requires approximately 20 amperes of reverse current to trip the generator. This current can only come from the battery. The battery-generator switch should be switched to BAT-EMERG position when the ammeter reads zero since the ammeter indicates only generator status.

1. THROTTLE - OFF.

2. TIGHTEN HELMET TIEDOWN CABLE.

3. PRESS-TO-TEST OXYGEN REGULATOR - if suit does not inflate at 42,000 feet cabin.

4. Unnecessary electrical equipment - off.

5. Establish glide airspeed for maximum range, or use fast descent procedure if required.

6. Dump fuel as necessary before landing.

7. Extend stall strips before landing.

TURBINE BUCKET FAILURE

If a high altitude flameout is accompanied by moderate engine roughness, it may possibly be caused by failure of a turbine bucket. A normal restart can be made. However, maximum altitude and power will be limited. Exhaust gas temperature should be monitored carefully and flight continued at reduced power. If maximum altitude cruise is resumed, a second flameout will probably occur. A landing should be made as soon as practicable, although the engine may operate at reduced power for many hours in an emergency situation.

> **NOTE**
>
> Do not confuse this type of failure with engine flameout due to other causes, when engine roughness is not present.

ENGINE VIBRATION

Severe engine vibration usually indicates internal engine failure and continued high power operation may result in complete failure. When severe engine vibration is encountered, proceed as follows:

1. ADJUST RPM FOR MINIMUM VIBRATION.

> **NOTE**
>
> The adjusted engine speed may not eliminate vibration but will reduce its severity and may permit continued use of the engine until a landing can be made.

2. MONITOR EGT.
   Closely watch EGT gage. If temperature rises above the maximum allowable inflight temperature, reduce power and descend to lower altitude. If temperature is still too high, shut off engine. Check for presence of fire in aft section.

   If a fire is suspected, turn the Ram air switch ON to preclude smoke entering the cockpit through the pressurization system. Be prepared for depressurization.

3. Land as soon as practicable.

> **NOTE**
>
> A form of airplane roughness, similar to that caused by engine vibration is sometimes encountered. This is caused by engine tailpipe misalignment and is easily corrected by ground adjustment.

AIRSTART

Airstart attempts have been very successful. The preferred altitude for airstarts is 35,000 feet or lower.

Before attempting an airstart, time and altitude permitting, a trouble check should be made to determine the cause of the flameout.

  a. If it was accompanied by an explosion, (other than the bang-bang of compressor stall), severe vibration, or fuel vapor in the cockpit, indicating mechanical failure in the engine, an airstart should not be attempted.

  b. If the fuel low level light is on, the flameout was probably due to fuel starvation and a restart cannot be made.

  c. If the fuel pressure has dropped to zero, indicating failure of both the hydraulically-driven fuel boost pump and electrically-driven auxiliary pump, a restart should still be successful and the flight then continued at any altitude providing fuel has had one hour to weather. After a flameout following a boost pump failure, vapor may form in the fuel control and cause restarting difficulties. This vapor can be flushed out by opening the throttle fully for the start. Several attempts may be required for a successful start.

3-7

SECTION III

> **CAUTION**
>
> When starting with throttle fully open, the engine will accelerate rapidly and the throttle must be retarded to avoid exceeding EGT, and RPM limits.

d. If compressor stall is the apparent cause of the flameout, a restart should be successful and flight may be continued as desired. The climb back to high altitude may be made if physiological factors are satisfactory. Avoid the conditions under which the flameout was encountered.

At 35,000 feet or below, airstarts can be made using the normal starting procedures and battery power (see figure 3-1). The engine will accelerate after a few seconds. Exhaust gas temperature lags slightly behind the RPM in restarting a cold engine.

AIR START PROCEDURE ON NORMAL FUEL SYSTEM

1. Airspeed and rpm - hold 160-180 knots and 15% rpm or more.

2. Bat & Gen switch - Bat-Emerg.

3. Inverter switch - Emerg.
   This provides EGT reading during start, using the inverter which requires the least battery power.

4. Compass switch - Comp (if required).

5. Ignition - Start and hold.

6. Throttle - Idle.

7. If start occurs generator-battery switch - Bat & Gen.
   Place the switch in Bat & Gen position and check generator output.

8. If start occurs - inverter switch - No. 1 or No. 2.

9. If start does not occur after 30 seconds of ignition - place throttle off, wait for 30 seconds to purge fuel, and repeat 1-6 above.

10. If start attempt is still unsuccessful - place throttle off for 30 seconds and accomplish procedures outlined for Airstart Procedures on Emergency Fuel System.

AIR START PROCEDURE ON EMERGENCY FUEL SYSTEM

The following procedure should be used when airstarts on the normal fuel system have been unsuccessful:

1. Airspeed and rpm - hold 160-180 knots and 15% rpm or more.

2. Bat & Gen switch - Bat-Emerg.

3. Inverter switch - Emerg.

4. Compass switch - Comp (if required).

5. Select emergency fuel system. Check for amber Fuel Sys Emerg light on.

6. Ignition - Start and hold.

7. Throttle - 1/4 to 1/3 open.

8. Adjust throttle as necessary for adequate acceleration and minimum chugging.

FLIGHT RELIGHT CHART

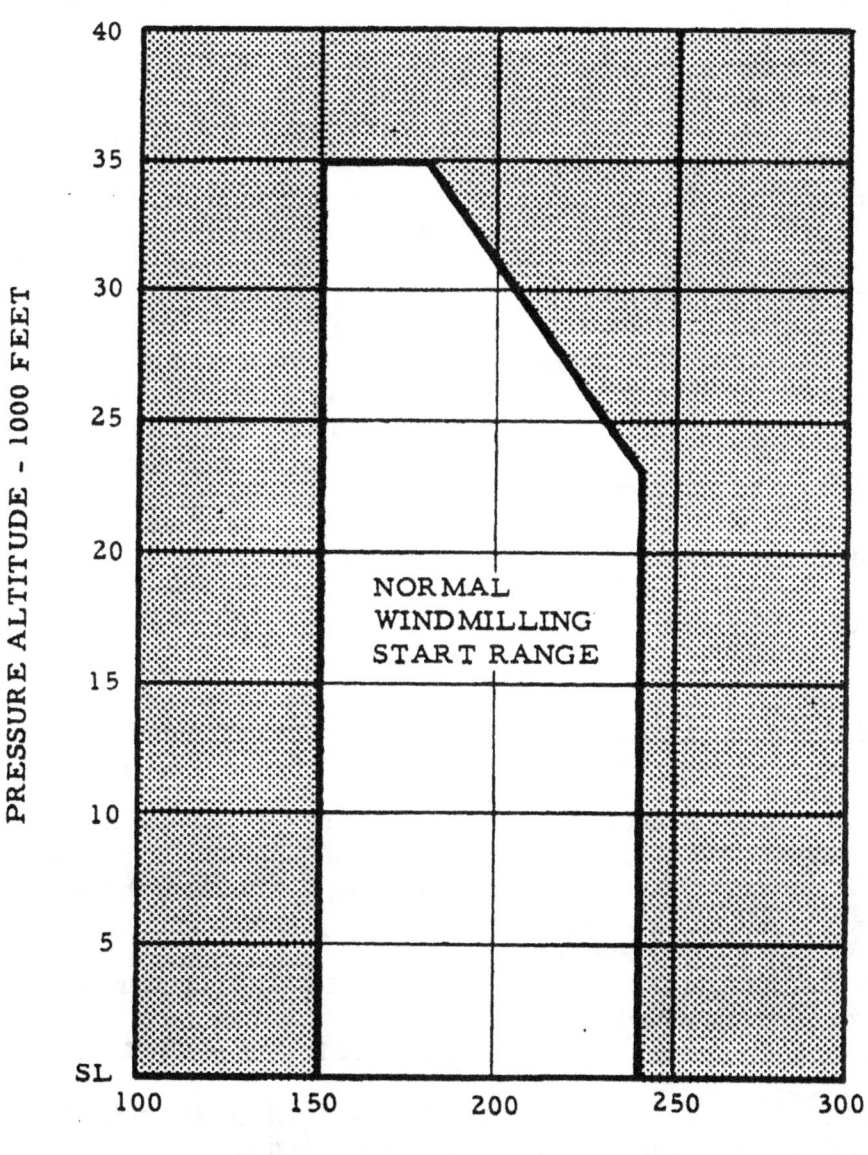

Figure 3-1

SECTION III

AF (C)-1-1

9. If start occurs - Generator-Battery switch - Bat & Gen.
   Check generator output.

10. If start occurs - inverter switch - No. 1 or No. 2.

WARNING

At lower airspeeds ignition will occur but acceleration is slow and a hot start may result. The airspeed should be increased before the start is attempted.

LOW ALTITUDE AIRSTART

If engine flameout occurs at low altitude where a quick airstart is of paramount importance, the following procedure should be used:

1. SELECT EMERGENCY FUEL SYSTEM.

2. THROTTLE - 1/4 to 1/3 OPEN.

3. IGNITION - START AND HOLD.

ENGINE FUEL SYSTEM MALFUNCTIONS

ENGINE OVERSPEED AT HIGH ALTITUDE

The most serious type of malfunction in the fuel control is runaway engine overspeeding. Corrective action must be taken quickly or complete failure will result. Complete the following procedure:

1. THROTTLE - OFF.
   Shut engine off immediately to prevent overspeeding.

2. TIGHTEN HELMET TIEDOWN CABLE.

3. PRESS-TO-TEST OXYGEN REGULATOR - if suit does not inflate at 42,000 feet cabin.

4. Head for nearest suitable landing field.

5. Make normal airstart.

6. If overspeed condition persists - switch to emergency fuel system with throttle in idle.

7. If overspeed condition still persists - throttle off and attempt airstart on emergency fuel system.

8. If overspeed still exists - attempt fuel control regulation with throttle between idle and off.

9. If overspeed cannot be controlled - throttle off and glide for flameout landing. Glide may be extended by starting and stopping engine as necessary.

10. Dump fuel as necessary before landing.

11. Extend stall strips before landing.

ENGINE OVERSPEED AT LOW ALTITUDE

If there is insufficient altitude to shut the engine off and attempt a restart, the following procedure should be used:

1. RETARD THROTTLE.

2. SELECT EMERGENCY FUEL SYSTEM.

3. THROTTLE OFF IF OVERSPEED PERSISTS.

EMERGENCY FUEL CONTROL

The emergency fuel control has been used up to 72,000 feet. All throttle movements must be made slowly and carefully when operating on this system, particularly at high altitude.

CLIMB

During climb, the throttle must be gradually retarded above 60,000 feet to keep the EGT and EPR within limits.

DESCENT

During descent, the throttle may be adjusted to give approximately the same RPM as when operating on the normal system.

The minimum EPR limits prescribed in Section V should be observed.

CHANGEOVER

The changeover from normal to emergency systems is easily made. The throttle position for a given RPM is not the same on the emergency system as on the normal system. Therefore, the RPM will change when the changeover is made. Up to an altitude of 45,000 feet, the changeover should be made at about 75% RPM. Above 45,000 feet the changeover becomes marginal and flameout may result. Once a transfer to the emergency fuel system has been made, the fuel control selector should not be returned to the NORMAL position unless the throttle has been placed in the cutoff position.

COMPRESSOR BLEED VALVE MALFUNCTION

If the bleed valves fail in the closed position, the engine EPR should be maintained in accordance with values listed in figure 5-1 for bleed valves closed. If the bleed valves fail to close when manual or auto switch position is selected, the bleed air circuit breaker on the cockpit circuit breaker panel may be pulled.

FLAMEOUT LANDING PROCEDURE

DEAD ENGINE GLIDE DISTANCE

Under zero wind conditions, the airplane will glide about 34 nautical miles per 10,000 feet of altitude in the clean configuration with the engine inoperative (windmilling or frozen). Optimum glide distance is obtained by descending at speeds which provide the maximum lift-to-drag ratio. The speed for maximum lift-to-drag ratio (and therefore optimum glide speed) is a function of gross weight. The glide distances shown in figure 3-2 are valid for any gross weight if the proper glide speeds per Notes 1 and 2 are used. The glide speed can be varied plus minus 5 knots from optimum without significantly affecting glide distance. A glide from 70,000 feet to sea level will take about 73 minutes and from 35,000 feet to sea level will take about 48 minutes. Glide distances shown in figure 3-2 are decreased, if drag items are used. Reduce the glide distances by the following percentages for other than clean configurations:

Gear down -	19%
Speed brakes out -	39%
15 degrees flaps -	46%
Gust position -	11%
Slipper tanks -	5%

Drag items should not be used until necessary if maximum glide distance is important.

DEAD ENGINE GLIDE DISTANCE

CLEAN CONFIGURATION ONLY

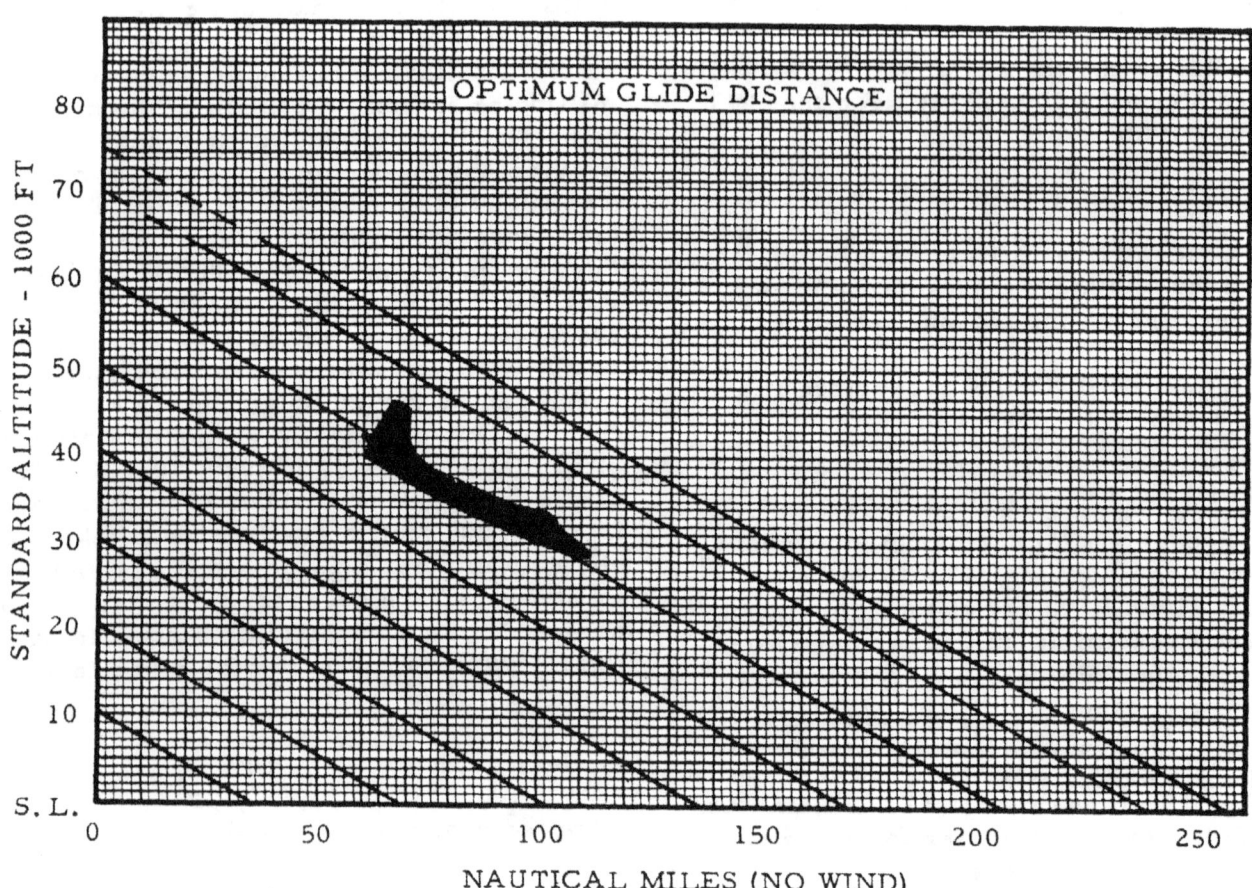

NOTE

1. WITH LESS THAN 600 GALLONS MAINTAIN STANDARD SPEED SCHEDULE UNTIL REACHING 115 KNOTS, THEN CONTINUE DESCENT AT 115 KNOTS.

2. WITH MORE THAN 600 GALLONS MAINTAIN STANDARD SPEED SCHEDULE UNTIL REACHING 130 KNOTS, THEN CONTINUE DESCENT AT 130 KNOTS.

Figure 3-2

FLAMEOUT LANDING PATTERN

Flameout landings should be made with flaps extended. Landings with flaps retracted will be avoided if possible.

After a flameout not associated with electric or hydraulic failure or engine seizure, the engine-driven hydraulic pump will furnish sufficient hydraulic pressure to actuate the wing flaps. At engine windmill speeds as low as 7% RPM, and with an IAS of 110 knots, the flaps can be lowered to 25 degrees in approximately 30 seconds. At higher airspeeds not accompanied by increased windmilling RPM, the flaps may lower at a slower rate due to increased air loads. More than 25 degrees of flap can be obtained but the rate of extension will be progressively slower. Flaps may be retracted as required to extend the glide. However, care must be exercised to avoid overshooting the runway.

If glide distance is not a factor, the landing configuration should be established at an altitude above the high key point. When conditions permit, this altitude should be high enough to make a series of orientation turns prior to entering the high key point as defined in figures 3-3 and 3-4. Each descending 360° turn should be flown to lose an increment of altitude equal to the altitude of the high key point for the configuration involved and to enter the high key point at the prescribed position.

When unable to start the pattern at the high key point or above, it is feasible to enter the pattern at one of the lower key points.

The recommended traffic pattern for a flameout landing is the 360° overhead as shown in figures 3-3 and 3-4. The point over the touchdown end of the runway is defined as the high key point. The downwind point opposite the end of the runway is the low key point and the base leg is the 270° point.

Plan to arrive at the high key point with the landing gear extended, speed brakes retracted, stall strips extended, and 25° flap extension if flaps are available. It is important to establish the high and low key points as outlined below so that proper threshold speed and altitude is reached.

FLAPS EXTENDED PATTERN
(See figure 3-3.)

HIGH KEY POINT:

1. Altitude - 3000 feet above the terrain.
2. Airspeed - 100 knots. Add 2 knots/100 gallons.
3. Landing gear - extended.
4. Wing flaps - 25°.
5. Speed brakes - retracted.
6. Cockpit canopy seal valve - OFF.
7. Stall strips - extended.

LOW KEY POINT:

1. Altitude - 1500 feet above the terrain.
2. Airspeed - 100 knots. Add 2 knots/100 gallons.
3. Landing gear - extended.
4. Wing flaps - 25°.
5. Speed brakes - retracted.
6. Stall strips - extended.

SECTION III AF (C)-1-1

TYPICAL FLAMEOUT LANDING PATTERN - FLAPS EXTENDED

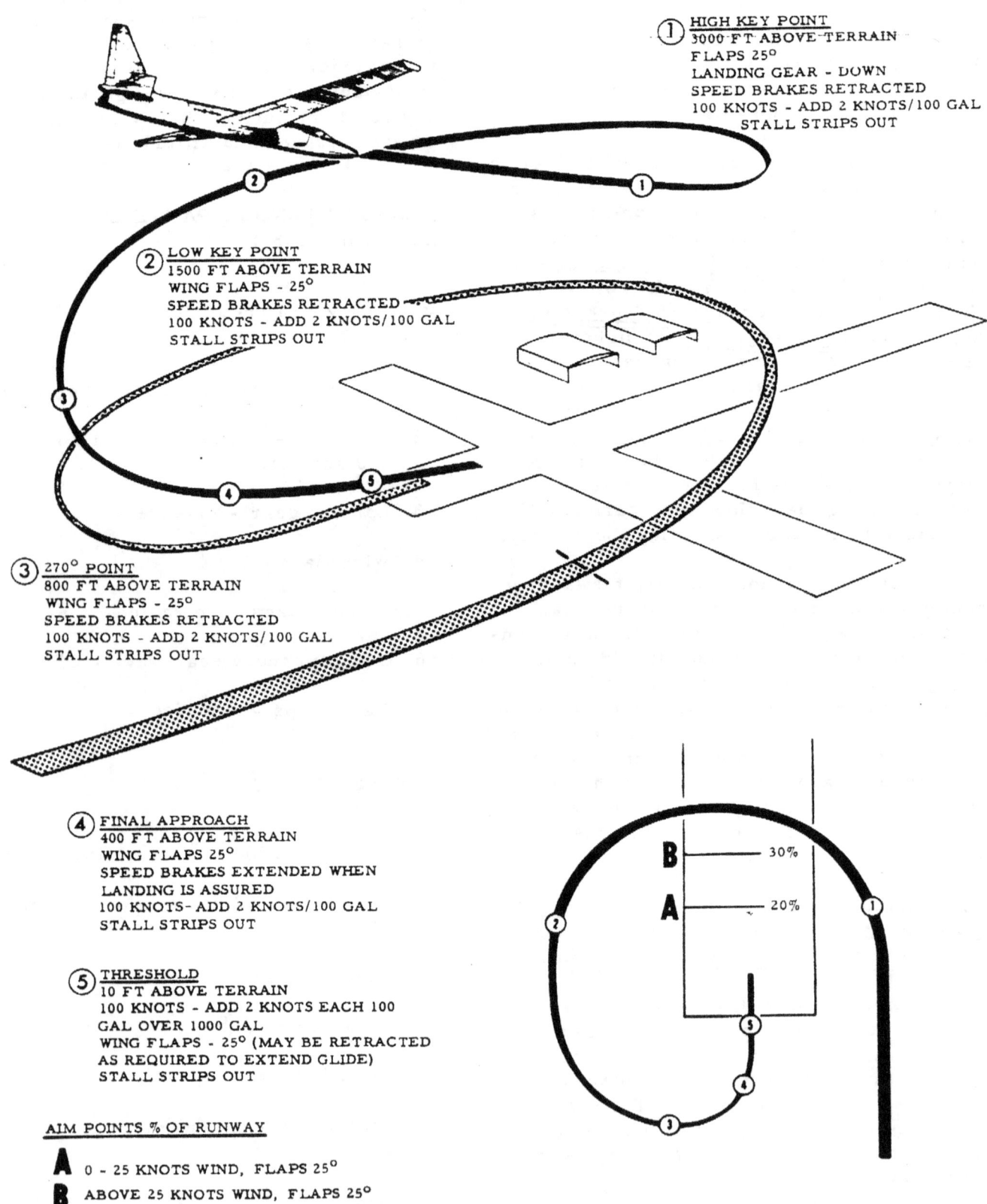

① **HIGH KEY POINT**
3000 FT ABOVE TERRAIN
FLAPS 25°
LANDING GEAR - DOWN
SPEED BRAKES RETRACTED
100 KNOTS - ADD 2 KNOTS/100 GAL
STALL STRIPS OUT

② **LOW KEY POINT**
1500 FT ABOVE TERRAIN
WING FLAPS - 25°
SPEED BRAKES RETRACTED
100 KNOTS - ADD 2 KNOTS/100 GAL
STALL STRIPS OUT

③ **270° POINT**
800 FT ABOVE TERRAIN
WING FLAPS - 25°
SPEED BRAKES RETRACTED
100 KNOTS - ADD 2 KNOTS/100 GAL
STALL STRIPS OUT

④ **FINAL APPROACH**
400 FT ABOVE TERRAIN
WING FLAPS 25°
SPEED BRAKES EXTENDED WHEN
LANDING IS ASSURED
100 KNOTS - ADD 2 KNOTS/100 GAL
STALL STRIPS OUT

⑤ **THRESHOLD**
10 FT ABOVE TERRAIN
100 KNOTS - ADD 2 KNOTS EACH 100
GAL OVER 1000 GAL
WING FLAPS - 25° (MAY BE RETRACTED
AS REQUIRED TO EXTEND GLIDE)
STALL STRIPS OUT

AIM POINTS % OF RUNWAY

A 0 - 25 KNOTS WIND, FLAPS 25°

B ABOVE 25 KNOTS WIND, FLAPS 25°

Figure 3-3

TURN TO 270° POINT:

1. Altitude - 800 feet above the terrain.
2. Airspeed - 100 knots. Add 2 knots/100 gallons.
3. Landing gear - extended.
4. Wing flaps - 25°.
5. Speed brakes - retracted.
6. Stall strips - extended.

FINAL APPROACH:

1. Airspeed - Reduce to arrive at threshold at 100 knots. Add 2 knots/100 gallons of fuel.
2. Wing flaps - 25° (may be milked up to extend glide if required).
3. Speed brakes - extended when landing is assured.
4. Stall strips - extended.

THRESHOLD:

1. Airspeed - 100 knots. (Add 2 knots for each 100 gallons over 1000 gallons.) Flare for normal landing.
2. Drag chute - deployed (if installed).
3. Stall strips - extended.

SIMULATED FLAPS EXTENDED FLAMEOUT LANDING

To simulate a flameout landing pattern, proceed as follows:

1. Arrive at high key point at 100 knots (add 2 knots/100 gallons) with landing gear down, speed brakes extended, 35° wing flaps, stall strips extended, and with 67% engine RPM. Maintain this configuration and speed to final approach.
2. On final approach, reduce airspeed to 100 knots. Add 2 knots/100 gallons for fuel above 1000 gallons.
3. If landing from this pattern, reduce throttle to idle. Otherwise, float distance is exaggerated.

FLAPS RETRACTED PATTERN
(See figure 3-4.)

HIGH KEY POINT:

1. Altitude - 1500 feet above the terrain.
2. Airspeed - 100 knots. Add 2 knots/100 gallons.
3. Landing gear - extended.
4. Stall strips - extended.
5. Speed brakes - retracted.
6. Cockpit canopy seal valve - OFF.

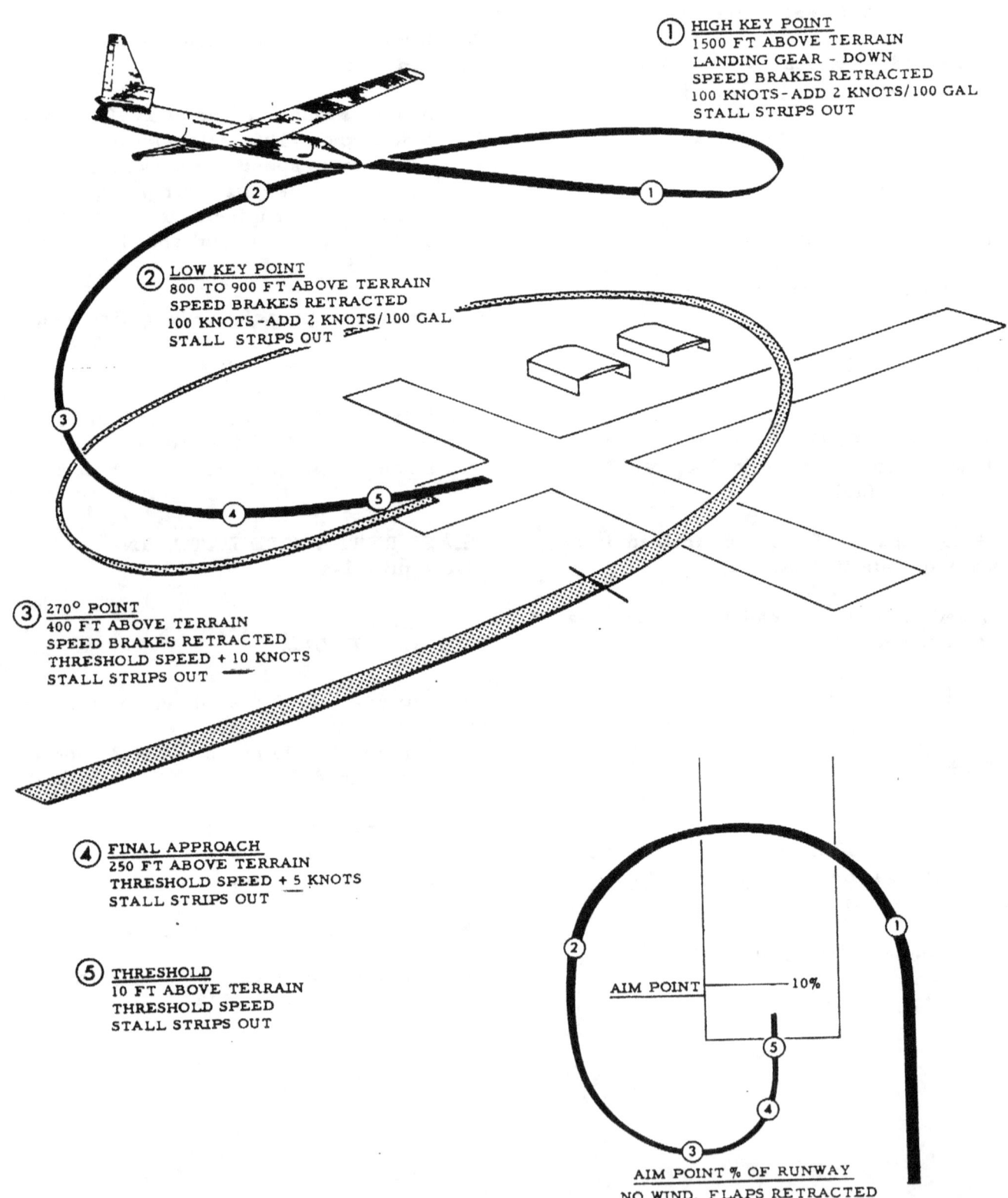

Figure 3-4

LOW KEY POINT:

1. Altitude - 800 to 900 feet above the terrain.
2. Airspeed - 100 knots. Add 2 knots/100 gallons.
3. Landing gear - extended.
4. Stall strips - extended.

TURN TO 270° POINT:

1. Altitude - 400 feet above the terrain.
2. Airspeed - threshold speed plus 10 knots. (See figure A8-2.)
3. Landing gear - extended.
4. Stall strips - extended.

FINAL APPROACH:

1. Altitude - 250 feet above the terrain.
2. Airspeed - reduce to threshold plus 5 knots. (See figure A8-2.)
3. Stall strips - extended.

THRESHOLD:

1. Airspeed - threshold speed. (See figure A8-2.)
2. Drag chute - deployed (if installed).

Do not exceed threshold speed as outlined in figure A8-2 for zero flaps otherwise the float distance will be excessive.

SIMULATED FLAPS RETRACTED FLAMEOUT LANDING

1. Arrive at high key point at 100 knots (add 2 knots/100 gallons) with landing gear down, speed brakes extended, 15° wing flap, 67% engine RPM and stall strips out.
2. Between low key point and 270° point airspeed to threshold plus 10 knots, wing flaps 15°, and with 67% engine RPM.
3. Between 270° point and final approach reduce airspeed to threshold plus 5 knots, power at 67% RPM, 15° flaps.
4. On final approach, gradually reduce airspeed to threshold speed.
5. If landing from this pattern reduce throttle to idle. Otherwise, float distance is exaggerated.

LANDING EMERGENCIES

WARNING

Prior to landing with any abnormal configuration or external damage to the aircraft, stall speed should be checked at a safe altitude, if conditions permit. Extend stall strips.

LANDING WITH WING FLAPS OF LESS THAN 15°

If the wing flaps cannot be extended, 15° or more, a satisfactory landing can be made if caution is exercised. With less than 15° of flaps the drag is low and the airplane will float farther. In order to overcome this characteristic, the proper threshold speed and altitude must be observed. If the approach is too high or

SECTION III   AF (C)-1-1

too fast, go around and make proper corrections on the next approach. Threshold speed is 5 knots higher than for a flaps extended landing. This speed will still give adequate margin over the stall for this emergency condition. Refer to figure A8-4.

On short runways, the drag chute is of great advantage for this type of landing; however, it is not necessary if other variables are properly controlled. Another aid to shorten the float and ground roll is to shut off the engine at the threshold. The following procedures should be used:

1. Speed brakes extended (if available).

2. Threshold altitude properly controlled.

3. Threshold speed properly controlled.

4. Deploy drag chute.

5. Throttle off if necessary after landing is assured.

NOTE

Use minimum run landing procedures if landing on short runways.

LANDING WITH LANDING GEAR UNSAFE

These instructions apply only to landing on prepared hard surfaces. If either landing gear indicates an unsafe condition when extended for landing, make a low pass by the control tower and ascertain if the gears appear to be extended. Whenever either of the gear cannot be extended with the normal hydraulic system, the emergency system will be used in an attempt to lower the remaining gear. (Refer to landing gear emergency extension, this section.) A gear down landing should be accomplished whenever the main gear can be extended to any degree. The techniques for main gear landings and a wheels up landing on prepared surfaces differ considerably and are discussed separately in the following paragraphs.

Both Gears Down - Show Unsafe

If both main and tail gear appear to be in the down position, but an unsafe condition is indicated on one or both, make a normal landing. The engine must be left running in order to maintain hydraulic pressure until ground safety pins are installed by the maintenance crew.

Main Gear Only Fully Extended

Make a normal pattern using speed brakes and the desired amount of flaps if normal hydraulic pressure is available.

WARNING

Do not full stall the aircraft. This would probably result in hitting the tail first and cause considerable damage when the main gear hits.

Make a normal landing, using drag chute. After aircraft has landed, shut the engine OFF, flaps FULL DOWN, main fuel shutoff to the OFF position, battery-generator OFF, and use the brakes as much as possible without sliding the tires.

SECTION III

Wheels Up Landing

Fly aircraft onto ground at slightly higher speed than for normal touchdown so that the aircraft contacts the ground in a level attitude. To help attain level attitude, utilize full flaps during approach to landing. After aircraft has landed, shut engine off, main fuel shutoff to the OFF position, battery-generator OFF.

Hung Pogo Landing

In the event that efforts to release a hung pogo have failed, a normal landing can be accomplished without difficulty. The drag created by the hung pogo will usually result in a heavy wing developing on the side with the pogo. The pogo drag will also create an illusion of a slightly heavy wing even when the fuel is properly balanced, therefore use extra caution in establishing desired fuel balance prior to landing. The best balance for landing under otherwise normal conditions will be attained when the fuel is estimated to be evenly balanced and a slight yoke pressure is required to overcome the drag of the pogo. As speed decreases the drag will also decrease and the yoke force required to maintain a wings level attitude will gradually dissipate. A slightly heavy downwind wing will give the pilot added controllability.

LANDING ON UNPREPARED SURFACE

Because of the low landing speeds of the U-2, safe landing can be made on any reasonably smooth unprepared surface. However, because of structural limitations, a landing on extremely rough terrain or on terrain with many obstacles is considered hazardous, and to eject is normally preferable. For an emergency landing on an unprepared surface, the following procedures will be used:

1. All dump switches - dump.
2. Ejection safety pin - installed.
3. Facepiece - removed and stowed.
4. Oxygen - off.
   Turn both valves off.
5. Parachute and suit connections - unfasten.
6. Seat belt - tightened.
7. Shoulder harness - locked.
8. Stall strips - extended.
9. Landing gear - extended.
10. Speed brakes - as required.
11. Wing flaps - as required.
12. Make a normal approach.
13. Dump switches - closed.
14. Throttle off when landing is assured.
15. Fuel shutoff switch - closed.
16. Battery switch - OFF.
17. Open or jettison canopy after aircraft comes to a stop.
18. Evacuate aircraft.

NOTE

If it is desired to jettison the canopy prior to landing, it should be jettisoned just prior to touchdown.

Changed 1 September 1968

SECTION III   AF (C)-1-1

DITCHING PROCEDURES

In general, ejection is preferable to ditching if the choice exists. If actual conditions preclude a safe ejection or an immediate rescue is more probable with ditching, the following procedure will be used:

1. ATC - Code 77.
2. Radio Call - Mayday.
3. Gear, speed brakes and flaps - retracted.
4. Stall strips - extended.
5. All dump switches - dump.
6. Remove facepiece.
7. Seat pack quick disconnect - disconnect.
8. Shoulder harness - locked.
9. All dump switches - closed.
10. Bat & Gen switch - off.
11. Jettison canopy just prior to touchdown.
12. Throttle - off.

Plan touchdown on the down side of a wave crest, parallel to the crest pattern. Touch down slightly faster than normal - do not stall in. Wait until aircraft has come to a complete stop, then leave as soon as possible.

EMERGENCY DESCENT

If it becomes necessary to reach lower altitudes in the minimum time, the descent can be made faster by remaining in a banked spiral and increasing the speed to maximum allowable IAS limits. The bank angle and stick forces should be varied as necessary to avoid exceeding G limits and to stay out of heavy high speed buffet. The only disadvantage is an increase in general roughness. Shutting the engine down will increase the rate of descent but may result in windshield and canopy frosting.

WARNING

This maneuver will require a constant acceleration and does not leave much margin for recovery from an overspeed condition.

If at any time the pilot should feel he might lose consciousness or be incapacitated in some way, it would be advisable to utilize the autopilot and Mach sensor to control the descent. There are several actions which should be completed if at all possible; however, the aircraft will descend if only the landing gear is extended and the power lever is retarded assuming that the autopilot and Mach sensor are already engaged. The following procedures should be followed, in order, for both a manual or an autopilot controlled descent:

1. PULL GREEN APPLE.
   (If there is any question of oxygen supply.)
2. EXTEND LANDING GEAR.
3. EXTEND SPEED BRAKES.
4. BLEED VALVE SWITCH - OPEN.
5. RETARD THROTTLE.
   Maintain minimum EPR.
6. PLACE GUST CONTROL IN GUST.

If losing consciousness -

7. ENGAGE AUTOPILOT.
8. ENGAGE MACH HOLD.

WARNING

The airplane will reach airspeed placard limitations as it descends. If the gust control is FAIRED, the limitation will be reached fairly early. This will require action by the pilot when the speed warning system is actuated. Therefore, it is very important to get the gust control in GUST position before descent since more structural margin exists for this configuration.

The autopilot descent can also be used with a dead engine if absolutely necessary. This will require heavy usage of the battery, and all nonessential electrical loads should be turned off if possible.

HIGH MACH RECOVERY

If the speed is inadvertently allowed to increase beyond normal allowable limits into the Mach buffet region, the aircraft will be subjected to one or more of the following conditions depending on the <u>degree of penetration:</u> wing buffet, nose tucking, high stick force, wing rolloff, aileron buzz and rudder buzz.

Depending upon the Mach acceleration rate and the <u>degree of penetration</u> into Mach buffet the following action should be taken. If the penetration is only moderate it may be unnecessary to shift into Gust to effect recovery. However, if the Mach acceleration rate is rapidly increasing and if the nose tuck counteracting stick forces are becoming extremely high, it may be necessary to shift to Gust immediately in order to relieve the tucking or pitchdown forces and to effect recovery.

1. AUTOPILOT - DISCONNECT (hands on wheel).
2. WINGS - LEVEL (if in a turn).
3. IAS - REDUCE (apply positive back stick pressure smoothly).
4. SPEED BRAKES - EXTEND (controllable porpoise may occur if speed is at or near .8 Mach).
5. THROTTLE - RETARD.
6. GUST CONTROL - GUST (if stick force excessive).
7. Gust control - faired (when recovery completed).

FIRE

There is no fire extinguishing system installed in the aircraft.

PROBABLE CAUSES OF FIRE OR OVERHEAT INDICATION

Illumination of the red O'HEAT warning light is generally caused by the following:

a. Fire (in the aft section).

b. A leaky tailpipe, tail cone or connections.

c. Improper installation of, or defective warning light circuit.

Illumination of the red FIRE warning light is generally caused by:

a. Fire (in the forward engine section).

b. Hot spots in the combustion chamber.

c. Defective warning light circuit.

GROUND FIRE

In the event of visible or indicated fire on the ground:

1. THROTTLE - OFF.
2. FUEL SHUTOFF VALVE - OFF.
3. BAT & GEN SWITCH - OFF.
4. DISCONNECT SEAT PACK QUICK DISCONNECT.
5. Evacuate aircraft.

SECTION III   AF (C)-1-1

FIRE OR OVERHEAT LIGHT DURING TAKEOFF ROLL

If either the overheat or fire warning light illuminates during takeoff roll, and there is sufficient runway left to abort the takeoff, proceed as outlined under ABORT BEFORE LEAVING GROUND, this section. If the warning occurs too late to abort the takeoff, proceed as outlined under FIRE OR OVERHEAT LIGHT WHILE AIRBORNE, below.

FIRE OR OVERHEAT LIGHT WHILE AIRBORNE

The high flash point of the fuel combined with the lack of oxygen at the operating altitudes of this aircraft greatly reduce the probability of fire at high altitude. However, if the overheat or fire warning light illuminates when airborne or too late to abort the takeoff, proceed as follows:

1. **RETARD THROTTLE.**
   Use minimum practical power.

2. **CHECK FOR FIRE.**

Check for evidence of:

   a. Fluctuating fuel pressure and RPM.

   b. Loss of engine RPM.

   c. Smoke entering cockpit through pressurization ducts.

   d. Smoke trail behind aircraft.

3. **THROTTLE OFF IF FIRE IS EVIDENT.**

4. **FUEL SHUTOFF VALVE - OFF.**

5. Make a decision to eject or execute emergency deadstick landing.

6. If warning light is only indication of fire, continue at reduced power and land as soon as practicable.

COCKPIT FOG OR SMOKE ELIMINATION

COCKPIT FOG

Under certain atmospheric conditions, the cabin air cooler will create a vapor which resembles smoke. This can be eliminated by moving the cabin temperature rheostat to a warmer air position.

COCKPIT SMOKE

If smoke occurs in cockpit:

1. Defroster - off.

2. Temperature control - cold.
   This will reduce the quantity of smoke.

3. Generator-battery switch - off.
   Turn the generator-battery switch to the OFF position until it is determined that the smoke is not caused by an electrical short.

If smoke is so bad that the mission cannot be continued:

4. Throttle - reduce power and make descent.
   The smoke may clear up at lower altitudes and lower engine power.

5. Ram air - on, if smoke persists.
   Be prepared for cockpit depressurization.

6. Throttle - off, if smoke condition remains hazardous.

If the condition persists and is too bad to continue the descent and landing, place the throttle off and switch main fuel shutoff to off.

7. Canopy - jettison if necessary.

 As a last resort, jettison the canopy to obtain visibility. Be prepared for any necessary action due to aircraft damage if the canopy is jettisoned.

CAUTION

Any time smoke is present in cockpit, insure that facepiece is worn and oxygen is on. Leave facepiece on if canopy jettison is contemplated to protect against air blast.

EMERGENCY CANOPY REMOVAL

There is a normal canopy lock and release handle located on the right cockpit sill and a yellow emergency canopy jettison handle on the left cockpit sill. In addition, an emergency canopy release handle for the left-hand latch is stowed on the right-hand console.

EMERGENCY CANOPY JETTISON

Press handle release button and pull yellow T-handle on left sill.

WARNING

If in flight, be prepared to leave the aircraft immediately as the canopy may strike the tail surfaces after it is jettisoned.

NOTE

Retain facepiece for wind protection after canopy removal.

MANUAL CANOPY REMOVAL

1. Turn canopy seal valve off.

2. Insert stowed handle in receptacle on left sill with hand grip below the receptacle.

3. Push left and right canopy release handles forward simultaneously.

EXTERNAL CANOPY OPENING

Depress the spring-loaded retaining button at the base of the canopy locking lever on the right side of aircraft below the canopy rail. The canopy locking lever will spring free from its flush mounted position. Rotate the lever in a clockwise direction to release canopy lock. Lift canopy from rail and open toward left side of aircraft.

EJECTION

A very high altitude ejection should be avoided except in an extreme emergency. A bad cockpit fire or uncontrollable maneuvers which could cause serious bodily injury are probably the only situations that would cause an immediate ejection at high altitude. With aerodynamic or structural trouble, the pilot may be able to remain in the cockpit until reaching a lower altitude.

SECTION III　　　　　　　AF(C)-1-1

PRIOR TO EJECTION

be accomplished prior to ejection:

1. ATC - Code 77.

2. Radio Call - Mayday.

3. Bat & Gen switch - Bat emerg.

4. Helmet tiedown - tightened.

5. Mike and headset jack - disconnect. (If radio bypass cord is in use.)

6. Pull green apple.

7. Destruct system - arm and actuate. (If such action is required.)

NOTE

Generator-battery switch must be in either the BAT EMERG or OFF position to operate destruct system. However, in order to maintain Birdwatcher system operation the switch must be placed in the BAT EMERG position. (This should be the last step before pulling the D-ring or leaving the airplane without using an ejection seat.)

EJECTION PROCEDURES
(With Ejection Seat)

This is the preferred method. If time permits, complete the applicable Prior to Ejection items before proceeding with the following:

1. ASSUME PROPER BODY POSITION. Normal ejection position is taken with feet in stirrups, head back, and hands on D-ring between knees.

2. PULL SEAT EJECTION D-RING. Raise hands, leaving elbows in lap.

NOTE

The D-ring should be kept in hands during ejection to restrain arms.

WARNING

The delay which is experienced between the time the canopy is jettisoned and the time when the seat ejects, although only one second, may seem unreasonably long. Pilots are warned not to lower head until sufficient time has elapsed for the delay initiator to function. The D-ring should be kept in the hands during ejection to restrain the arms.

AFTER EJECTION

1. Release ejection D-ring.

WARNING

Immediately after ejection, attempt to manually open the seat belt. This is strictly a precautionary measure in case the belt fails to open automatically. If the belt is operating normally, it will be impossible to beat the automatic opening action.

As soon as the belt releases, a determined effort must be made to separate from the seat to obtain full parachute deployment at maximum terrain clearance - this is extremely important for low altitude ejections. If the seat belt is opened manually, the automatic feature of the parachute is eliminated. Therefore, under these circumstances, the parachute arming lanyard must be pulled if above 14,000 feet, or the parachute D-ring must be pulled if below 14,000 feet.

2. Manually pull parachute D-ring immediately following seat separation for all ejections below 14,000 feet. This is strictly a precautionary measure since the parachute should deploy automatically.

3. Facepiece - remove after normal parachute deployment at or below 14,000 feet.

BAILOUT PROCEDURES
(Ejection Seat Failure)

1. PULL GREEN APPLE.

2. DISCONNECT SEAT PACK QUICK DISCONNECT.

3. JETTISON CANOPY.

WARNING

Be prepared to leave the aircraft immediately as the canopy may strike the tail surfaces.

4. DISCONNECT SEAT BELT AND HARNESS.

WARNING

If the ejection seat D-ring has been pulled below 14,000 feet and ejection does not occur, the seat belt must be unfastened manually (even though the initiator may have blown the seat belt loose) in order to disengage the parachute arming lanyard before bailing out over the side. If this is not done, the lanyard will deploy the parachute in the cockpit when the pilot stands up.

5. BAIL OUT.

6. PARACHUTE ARMING LANYARD OR RIPCORD HANDLE - PULL.

7. Facepiece - remove AFTER normal parachute deployment at or below 14,000 feet.

EJECTION ATTITUDE

During any low altitude ejection, the chances for successful ejection can be increased by zooming the aircraft (if airspeed permits) to exchange airspeed for altitude. <u>Ejection should be accomplished while the aircraft is in a positive climb and airspeed is above 120 KIAS.</u> This will result in a more nearly vertical trajectory for the seat and crew member, thus providing more altitude and time for seat separation and parachute deployment.

The use of the automatic seat belt and the low altitude escape lanyard make possible successful ejections at very low altitude. The automatic seat belt is an integral part of the ejection seat system and is operated by gas pressure from an initiator after a one-second delay. The automatic or aneroid controlled parachute has a lanyard which is attached to the seat belt. After ejection, this lanyard is pulled by the separation of the pilot from the seat. This actuates the parachute aneroid which will open the parachute at a preset altitude or after one second if below the preset altitude.

The low altitude escape lanyard is connected directly to the parachute D-ring. Therefore, separation from the seat will pull the parachute D-ring. This connection must be made by the pilot at low altitudes only, since it would be disastrous at high altitudes.

The use of this low altitude escape lanyard, which pulls the D-ring directly, eliminates the one-second aneroid timer delay and ensures prompt release of the parachute upon separation from the seat. This system of one second seat belt opening and immediate parachute deployment will make possible a successful bailout at very low altitude. However, since aircraft forward speed is also a factor, no definite minimum ejection altitude can be given. It is very doubtful that a bailout below 500 feet would be successful. Above 500 feet the chances improve rapidly.

As a general rule bailout altitudes should be as follows:

1. UNDER LEVEL FLIGHT CONDITIONS, EJECT AT LEAST 2,000 FEET ABOVE TERRAIN WHENEVER POSSIBLE.

2. UNDER SPIN OR DIVE CONDITIONS, EJECT AT LEAST 10,000 FEET ABOVE THE TERRAIN WHENEVER POSSIBLE.

EJECTION SPEED

EJECT AT THE LOWEST PRACTICAL AIRSPEED ABOVE 120 KIAS AT LOW AND MEDIUM ALTITUDES.

Below 120 KIAS, airflow is not sufficient to assure rapid parachute deployment. Therefore, it becomes extremely important during low altitude ejection to obtain at least 120 KIAS, if possible, to ensure complete parachute deployment at the greatest height above the terrain. During medium altitude ejection, observing this minimum airspeed (120 KIAS) becomes less important since there is adequate time (altitude) for parachute deployment.

The need to be at the lowest possible airspeed down to 120 KIAS prior to ejection is predicated on many factors such as avoiding bodily injury, precluding parachute or seat structural failure, and providing adequate tail clearance. If the aircraft is controllable, airspeed will be reduced to as near 120 KIAS as practical, which eliminates high speed problems.

If the aircraft is not controllable, ejection must be accomplished at whatever speed exists since ejection offers the only opportunity for survival.

OXYGEN SYSTEM MALFUNCTION

EXCESSIVE OXYGEN CONSUMPTION

A running check should be maintained of the oxygen system pressure during any mission above 20,000 feet to preclude the possibility of running out of oxygen above a safe altitude if a leak should occur. The oxygen pressure should be checked at frequent intervals and plotted on the oxygen consumption chart every 30 minutes. Due to the cooling of the oxygen supply, the oxygen pressure drop will be fairly rapid at first. This is explained in detail in Section IV. Generally in the first hour of flight, the pressure will drop about 500 to 600 psi and should begin to fall off at a reduced rate after this. If on the U-2C, the rate of consumption after two hours of flight at cruise altitude exceeds the values listed below, project the oxygen plots to determine if the mission can be completed with at least 500 psi remaining at destination. If this is impossible the mission shall be discontinued. Since the U-2F has a larger oxygen supply, the maximum desired consumption rates in psi/hr are 75% of the U-2C rates.

If the oxygen consumption is excessive and the oxygen pressure has dropped below the minimum shown for the altitude being flown, the pilot will descend so as to maintain the minimum pressure/altitude ratios.

U-2C OXYGEN PRESSURE TABLE

CRUISE ALTITUDE	MAX. DESIRED OXY CONSUMPTION	MIN. OXY PRESS.
60 M +	100 psi/hr	500 psi
50 M	150 psi/hr	250 psi
40 M	175 psi/hr	-
30 M	200 psi/hr	-

If the oxygen pressure has fallen below 250 psi the pilot will descend to 10,000 feet cabin altitude.

WARNING

Serious consideration must be given to the type of mission being flown, as well as the availability of suitable alternates, before continuing a mission when oxygen consumption is even slightly in excess of that desired.

NOTE

If at high altitude with no ship oxygen, pull the green apple and make a fast descent to a 10,000 foot cabin altitude, and remove facepiece.

WARNING

If the pilot is unable to regain an oxygen supply, he will have to open the facepiece in flight. Use the autopilot emergency descent and exercise extreme caution to avoid possible flameout and accompanying loss of pressurization.

OXYGEN DIFFICULTY

If the pilot suspects oxygen difficulty due to symptoms of hypoxia or by noting the helmet bladder collapsing, he should take the following action:

1. PULL GREEN APPLE.

2. ACTUATE PRESS-TO-TEST.

3. INSPECT SUIT T-FITTINGS, QUICK DISCONNECT, AND FACEPIECE.

If difficulty is not corrected:

4. MAKE EMERGENCY DESCENT (to cabin altitude of 10,000 feet and remove facepiece).

NOTE

If at any time during flight the pilot finds it necessary to open the facepiece, other than for complete oxygen depletion, an immediate descent will be made to an aircraft altitude of 45,000 feet prior to opening the facepiece. The facepiece will then be checked for proper seating by using the mirrors in the aircraft and will also be subjected to Press-to-Test pressure. If these checks are satisfactory, a re-ascent may be made if necessary.

In case the bends are experienced, descend until the symptoms disappear.

SECTION III

AF (C)-1-1

AIRCRAFT OXYGEN SYSTEM DIFFICULTY

In the event that the oxygen warning primary off light comes on:

1. Turn primary valve OFF to permit it to warm up and thaw out.

2. Turn cabin heat up as high as practical to aid in thawing primary valve.

3. Descend to a safe altitude if operational requirements permit. If it is not possible to descend to a safe altitude:

   Every ten minutes turn primary valve ON momentarily to determine whether the blockage has cleared. When this check results in the PRIM OFF light going out, the primary valve should be left ON.

If when operating on the secondary valve only, the oxygen warning LOW PRESS light comes on:

1. Turn primary valve ON.

2. Recheck lights. Both lights should be out if primary valve is clear. If both lights remain on, pull green apple and descend to a safe altitude and remove facepiece.

HYDRAULIC SYSTEM EMERGENCY OPERATION

COMPLETE HYDRAULIC PRESSURE LOSS

There is no action which can be taken to restore a hydraulic system after failure; however, its effects can be minimized.

When range is critical, subsequent descent should be made carefully reducing engine power to minimum EPR limit and descending with the aircraft in the clean configuration.

NOTE

The landing gear must not be extended to expedite descent. The landing gear cannot be retracted without hydraulic pressure and, if extended, will cause a considerable loss in range.

When range is not a critical factor the descent may be expedited by first pulling the Landing Gear Control Circuit Breaker and then extending the gear by means of the manual emergency system.

If the hydraulic pressure gage indicates zero, or very low pressure, check the fuel pressure gage for a supporting indication, operate the speed brakes to determine if the hydraulic pressure gage is faulty. In case the pressure is actually lost, the wing flaps and speed brakes will be inoperable in addition to the fuel boost pump. The landing gear can be extended by use of emergency procedures. The landing will be made as outlined under Landing With Flaps Retracted.

WARNING

Since the wing flaps cannot be moved to the GUST position, the placard speed for FAIRED must be observed.

The auxiliary boost pump, if on, will take over for the normal boost pump in the event of hydraulic failure. During high altitude cruise with the auxiliary boost on, the auxiliary pump pressure is above 10 psi but below the normal fuel pressure. Failure of the hydraulic system will show zero hydraulic pressure and a decrease in fuel pressure. A failure of this type does not require a descent, and the mission may be completed.

3-28

PARTIAL HYDRAULIC PRESSURE LOSS

A case of partial pressure loss may be due to pump compensator trouble or to an internal leak in the system. All services will operate but at reduced speed. If the pressure is too low, the landing gear may not fully retract.

LANDING GEAR EMERGENCY EXTENSION

The emergency landing gear extension system is a manual free fall system. Neither electrical power nor hydraulic pressure is required to operate this system.

1. Gear handle - down.

2. Emergency gear release - pulled.

3. Gear down and locked indication - checked.

   (In case of an electrical failure, no indication is available.)

4. Airspeed - increase if necessary to assist in locking the gear.

   Do not exceed limiting airspeed for existing configuration and condition.

LANDING GEAR CONTROL LEVER LATCH MALFUNCTION

If there is a malfunction in the landing gear control lever latch it may not be possible to move the landing gear control lever to the down position. The landing gear can be extended to the down and locked position by the following procedure.

1. Landing gear control circuit breaker - pulled.

2. Uplock release - pulled.

3. Gear down and locked indication - checked. (In case of electrical failure, no indication is available.)

4. Airspeed - increase if necessary to assist in locking the gear.

   Do not exceed limiting airspeed for existing configuration and condition.

After the landing gear has been extended by this procedure, resetting the landing gear control circuit breaker will cause the gear to retract.

AIRCRAFT FUEL SYSTEM MALFUNCTION

FUEL BOOST PUMP FAILURE

Failure of the fuel boost pump or its hyddraulic drive motor is indicated on the fuel pressure gage in the cockpit. The auxiliary boost pump will take over in the event of boost pump failure provided the boost pump and continuous ignition switch is on and the AC generator is on and operating. A failure of this nature will be evidenced by a decrease in fuel pressure. The mission may be completed with this condition.

FUEL CROSS TRANSFER PUMP FAILURE

If wing heaviness due to uneven fuel flow cannot be corrected by actuation of the cross transfer pump switch, another method may be used. The wing heaviness can usually be corrected by flying in a very slight continuous skid with the heavy wing high.

SECTION III   AF (C)-1-1

FUEL COUNTER MALFUNCTION

A fuel totalizing system malfunction may be evidenced by the stopping of the fuel counter or by erratic movement of the counter. If a fuel versus time plot was being made up to the time of the malfunction, this curve can be extended to determine approximate fuel remaining at any later time. **Plan the landing with ample estimated reserve.** The fuel counter operation should be checked frequently during the climb, since a temporary stoppage during this period of high fuel flow can result in a large error in the fuel remaining reading. A malfunction of this type should be suspected if the reading is higher than predicted at the beginning of the cruise.

FUEL LOW LEVEL INDICATION DURING CLIMB

If the low level light comes on during the climb, it must be assumed that the fuel is not transferring into the sump tank and only 50 gallons of fuel is available to continue flight. When this occurs, reduce power to level off and then proceed to the nearest suitable landing field. When landing field is within gliding range, descend and land as soon as practicable from a flameout pattern.

1. Declare emergency.

2. Proceed to closest suitable landing field.

3. Dump fuel as necessary.

4. Land as soon as practicable from a flameout pattern.

NOTE

If the fuel low level light comes on during or shortly after takeoff, it may be due to the fact that at high power, and low altitude, the engine can use fuel out of the sump tank faster than it can be replenished from the wing tanks. Note the totalizer reading, throttle back, stay in the area of the field. If the fuel low level light does not go out after 20 more gallons have been used, execute the above procedure.

FUEL LOW LEVEL INDICATION IN LEVEL FLIGHT AT ALTITUDE

Since there is no fuel quantity gage in the airplane, the fuel low level warning light is very important. When the light comes on during normal level flight, it must be taken as an indication that only 50 gallons of fuel remain in the aircraft to complete the flight. The engine should not be shut down immediately. If landing field is not within normal descent range, the throttle should be retarded to idle or minimum EPR and a clean configuration descent to 65,000 feet should be made on speed schedule. Add power at 65,000 feet and maintain altitude.

If field was approximately 300 miles or less distance (zero wind) when warning light came on, the engine may be shut off after 30 gallons of fuel is expended. Establish maximum distance glide, see figure 3-2. Restart the engine over the field and enter the flameout pattern for landing.

If maximum range is desired, the flight should be continued at 65,000 feet until engine flames out and then the maximum range glide configuration should be established.

FUEL LOW LEVEL INDICATION DURING DESCENT

If the FUEL LOW LEVEL light comes on during descent, it is probably caused by one of two conditions: a) Wing tanks contain fuel which is not being fed into the sump tank because of iced-over ram air scoops on the fuel tank suction relief valves, or b) there are actually only 50 gallons of fuel remaining.

Low Level Indication - Iced Scoops Suspected

If the totalizer reads considerably more than 50 and if icing conditions have been encountered on the flight, there probably is fuel in the wing tanks which is not feeding into the sump tank. In this case, fuel flow into the sump tank from the wing tanks can be restored by increasing engine RPM and maintaining it as high as practicable until the light goes out.

Actual Low Level

If the totalizer reads near 50 and the FUEL LOW LEVEL light comes on during descent, it should be assumed that only 50 gallons of fuel remain to continue flight. Proceed as follows:

1. Declare an emergency.

2. Continue descent to closest suitable landing field.

3. Land as soon as practicable from a flameout pattern.

DROP TANK SYSTEM MALFUNCTIONS

Drop Tank Fuel Does Not Feed

Evidence of this condition is indicated by illumination of either or both low pressure warning lights before 215 gallons of fuel have been used; or by failure of empty and low pressure lights to illuminate by the time 250 gallons of fuel have been used. If either of the above conditions exists, it must be assumed that the drop tank fuel is not feeding - proceed as follows:

1. Pull left and right drop tank circuit breakers.

2. Proceed to a suitable drop area.

3. At 50,000 feet or lower, establish an airspeed of 150 knots (IAS) in level flight.

4. Drop tanks - arm and release over drop area.

WARNING

Be prepared for a lateral trim change when drop tanks release because one drop tank may have been heavier than the other.

NOTE

If operational considerations dictate landing with drop tanks, proceed as follows:

1. Pull left and right drop tank circuit breakers.

2. Approaching destination, dump fuel selectively to obtain optimum landing configuration.

3. Check lateral trim. Transfer fuel as necessary to obtain the same wheel position noted after takeoff.

4. Check stall characteristics.

5. Land using normal pattern.

Drop Tank(s) Does Not Release

If one or both drop tank(s) does not release proceed as follows:

1. Approaching destination, dump fuel selectively to obtain optimum landing configuration.

2. Check lateral trim. Transfer fuel as necessary to obtain the same wheel position noted after takeoff.

3. Check stall characteristics.

4. Land using normal pattern.

ELECTRICAL SYSTEM MALFUNCTION

DC SYSTEM

High Ammeter Reading

The maximum design generator output corresponds to an ammeter reading of 225 amperes. If the ammeter shows a continuous reading above the maximum value allowable for the equipment being carried proceed as follows:

Check all electrically operated equipment such as radios, light switches, etc., by turning them OFF one at a time to determine whether any of these units is the cause of the high ammeter reading. If one of these units is the cause of the high reading and is not vital to the completion of the mission, turn it off. If it is vital, land as soon as practicable, operating without it unless it is necessary to the completion of the landing.

1. Check electrical equipment individually to isolate cause.

2. Turn off malfunctioning equipment, if unnecessary for further flight.

3. Land as soon as practicable if mission cannot be successfully completed.

Zero Ammeter Reading

If the ammeter indicates zero, proceed as follows:

Operate an electrical device, such as the radio transmitter, to ascertain that the gage is indicating an erroneously low reading. The indication normally increases when an electrical device is placed in operation. Press the warning light test button to ensure that the low indication is not caused by a failed generator. If the reading fails to increase, it indicates the ammeter is inoperative. This malfunction cannot be corrected while airborne. Closely monitor electrical equipment during remainder of flight.

Generator Failure

Failure of the generator will be evidenced in the following ways:

a. Generator out light on.

b. Ammeter indicating zero.

c. Battery discharge light on.

In case of generator failure, the battery-generator switch should be moved to the BAT EMERG position. All electrical loads not absolutely needed should be turned off. Examples of the loads with large power requirements are: all radios, AC inverter, aileron shifter, trim tabs, camera equipment, etc.

Flight can be continued in case of generator failure so long as battery power remains for indispensable items such as facepiece heat. However, it should be realized that depletion of the battery will make all electrical items inoperable. If possible, descend to 45,000 feet to descrease the likelihood of flameout and to conserve the battery.

All AC powered instruments, including flight instruments, will be available in case of generator failure since they are powered from the No. 1 or No. 2 main inverter. However, the emergency inverter should be used in lieu of the main inverter, in order to keep battery drain to a minimum. The emergency inverter will supply power to the remote compass, attitude indication, EGT, and the ADF/VOR needle.

1. BAT & GEN switch - BAT EMERG.

2. Inverter - EMERG.

3. Unnecessary electrical equipment - off.

4. Compass switch to COMP.

5. Battery - use only as required.

6. Descend to 45,000 feet if possible.

7. Land as soon as practicable.

Complete Electrical Failure

Flight can be continued without electrical power and all fuel, except drop tank fuel, will feed. This would be caused by generator failure followed by complete battery depletion or some other factor. The important items to remember are:

a. If the battery suffers a rapid discharge, all the electrical loads should be disconnected. This will give the battery an opportunity to build up a charge. This charge would not support a heavy load such as a main inverter for very long, but if used intermittently on small loads such as EMERG inverter it might last considerably longer.

b. Facepiece heat will not be available. This will probably make a descent necessary since vision will eventually become obscured. High cabin temperature will help to keep the facepiece clear. If MANUAL heat is selected, the temperature control will impose a load on the battery only when in the COLD or HOT position.

c. Flight can be continued without electrical power and all fuel except drop tank fuel will feed. The fuel counter will be inoperative and fuel quantity remaining must be estimated.

d. All electrical instruments will fail with the pointers in operating range and the landing gear indicators will show barber pole.

e. The landing gear can be extended with the emergency release. There will be no indication of gear position.

SECTION III   AF (C)-1-1

f. Trim tabs will remain as set prior to electrical failure. Prior to landing it may be possible to retrim the aircraft if battery charge rebuilds.

g. The turn needle of the turn and slip indicator will not be operable.

h. An air start usually is impossible. If conditions make an air start attempt imperative, all load should be removed from the battery during descent to allow it to obtain the greatest charge possible. The start should be attempted at 35,000 feet or lower, as only one attempt could be made before the battery would again be depleted.

If possible, descend to 45,000 feet to decrease the likelihood of a flameout. The flight should be terminated as soon as practicable.

1. Descend to 45,000 feet if possible.

2. Land as soon as practicable.

AC SYSTEM

No. 1 and/or No. 2 Inverter Failure (Main Inverters)

Inverter failure is indicated by the glowing of the INVERTER OUT warning light. When the No. 1 inverter fails, the No. 2 inverter is selected by placing the inverter switch in the No. 2 position. There will be a loss of the navigation system when the No. 2 inverter is selected. All other equipment will continue to operate from the No. 2 inverter. If the No. 2 inverter also fails, flight will be handicapped by loss of the autopilot, remote compass, ADF, VOR, attitude indicator, and most of the engine instruments including the EGT. Placing the inverter switch in the EMERG position will restore power to the remote compass, ADF. VOR needle, attitude indicator and EGT.

NOTE

Compass control must be placed in COMP position for operation on the emergency inverter.

1. No. 1 inverter failure - inverter switch - No. 2.

2. If inverter out light continues to glow - inverter switch - emerg.

3. Switch compass to comp if on emerg inverter.

No. 1, No. 2, and Emergency Inverter Failure

If the No. 1, No. 2, and Emergency inverters fail the EGT indication will not be available. It will then be advisable to level off to prevent engine overtemperature. If both the main inverters and the EMERG inverter have failed, flight should be terminated as soon as practicable.

1. Level off and maintain altitude.

2. Land as soon as practicable.

AC Generator Failure

Failure of the engine-driven AC generator is evidenced by the glowing of the AC GENERATOR OUT warning light. Normal flight will not be affected. Turn switch to the OFF/RESET position and return to AC GENERATOR position. If light does not extinguish, turn switch to OFF/RESET position.

3-34

NOTE

If light comes on due to underfrequency (below 80% engine RPM) no reset is required. If light does not go out after returning engine speed to above 80% RPM, reset should be attempted.

When the AC generator is switched off or fails, a relay automatically transfers Systems 9, 12, and the continuous ignition system to the inverter bus. System 13 is automatically transferred to the No. 2 inverter and power to the Dead Reckoning Navigation system is automatically removed, making it inoperative. All other AC generator-powered equipment becomes inoperative.

NOTE

Under the foregoing conditions, with the No. 1 inverter carrying the inverter bus loads, the No. 2 inverter automatically operates and assumes only the System 13 load. No. 2 inverter will also assume the System 13 load if the inverter switch is in the OFF position.

If the No. 2 inverter has been selected (because of No. 1 inverter failure) at the time of ac generator failure, the No. 2 inverter will assume System 9, 12, and continuous ignition loads, but System 13 will be rendered inoperable.

ENGINE OIL SYSTEM MALFUNCTION

The normal oil pressure operating range is 40 to 55 psi, with an allowable fluctuation of plus or minus 2 psi. If the pressure fails below 40 psi or exceeds 55 psi, the mission should be discontinued and the oil pressure carefully monitored. If the pressure falls below 35 or exceeds 60 psi, an immediate landing should be made if operational conditions permit. The engine should be operated at the lowest practical RPM since continued operation with an oil pressure below 35 or above 60 psi may result in eventual engine failure.

1. Oil pressure below 40 or above 55 psi - abort mission.

2. Oil pressure below 35 or above 60 psi - operate at lowest possible RPM and land as soon as practicable.

TRIM TAB MALFUNCTION

RUNAWAY TRIM ON MANUAL FLIGHT

A fault in the trim circuit may cause the trim tab to move beyond the desired setting. Countering the motion by actuation of the trim switch may not correct the out of trim condition.

The amount of control force necessary to overcome the runaway trim effect will vary with airplane configuration, and will materially increase at high airspeeds.

NOSE UP TAB

1. YOKE - APPLY COUNTERING FORCE.
2. TRIM POWER SWITCH - OFF.
3. AIRSPEED - AS SLOW AS PRACTICAL.
4. Gust Control - faired.
5. Speed Brakes - retract (unless required to reduce speed).

NOTE

Manual trim operation may be rechecked by momentarily moving the trim power switch on and noting the tab position indicator. If no movement is detected actuate the trim switch to relieve control forces. When control forces are relieved turn trim power switch off. Repeat above as required during remainder of flight.

CAUTION

Extreme care must be exercised to assure that tab is moving in proper direction to alleviate control forces. Turn trim power off immediately if tab moves in direction of runaway.

NOSE DOWN TAB

1. YOKE - APPLY COUNTERING FORCE.
2. TRIM POWER SWITCH - OFF.
3. AIRSPEED - AS SLOW AS PRACTICAL.
4. Gust Control - gust.
5. Speed Brakes - extend.

SECTION III

RUNAWAY TRIM ON AUTOPILOT

The automatic trim failure system will activate and will automatically stop a runaway pitch trim when a pitch trim coupler fails. In this event the Auto Trim Failure light on the auto trim control will illuminate and the pitch trim index will deflect up or down, depending upon the direction the runaway started. Other runaway trim conditions can develop with no warning to the pilot other than an indication on the elevator trim tab position indicator.

If Auto Trim Failure Light Illuminates

1. Push the auto trim control. This will shift control of the trim tab to manual trim (wheel switch) and will light the manual trim light. The auto trim fail light will go out.

2. Use manual trim switch to center pitch trim index to a near neutral position for remainder of the flight while on autopilot.

If Runaway Trim Occurs With or Without Warning

1. YOKE - APPLY COUNTERING FORCE.
2. AUTOPILOT - DISCONNECT (WHEEL SWITCH).
3. MANUAL TRIM POWER SWITCH - OFF.
4. Proceed as outlined for runaway trim on manual flight.

GUST CONTROL MALFUNCTIONS

There are several types of malfunctions of the gust control. Since the gust control is so important to the structural capabilities of the airplane, the pilot must always be positive that both wing flaps and ailerons have operated properly. There are several indications to show whether these surfaces are in the GUST or FAIRED position:

a. The gust control lights.
   These lights are illuminated for GUST position.

b. The wing flap position indicator.
   The indicator shows minus 4° for GUST position.

c. Change in pitch trim.
   GUST position gives a nose up change.

d. Amount of aileron travel.
   Wheel travel is reduced and forces are lighter in GUST.

e. Aileron faired light.

All of these indications should be observed before exceeding the gust control FAIRED limitations.

1. If either flaps or ailerons malfunction - check circuit breakers while observing Faired limitations.

2. Ailerons only in gust - observe Faired limitations. Return ailerons to Faired when desired.

   Ailerons only in gust will give wing bending alleviation.

3. Ailerons stuck in gust - ensure even fuel load and land from straight in approach.

   Aileron control will be limited.

4. Flaps only in gust - return to Faired.

5. Flaps stuck in gust - reduce air speed to 130 knots. Plan no-flap landing. Add 5 knots to threshold speed.

   An adverse wing load distribution is created if only the wing flaps are in gust position.

6. Flaps and ailerons stuck in gust - balance fuel load and make straight in approach. Plan no-flap landing. Add 5 knots to threshold speed.

   Aileron control will be limited and "no-flap" drag characteristics must be considered for landing.

7. Flaps continue down - return to gust.

In the event that the wing flaps continue past the Faired position while returning to Faired from Gust, return the gust control switch to gust. A mission may be flown in gust with only a 1.5% loss in range. Normal landing can be made with flaps.

WING FLAPS MALFUNCTION

There is no emergency method of extending the wing flaps. If the flaps cannot be extended due to hydraulic system failure or other cause, a landing must be made as outlined in Landing With Wing Flaps Retracted.

ASYMMETRICAL FLAP CONDITION

In the event that the left and right flap sections do not remain synchronized, they will be automatically stopped. This will occur after the two sides have reached a maximum difference of 5° in angle. They will remain in the stopped position and no cockpit action can be taken.

RETRACTABLE STALL STRIP MALFUNCTION

A malfunction of this system is not likely, since it is a reliable, cable-operated system. Possible malfunction and proper procedure to follow is as follows:

Either one or both stall strips fail to extend:

1. Return T-handle to the retract position.

2. At a safe altitude, stall or slow-fly the aircraft close enough to a stall to establish characteristics to expect on landing.

NOTE

The stall characteristics of the airplane are not as good with the stall strips retracted - there may be a tendency for a wing to drop.

3. Land with stall strips retracted.

FUEL DUMP SYSTEM MALFUNCTION

If control wheel offsets 10° from original position after dumping is started, turn off dump switches on light side and check for continued dumping on heavy side. If wheel returns to the original wing balanced position, resume dumping from both wings. If it is apparent that one dump valve is malfunctioning, dumping should be discontinued and fuel balanced with the transfer pump.

Dumping under an emergency situation when an immediate landing is necessary, particular attention must be paid to the wing balance condition. With a malfunctioning dump valve, a wing-heavy condition can occur so rapidly that not enough time would be available before landing to balance the wings with the transfer pump.

1. If one wing does not dump, close fuel dump valves.

2. Balance fuel load with transfer pump.

3. Do not attempt further dumping.

AIR CONDITIONING MALFUNCTION

AUTOMATIC TEMPERATURE CONTROL FAILURE

If the cockpit air becomes too hot or too cold and does not respond to the automatic temperature control rheostat, try the manual system. Turn the cabin heat selector to manual HOT or manual COLD as needed.

Changed 15 August 1968

SECTION III AF (C)-1-1

CABIN COOLER FAILURE

If air remains hot after manual adjustment, it indicates failure of the cabin cooler.

1. Reduce engine power. This lowers the cooler inlet air temperature.

2. Check defroster - off.

3. Ram air - on if necessary. Note that the cockpit will be depressurized.

COOLER BYPASS VALVE FAILURE

If air remains cold after manual adjustment, it indicates failure of the cabin cooler bypass valve in a position which diverts all or most of the cabin air supply through the cooler unit.

1. Reduce engine power. This lowers the cold air flow.

2. Defroster - on.

3. Descend to a warmer altitude if necessary.

4. Ram air - on if necessary. Note that the cockpit will be depressurized.

PRESSURIZATION SYSTEM EMERGENCY OPERATION

The most frequent cause of loss of cockpit pressurization is engine flameout. There is no corrective action except to descend for an engine restart. If cockpit pressure is partially or completely lost for reason other than engine flameout:

1. TIGHTEN HELMET TIEDOWN CABLE.

2. PRESS-TO-TEST OXYGEN REGULATOR - if suit doesn't inflate at 42,000 feet cabin altitude.

3. Seal valve - on.

4. Cockpit air temperature - hot.

5. Defroster - on.

6. Descend to a safe altitude if possible. If it is necessary to remain at maximum altitude and the pressure suit is operating properly, do so for as long as physiological factors permit.

NOTE

If it is necessary to remain at altitude or descend to low altitudes without pressurization, the windshield may become frosted. The cockpit fan and defroster should be used to minimize this condition.

7. Inverter and command radio - as required.

With equipment bay pressure of 40,000 feet or above, the normal inverter may fail. In the event of loss of Q-Bay pressure switch to emergency inverter.

Since ambient enviromental limits may be exceeded at low Q-Bay pressure the UHF radio transmitter should not be used above 50,000 feet Q-Bay pressure, except in emergency.

3-38

An altitude switch will cut off the 618T-3 transmitter and illuminate the HF transmitter disabled light when the cabin altitude exceeds 39,000 feet. Normal receiver operation will continue. However, transmission will not be possible until the altitude decreases below 35,000 feet.

An override switch is provided to permit transmissions at any altitude when required.

FACEPIECE HEAT FAILURE

Failure of the facepiece heat is evidenced by fogging of the facepiece. A completely separate system is attached to the left side of the facepiece. To obtain emergency face heat, remove normal system wires on the right side of the facepiece and push in EMER FACE HEAT circuit breaker. There is no automatic temperature control in the emergency system. Limited temperature control can be obtained by pulling and resetting EMER FACE HEAT circuit breaker.

1. Normal face heat wires - disconnect.

2. Emergency face heat circuit breaker - in.

3. Normal face heat circuit breaker - pulled.

NOTE

In an emergency requiring more than the normal system can provide (such as canopy loss in flight) using both normal and emergency systems simultaneously, provide approximately twice the heat otherwise available. The face heat control can still be used but the full cold position is hotter than normal full hot alone.

COMMUNICATIONS FAILURE

Some communications failures are traceable to the pilots personal equipment wiring. In these cases, service can sometimes be restored by using the bypass system, which is on the left radio console. This system is plugged directly into the helmet plug.

POGO RELEASE FAILURE

If a pogo fails to release after takeoff, make a straight-in low approach. After the aircraft is slowed to threshold speed, the pogo will normally release. If the pogo does not release, dump fuel before landing.

WARNING

Do not fly over populated areas with a hung pogo as it may drop at any time.

MINIMUM RUN TAKEOFF

Since the airplane has excellent takeoff performance using normal procedures, the need for a special minimum run takeoff run can be shortened by use of the following techniques:

1. The gross weight of the aircraft should be held to a minimum by accurately controlling the fuel load at the bare minimum required for the mission.

2. Place wing flaps at 15-degree position.

3. Advance throttle to 80% prior to releasing brakes.

4. Release brakes and throttle-burst engine to maximum power.

SECTION III   AF (C)-1-1

5. Hold control column neutral at the start of takeoff run.

6. Fly aircraft off runway in a two-point attitude, using care not to stall after becoming airborne.

FULL POWER TAKEOFF

Full power takeoff technique is the same as for the gate power takeoff. However, the acceleration to takeoff airspeed will be much faster. The possibility of engine overspeed and/or overtemp is quite likely with the throttle at the full open position. Since the pilot's attention during the takeoff roll must be devoted mainly to the control of the aircraft, it is recommended that when the throttle is advanced to the full power position, that it then be backed off slightly to insure that an engine overspeed or overtemp condition will not occur. After climb airspeed is established, the throttle can then be adjusted to the desired EGT.

EPR SYSTEM MALFUNCTION

If the EPR system is suspect due to obvious indicator error, the flight may be continued by using the EGT versus OAT schedule.

Pages 3-41 through 3-66 Deleted   Changed 6 November 1967

AF (C)-1-1

SECTION 4A
DESCRIPTION AND OPERATION OF AUXILIARY EQUIPMENT

TABLE OF CONTENTS

	PAGE		PAGE
AHRS (COMPASS) SYSTEM	4-1	AN/ARC-59 RADIO COMPASS	4-42
AFCS AUTOPILOT	4-4	ARC TYPE 15F VHF NAVIGATION EQUIPMENT	4-42
AIR CONDITIONING AND PRESSURIZATION SYSTEM	4-7	ATC SYSTEM	4-43
WINDSHIELD AND CANOPY DEFROSTING	4-11	AN/APN-135 RENDEZ BEACON	4-46
HATCH WINDOW HEATER	4-12	ARC TYPE 12 VHF TRANSCEIVER	4-47
PITOT HEAT SWITCH	4-13	AN/ARC-3 VHF RADIO SET	4-47
OXYGEN SYSTEM	4-13	AUTOMATIC OBSERVER	4-48
PERSONAL EQUIPMENT	4-15	F-2 FOIL SYSTEM	4-48
LIGHTING EQUIPMENT	4-18	B/400 RATE RATE METER	4-50
DRIFTSIGHT SYSTEM	4-19	P-3 PLATFORM	4-51
SEXTANT SYSTEM	4-23	MISCELLANEOUS EQUIPMENT	4-53
AERIAL REFUELING SYSTEM	4-29	SUNSHADE	4-53
COMMUNICATION AND ASSOCIATED ELECTRONIC EQUIPMENT	4-39	DEFROSTER FAN	4-53
MICROPHONE AND HEADSET	4-39	MAP CASE	4-53
INTERPHONE CONTROL	4-39	RELIEF BOTTLE	4-53
COMM SELECTOR PANEL	4-39	INTERNAL REAR VIEW MIRROR	4-53
AN/ARC-34 UHF RADIO	4-40	WINDSHIELD SWAB	4-53
618T-3 HF TRANSCEIVER	4-40	RUBBER CONE	4-54
		PINGER	4-54
		DEAD RECKONING NAVIGATION	4-54

ATTITUDE HEADING REFERENCE SYSTEM (COMPASS)

The compass system is a remote indicating, gyro-stabilized system designed for use in all latitudes. It is part of the Attitude Heading Reference System (AHRS). It consists of a magnetic flux valve located in the left wingtip of the airplane; a gyro platform, adapter, and amplifier located in the nose compartment; and a pilots directional indicator, and a compass control panel located on the right side console. See figure 4-1.

INDICATORS

Directional Indicator - During fast slaving the directional indicator should be observed to establish the proper heading. The indicator card will show immediately whether the set heading switch is being turned in the proper direction.

Synchronization Indicator - The SYNC IND is located above the SET HDG switch on the compass control panel. The position of the needle indicates to the pilot how well the gyro signal is synchronized with the flux valve. The indicator is very sensitive and will deviate from centered for short periods especially during turns. The SYNC IND is operative only in the SLAVE mode.

CONTROLS

Compass Power Switch

There is one magnetically-held compass power switch on the right-hand console. This switch only controls compass power when external power is applied to the receptacle. If DC external power is applied, a relay opens the compass power ac circuit and turns the compass off. When external power is removed, the relay relaxes and switches the compass power back on to the inverter bus. For ground checks of the compass, this automatic cutoff feature can be overridden by placing the compass power switch in EXTERNAL. With the switch in AIRCRAFT position and ground power applied, the compass will be off. When no ground power is provided, the compass will be on when the inverter is on, regardless of switch position.

Compass Control Panel

This panel, figure 4-1, is located on the right forward side console.

The panel contains the following:

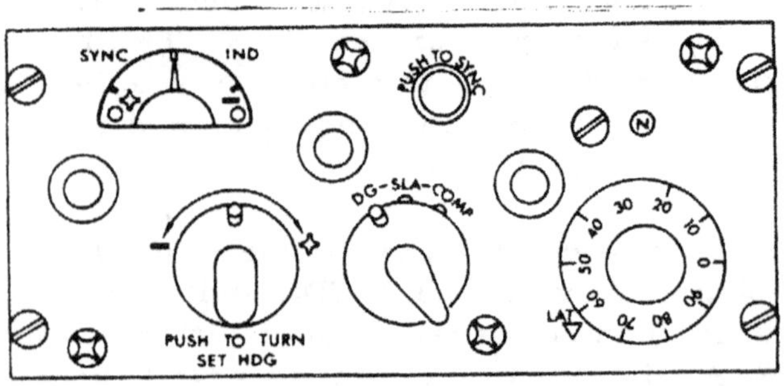

Mode selector switch - positioning of this switch determines which source of heading information is presented on the heading indicator (BDHI) Card: DG - free directional gyro with correction for the effect of the earths rotation; SLA - magnetically slaved, gyro stabilized; or COMP - magnetically slaved, but not gyro stabilized (bypasses the gyro platform).

SET HDG (set heading) switch - in DG mode, this switch can be used to slew the BDHI card left or right to any desired heading for grid navigation purposes. It also can be used to manually synchronize the system in the slaved mode.

PUSH TO SYNC button switch - momentarily pushing this button fast synchronizes the system.

LAT - this control is used only when operating as a free directional gyro. The latitude control knob is turned to set the existing latitude to the index mark on the panel. This provides electrical compensation for drift caused by the rotation of the earth. When set properly, the gyro spin-axis stays fixed relative to the earth and provides a stable directional reference. When operating as a magnetically slaved system, the latitude control is disconnected and may be left at any setting.

The proper hemisphere is set by switching to N for northern or S for southern hemisphere. This is done by turning a stem slotted for a screwdriver. This selector is located adjacent to a window displaying either the N or S.

Figure 4-1

OPERATION

The system may be operated in three modes; Slaved, Directional Gyro, and Comp. The system requires both DC and inverter power.

SLAVED MODE

Operation as a magnetically slaved compass (SLA mode) may be used in any locality except latitudes in excess of 70° N or S, or in areas where severe magnetic distortion occurs. When operated in this mode, the flux valve defines the direction of magnetic north. The electrical signals from the flux valve are supplied to the gyro platform and amplifier of the AHRS. These components, in turn, furnish the magnetic heading information to the autopilot and to the directional indicator (BDHI Card) in the cockpit. The compass heading signal errors will not exceed plus or minus 0.25 degree in this mode.

DIRECTIONAL GYRO MODE

Operation as a free directional gyro (DG mode) may be used in any latitude but is especially useful where the magnetic field is weak or distorted. Distortion occurs at any latitude when in close proximity to large masses of iron. Above approximately 70° latitude the declination of the earths magnetic lines of force creates errors in a magnetically slaved system, and the DG mode should be used. In this mode the heading information is supplied to the directional indicator (BDHI Card) and to the autopilot strictly by the orientation of the gyro platform. Heading errors in this mode of operation will not exceed 0.50 degrees per hour.

COMP MODE

In the COMP mode, the heading information is supplied by the flux valve through the AHRS amplifier, bypasses the gyro platform, and goes to the directional indicator (BDHI Card). The COMP mode should be used whenever the red flag on the BDHI appears or slave mode is inoperative. The red flag will appear whenever the gyro platform is not functioning properly. If the red flag appears, the autopilot will be inoperative.

NOTE

When operating on the EMERG inverter, it will be necessary to select COMP mode in order to maintain heading information on the BDHI.

The heading information in this mode should be accurate to within plus or minus 2 degrees but the card will not be steady because the flux valve signals are not gyro stabilized.

SLAVING

The system automatically synchronizes at 50 degrees per second when first turned on and then automatically assumes the normal slave rate of 1° to 2° per minute. Synchronization can be speeded up, if necessary, by momentarily pushing the PUSH TO SYNC button.

Operating Procedure

1. Inverter - No. 1 or No. 2.

2. Compass Power Switch - EXTERNAL (if ground power is connected). Switch is spring-loaded to AIRCRAFT position at all other times.

3. Compass Mode Selector - Set for desired mode.

4. Latitude - Set for proper latitude and hemisphere when operating in the DG mode.

5. SYNC-IND - Check needle centered.

SECTION IV A AF (C)-1-1

AUTOMATIC FLIGHT CONTROL (AUTOPILOT) SYSTEM

The autopilot is a Lear 201 Automatic Flight Control (AFCS) system. Attitude and heading reference information is provided by a two-gyro, all-attitude platform; heading information is synchronized to the magnetic flux valve signal. The autopilot effects control of the aircraft through electric servos connected to the ships control cable system.

The basic autopilot is engaged by actuating the three individual AXIS CUTOFF switches, the POWER switch and the ENGAGE switch located on the autopilot control panel. In this basic mode Roll and Pitch attitude, Heading Hold and Yaw damping is provided to the aircraft control system. Pitch attitude can be varied incrementally by rotation of the control panel PITCH DN-UP attitude wheel, and turns can be made by rotation of the control panel PUSH TO TURN knob. Maximum bank angles are limited to 30°.

The autopilot operates on either the No. 1 or No. 2 main inverter, but will not operate on the emergency inverter.

There are three potentiometer trim adjustment knobs located on the lower left-hand portion of the instrument panel: roll rate, pitch rate, and Mach. Adjustment of these pots in flight will not normally be necessary, but under certain conditions, adjustment of one or more of these pots may improve autopilot performance. Higher gain settings sensitize the response and lower settings desensitize the response.

Autopilot evaluation should be made only, at cruise altitudes.

The autopilot may be completely disengaged either by the disengage switch on the right control wheel grip or by switching the ENGAGE switch to OFF on the control panel. Also, any of the three axes can be selectively disengaged. This enables continued use of good axes in case malfunctions occur in one or two axes only. The AUTOPILOT DISCONNECT light on the annunciator panel and the MASTER CAUTION light should illuminate. A dual frequency tone will be heard in the headset when the AUTOPILOT DISCONNECT light illuminates. The audio tone can be silenced by actuation of the A/P AUD CO pushbutton switch. The tone disabling circuit is removed by re-engaging of any one or all of the autopilot axes. The tone will not function with the autopilot POWER switch in OFF position or if subcomponent failures occur which do not illuminate the AUTOPILOT DISCONNECT light.

WARNING

Subcomponent failures can occur in the autopilot system which can render any one axis or all three axes inoperative without benefit of the MASTER CAUTION light, the AUTOPILOT DISCONNECT warning light, or the dual frequency audio tone in the headset. Do not rely exclusively on these indications to determine that the autopilot is engaged and controlling properly.

PITCH AXIS AND MACH HOLD

The pitch axis can be controlled in either attitude or Mach. The altitude hold mode is not installed in this aircraft at present except for the autopilot control panel and amplifier provisions. Mach hold is selected by pressing the MACH button on the control panel and engagement is indicated by a green light inside the MACH button. During Mach hold, the pitch axis is controlled to maintain whatever Mach number existed at

Changed 1 September 1968

the time Mach hold was selected. Incremental changes in Mach/airspeed (approximately plus or minus 5 knots) can be made by rotating the Mach trim wheel while in the Mach hold mode. The Mach trim wheel has no stops in either direction - it is electrically centered each time Mach hold is selected. If the Mach trim wheel has been used to the limit of its effectiveness, it is merely necessary to press the OFF button, then press the MACH button and once again incremental adjustments can be made about the new Mach trim wheel position.

Use Mach vernier; do not use pitch command knob while in Mach mode.

AUTOMATIC PITCH TRIM

This system automatically moves the elevator tab to minimize elevator control forces being applied by the autopilot servo motor.

If manual trim is desired, while on autopilot, or if the AUTO TRIM FAILURE light is illuminated, Manual trim may be selected by actuating the AUTO TRIM FAILURE switch.

The TRIM PWR switch must be in the ON position for either manual or automatic trim operation.

A split lamp and pushbutton switch assembly and a pushbutton switch are located above the annunciator panel on the lower right side of the instrument panel. The upper section of the light, AUTO TRIM FAILURE, is part of the master caution system and illuminates together with the CAUTION light whenever the autopilot attempts to trim the tab to increase control forces. Actuating the switch section of the light assembly transfers the pitch trim control from automatic to manual. The lower section of the light, MANUAL TRIM, illuminates whenever the autopilot is engaged and manual pitch trim control has been selected. The lower section of the assembly is an advisory light. The CAUTION light does not illuminate in manual trim operation. Both sections of the light assembly are dimmed when the INSTRUMENT LTS control is moved out of the OFF position.

The AUTO TRIM FAIL TEST switch is located to the right of the split light assembly. It is used for a system check. Pressing the button simulates a failure in the automatic pitch trim control circuit. If the AUTO TRIM FAILURE and CAUTION lights illuminate, proper operation of the auto trim failure indicating system is indicated.

ROLL AXIS - HEADING HOLD, HEADING SELECT

Selection of the HDG mode slaves the autopilot to whatever heading is selected by the "bug" on the BDHI instrument and engagement is indicated by a green light inside the HDG button. When in this mode, turns can be accomplished by using the preselect heading knob as a turn knob. If it is desired to select a heading to be used later, push the OFF button and select the desired heading by turning the preselect heading knob located on the BDHI instrument. When it is desired to take up the preselected heading, push the HDG button and the airplane will turn to the new heading. All turns made in the preselect heading mode are made at a bank angle of 12°. Use of the turn knob will automatically remove the HDG mode. If the HDG select mode is not engaged, the autopilot holds whatever heading the airplane was on at the time the roll axis was engaged or whatever heading existed at the time the HDG mode was turned off. There is a NAV mode button on the control panel, but it is not used on this aircraft.

On heading select mode, bank angle should not be referenced against the attitude indicator since this instrument is subject to precession during turns; also the performance of HDG select mode should be evaluated at cruise altitude only.

AILERON AND ELEVATOR TRIM INDICES

Aileron and elevator trim indices are located on the left-hand instrument panel. These indicate (by deflecting bars) the direction and magnitude of the signals being supplied by the AHRS to the respective autopilot servo clutch motors. The pilot should note the magnitude and direction of the bar deflection before engaging or disengaging the autopilot because they will show him what sort of transients to expect.

NOTE

The aircraft should be trimmed as close to hands off as possible to avoid transients upon engagement.

Trim Indices		At Engagement	At Disengagement
Roll	(AIL)	Roll Left	Roll Right
Pitch	(EL)	Nose Down	Nose Up

CONTROL PANEL

The control panel, figure 4-2, is located on the right console, just forward of the MK III driftsight hand control.

NOTE

Autopilot control panel has NAV and ALT mode buttons, but these are not not used in this airplane.

AFCS CONTROL PANEL

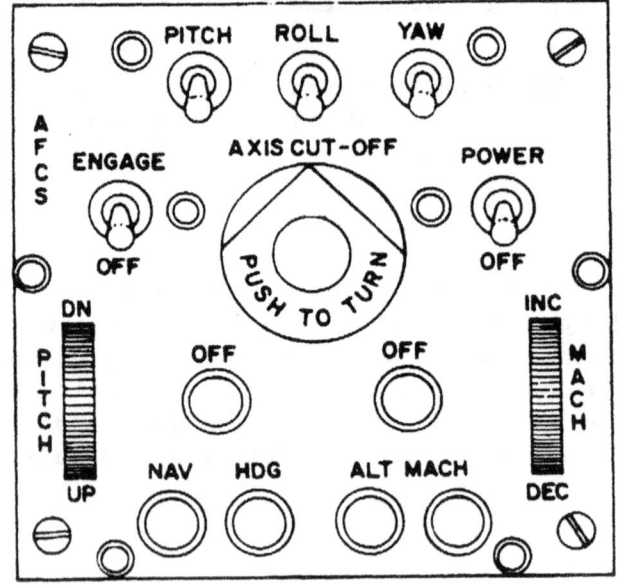

Figure 4-2

OPERATION

1. Inverter - ON No. 1 or No. 2.

2. AHRS operating (indicated on BDHI flag retracted).

3. Airplane trimmed in all three axes.

4. Power switch - POWER.

5. Pitch, Roll, and Yaw axis cutoff switches - checked ON.

6. PUSH TO TURN knob in the center (detent) position.

7. Check trim indices near neutral.

8. Engage switch - ENGAGE.

9. Select MACH hold mode if desired - green light on.

10. Select HDG SEL mode if desired - green light on.

11. When desired, disengage using control wheel button or control panel switch.

The autopilot should engage smoothly. If there is a displacement of the rudder pedals on engagement, it indicates that either the ground adjustable rudder tab was not properly set and the airplane was in yaw, or the yaw axis of the autopilot was not synchronized with the normal rudder trail position. If the rudder is displaced by the autopilot, the yaw introduced tending to make the airplane turn, will be compensated for by the roll axis holding a wing low in the direction opposite the rudder displacement. In this case, the YAW AXIS CUTOFF switch may be turned off. Any wing heaviness that may develop later in the flight as shown by the roll trim index will be the result of unbalanced fuel and should be corrected with the fuel cross transfer pump.

NOTE

The roll trim index may be used as an indication of proper fuel balance. Fuel can be transferred until the roll trim index is level. The control wheel position should be cross checked with the roll trim index during fuel transfer.

Should oscillations in pitch or roll be present after autopilot engagement, it is a probable indication that the rate gain pots are set too low and if chatter is encountered, that they are set too high. They should be adjusted to a higher or lower sensitivity as necessary to obtain stable conditions. If even light turbulence is present, there may be oscillations in the pitch and roll trim indices before and after autopilot engagement. This is a normal condition and should not concern the pilot.

The basic autopilot will hold the heading of the airplane at time of engagement and if a turn is required, it may be made by pressing the PUSH TO TURN knob and rotating it in the direction of the desired turn. The further the knob is turned, the greater the angle of bank up to 30 degree limit. As the desired heading is approached, start rolling the wings level by moving the turn knob back toward the center position. Hold the turn knob down until the airplane is on the desired heading and then release it in the center (detent) position. The airplane will hold the heading existing at the time the knob is returned to the detent position. Turns may also be made by use of the bug on the BDHI in one of two ways. Place the bug on the heading of the airplane, press the HDG select button and then rotate the bug around the BDHI to the desired heading. The second method is to place the bug on any desired heading and then press the HDG select button to initiate the turn to that heading. Turns made in the HDG select mode will be made at 12 degrees angle of bank.

The autopilot will turn the shortest way toward a heading selected by the bug, therefore, if a turn of more than 180 degrees is desired, move the bug to some heading less than 180 degrees in the direction the turn is to be made. As the turn progresses, move the bug on toward the desired heading. If the bug is moved more than 180 degrees from the existing airplane heading, the autopilot will reverse the turn.

The Mach hold function may be used at any time the autopilot is engaged but its use becomes more valuable at the higher altitudes since the airplane can be flown at the recommended climb speed of 160 KIAS up to 50,000 feet quite easily by controlling the attitude with the pitch control wheel. It is recommended that the Mach hold mode be engaged on the scheduled climb speed for the altitude of the airplane. To engage, push the MACH button and check for illumination of the green light inside the button. There should be little, if any, displacement of the control column on engagement. The climb schedule is not a constant Mach number and the Mach trim wheel should be used to keep the indicated airspeed on the climb schedule.

The Mach hold mode should control the pitch attitude to hold the Mach precisely constant. If pitch oscillations are present it is a probable indication that the Mach gain pot is set too low and if abrupt corrections or chatter in the control column is present, the setting is too high. The Mach gain pot should be adjusted for optimum performance but only after best performance has been achieved in pitch while on basic autopilot. Once the autopilot has been adjusted for peak performance, it should not require further adjustments in flight.

AIR CONDITIONING AND PRESSURIZATION SYSTEM

Air conditioning and pressurization air is bled from the compressor section of the engine as shown in figure 4-3. It is cooled by a conventional aircraft type refrigeration unit, ducted through a water separator, and directed into the cockpit through five outlets. Cockpit exhaust air is discharged into the equipment bay through a cockpit pressure regulator valve which automatically maintains the cockpit pressurization schedule. This schedule is maintained regardless of the equipment bay pressure level, as the valve senses only true cockpit-to-ambient differential. After circulating through the equipment bay, this air is exhausted through a second pressure regulator valve into the unpressurized landing gear bay. This valve maintains essentially the same pressurization schedule in the equipment bay as that of the cockpit. This schedule provides a cabin altitude of 29 to 30 thousand feet at an airplane altitude of 70 thousand feet as shown in figure 4-4.

In addition to the regulator valves, there is a safety relief valve in both the cockpit and the equipment bay to limit compartment pressures to structurally safe values in the event of a failed-closed regulator. To prevent excessive crushing pressures on the fuselage during high rates of descent, vacuum relief is provided by automatic opening of both the regulator and the safety relief valves.

CABIN HEAT SELECTOR

Located below the landing gear handle, the four position cabin heat selector is used to select either manual or automatic temperature control. Normally, the selector is left in the AUTO position, in which case it is only necessary for the pilot to set the automatic temperature control to a position which gives a satisfactory temperature. The cockpit thermostat will then control the electrical positioning of the refrigerator bypass valve so as to maintain the selected temperature. A separate thermostat in the duct downstream of the water separator acts as an automatic override to limit the duct temperature to a safe high value.

If the pilot desires more heat than can be obtained with the selector set on AUTO and the automatic temperature control adjusted full HOT, or if the automatic control system malfunctions, the cabin heat selector should be set to HOLD. This removes the automatic temperature control rheostat from the system. Temperature control is then accomplished by manually regulating the electrically-driven refrigerator bypass valve. To drive the valve further open for more heat, turn the selector to HOT for just a few seconds and then return it to HOLD. Cooler settings are obtained by momentarily selecting COLD and then returning to HOLD. A short time is required for the cabin inlet temperature to stabilize after repositioning the bypass valve.

Never leave the cabin heat selector in the HOT position. This will drive the refrigerator bypass valve full open and will result in extremely hot air to the cockpit. An appreciable amount of time will be required to cool the ducts sufficiently for the cabin inlet air to again be comfortable.

SECTION IV A

AIR CONDITIONING SYSTEM

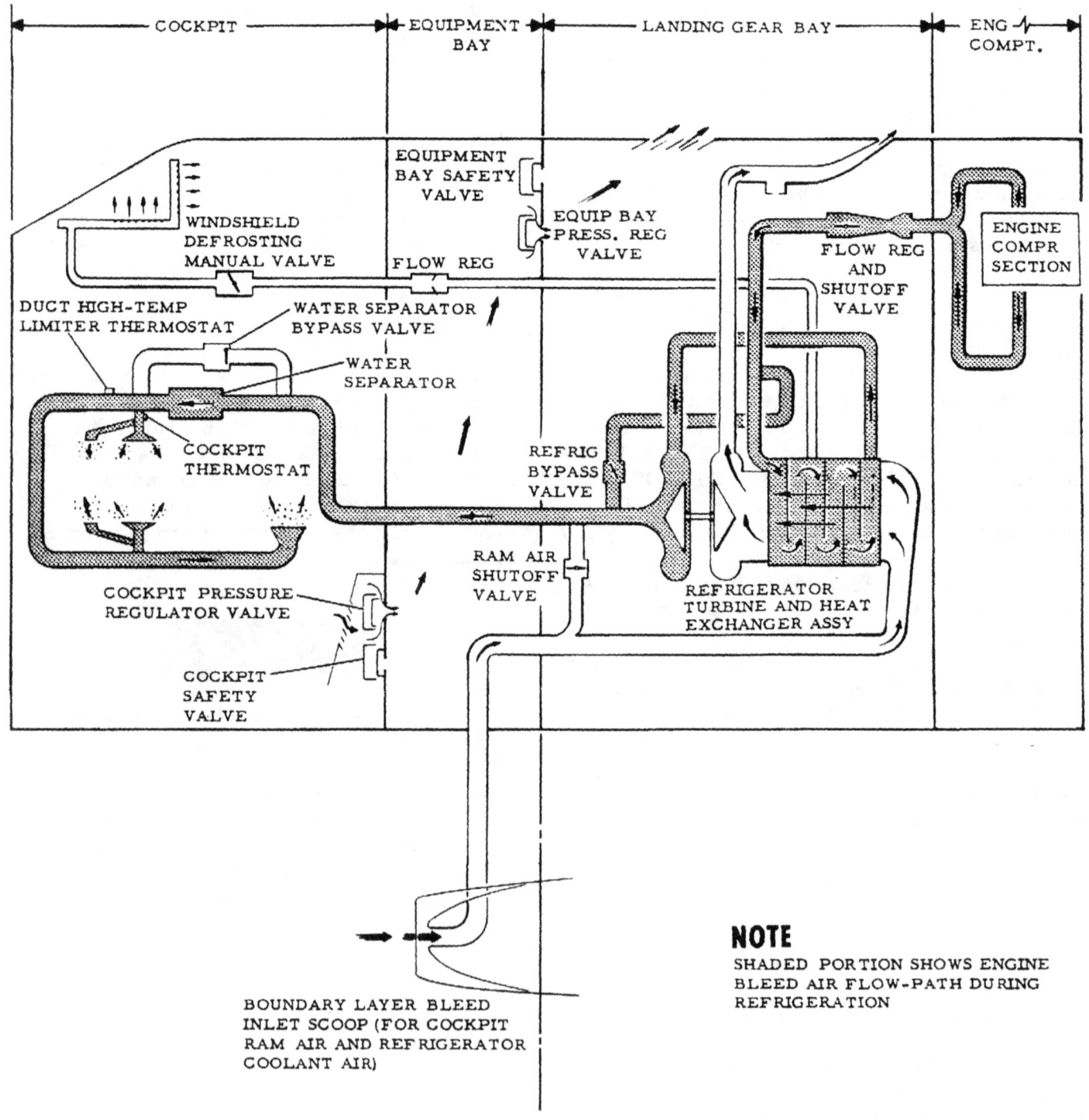

Figure 4-3

PRESSURIZATION SCHEDULE

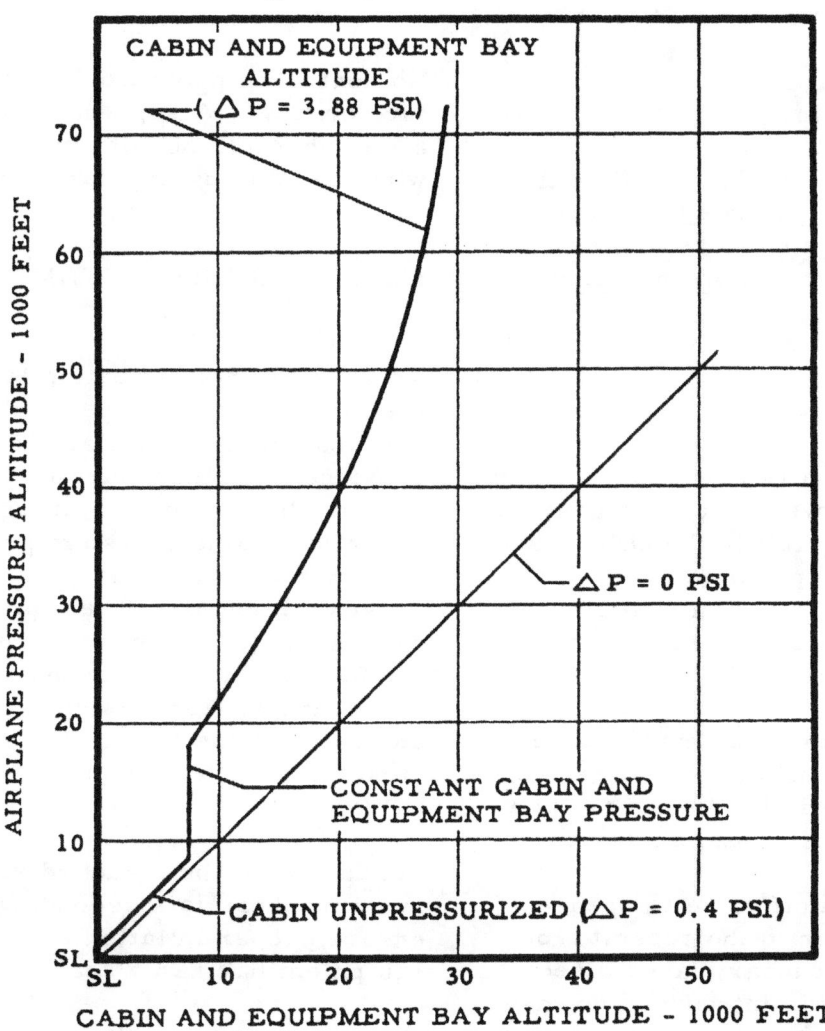

Figure 4-4

CABIN HEAT AUTOMATIC TEMPERATURE CONTROL

The cabin heat automatic temperature control rheostat is operated by a knob and shaft concentric with the cabin heat selector knob and shaft. This control is used to set the desired cabin temperature whenever the selector is in AUTO. When the selector is in any other position, the rheostat is not effective.

NOTE

It is recommended that the cabin temperature at cruise altitude be maintained on the high side to assist in keeping the camera equipment warm.

RAM AIR SWITCH

The ram air switch is a two position, guarded, toggle type located above the landing gear handle. It is normally left in the OFF position, allowing normal operation of the air conditioning and pressurization system. Three primary functions are performed by turning the ram air switch ON.

a. The flow regulator and shutoff valve is closed, blocking the flow of engine compressor bleed air into the air conditioning and pressurization system.

b. The ram air shutoff valve is opened, supplying ram air to the cockpit from the right-hand boundary layer bleed scoop. This air is introduced downstream of the turbine discharge, passing through the water separator and into the cabin.

c. The cockpit and equipment safety relief valves are opened, dumping all pressurization.

As a secondary function, turning the ram air switch ON runs the refrigerator bypass valve full closed, preparing the system for repressurization.

NOTE

It takes approximately three seconds for the ram air system to operate after turning the switch ON.

No control is provided for ram air temperature or flow. Also, since the engine bleed air is shut off, no hot air is available for windshield defogging during ram operation.

SEAL PRESSURE SYSTEM

To prevent pressurization leakage around the canopy and the upper and lower equipment bay hatches, inflatable seals are provided. Air pressure for the seals is supplied by the engine through a regulator which maintains a system pressure of approximately 17 to 18 psi. A check valve prevents bleed off of seal pressure in the event of engine flameout. Selection of ram air ON does not deflate the seals, since the engine air for this system is obtained upstream of the air conditioning and pressurization system regulator and shutoff valve.

In addition to the standard seal pressure system, there is a secondary system to ensure proper inflation of the canopy and equipment bay hatch seals.

The secondary system consists of a 57 cubic inch nitrogen bottle located in the main landing gear well and connected to the standard seal pressure line downstream of the check valve in this line. In the line from the nitrogen bottle are two regulators and a check valve. The bottle is normally serviced to about 1500 psi. The first regulator is a high pressure regulator which drops the pressure to about 400 psi. The second regulator is a low pressure regulator which drops the pressure to about 11 psi. When the pressure in the standard seal system drops below 11 psi, the nitrogen bottle will automatically supply regulated pressure to keep the seals inflated if the canopy and Q-Bay seal controls are on. No pilot action is required other than monitoring the pressurization altimeter. An unusually high equipment bay altitude indicates a probable loss in seal pressure and should be reported for corrective action. Loss of equipment bay pressure will not result in loss of cabin pressure.

CANOPY AND HATCH SEAL CONTROLS

Two manually operated seal pressure valves are provided on the left side console; one is to control the pressure to the canopy seal, the other to the hatch seals. Each of these valves is a three-port, two-position type. Selecting OFF connects the respective seal to the exhaust port (cockpit ambient) while blocking the pressure source; selecting ON connects the pressure source to the seal while blocking the exhaust port.

PRESSURIZATION ALTIMETER

An altimeter, located on the lower right side of the instrument panel and labled CABIN & EQUIP BAY ALTITUDE; is provided to monitor both cockpit and equipment bay pressures.

PRESSURIZATION ALTIMETER SELECTOR SWITCH

A two-position selector switch is provided on the upper right side console to connect either the cockpit or the equipment bay pressure to the pressurization altimeter.

NOTE

It is recommended that the pressurization altimeter selector switch normally be left in the EQUIP BAY ALTITUDE position since it is not possible to lose cabin pressure without losing equipment bay pressure.

WINDSHIELD AND CANOPY DEFROSTING

PRIMARY DEFROSTING SYSTEM

The windshield and canopy primary defrosting system is supplied by hot bleed air from the second stage of the air conditioning and pressurization system refrigeration unit. This air is ducted through a flow limiting regulator to a manifold along the bottom edge of the windshield and the forward edge of the canopy.

CANOPY DEFROST CONTROL VALVE

A manually operated valve adjusts the flow below the regulated maximum. The canopy defrost control is a push-pull lever located on the right console and may be set to any position.

AUXILIARY DEFROSTING SYSTEM

An auxiliary windshield defrost system is used when the primary engine bleed air system is incapable of clearing the windshield. This may occur during high altitude cruise or idle descent. The system uses a 1200 watt, DC powered heater-blower mounted forward of the driftsight. The blower exhausts over the front windshield panel and ducts are provided to direct air over the windshield side panels. A three-position HTR BLOWER-OFF-BLOWER switch is located on the upper right-hand console. The switch is moved down for blower only operation, and up for heater and blower operation.

An amber light glows when the HTR BLOWER position is selected. The light does not glow when the BLOWER position is selected.

NOTE

The amber light will cycle on and off as warning that an overheat condition exists. If this occurs, the switch should be moved either to the center OFF position or to the BLOWER position. Only the blower portion of the heater blower will operate in the event the DC generator is off the line.

The blower only can be used continuously if desired, but the heater-blower should be used only when absolutely necessary.

The heater-blower directs very hot air directly against the windshield and should be used very sparingly to prevent windshield crazing.

DEFROST FAN

A small two-speed, rubber-bladed fan mounted on the left side of the canopy constitutes a second defrosting system. The general circulation provided by this fan aids in keeping the canopy free of frost.

The fan is also useful during engine out descents as an aid to windshield defrosting.

DEFROST FAN SWITCH

A three position center OFF switch is installed on the left console for operating the defrost fan. The forward position is for HI and the aft position for LO speed.

HATCH WINDOW HEATER SYSTEM

With high humidity conditions, moisture can be deposited on the cold inner surface of the hatch windows. In order to prevent this, a heater-blower system is installed in the lower hatch. A small blower with a 1000-watt heater attached is ducted to each window so that the warm air flows directly across the inner surface.

CONTROLS

The blower is connected so that is is always running for normal flight. There is no control from the cockpit. However, in case of generator failure the blower will be disconnected and cannot drain the battery. The heater switch is on the lower instrument panel. This is a two position switch for heater ON or OFF.

OPERATION

In order to obtain maximum benefit from the heater-blower system, the heater should be turned ON at takeoff and remain on throughout the flight. It should be turned OFF after landing.

A red light on the annunciator panel glows when the hatch heater switch is in the ON position and an overheat condition exists. The heater switch should be turned off if this occurs. The light will also illuminate when the switch is turned on and no heater element installed.

PITOT HEAT SWITCH

A single switch on the lower instrument panel operates heater elements for both pitot masts.

OXYGEN SYSTEM

The high pressure dual oxygen system consists of three (four for U-2F) 514 cubic inch cylinders, a filler valve, check valves, two pressure reducers, two shutoff valves, a high pressure gage, two pressure switches and two warning lights as shown in figure 4-5. The high pressure storage cylinders are located in the left-hand cheek area with the filler on the left side of the fuselage just below the wing leading edge. The check valves and lines are arranged so that a minimum of oxygen will be lost to the pilot if a line becomes blocked or broken.

WARNING

Smoking in the cockpit is prohibited.

The system is normally serviced to 1800 psi and has the capacity to supply a normal flow of oxygen to the pilot for a period of time in excess of the maximum duration of a mission. See Section II for oxygen average consumption schedule for a maximum high altitude mission. The high pressure gage is located on the left forward panel.

U-2C OXYGEN DEPLETION RATES

CRUISE ALTITUDE	MAX. DESIRED OXY CONSUMPTION
60 +	100 psi/hr
50 M	150 psi/hr
40 M	175 psi/hr
30 M	200 psi/hr

NOTE

As the aircraft ascends to high altitudes where the temperature is normally quite low, the oxygen cylinders become chilled. As the cylinders grow colder, the oxygen gage pressure is reduced, sometimes rather rapidly. With a 100° F decrease in temperature in the cylinders, the gage pressure can be expected to drop 20%. This rapid fall in pressure is occasionally a cause for unnecessary alarm. All the oxygen is still there and as the aircraft descends to warmer altitudes, the pressure will tend to rise again so that the rate of oxygen usage may appear to be slower than normal. A rapid drop in oxygen pressure while in level flight or during descent, is not due to falling temperatures. If this occurs, leakage or loss of oxygen must be suspected.

OXYGEN SYSTEM

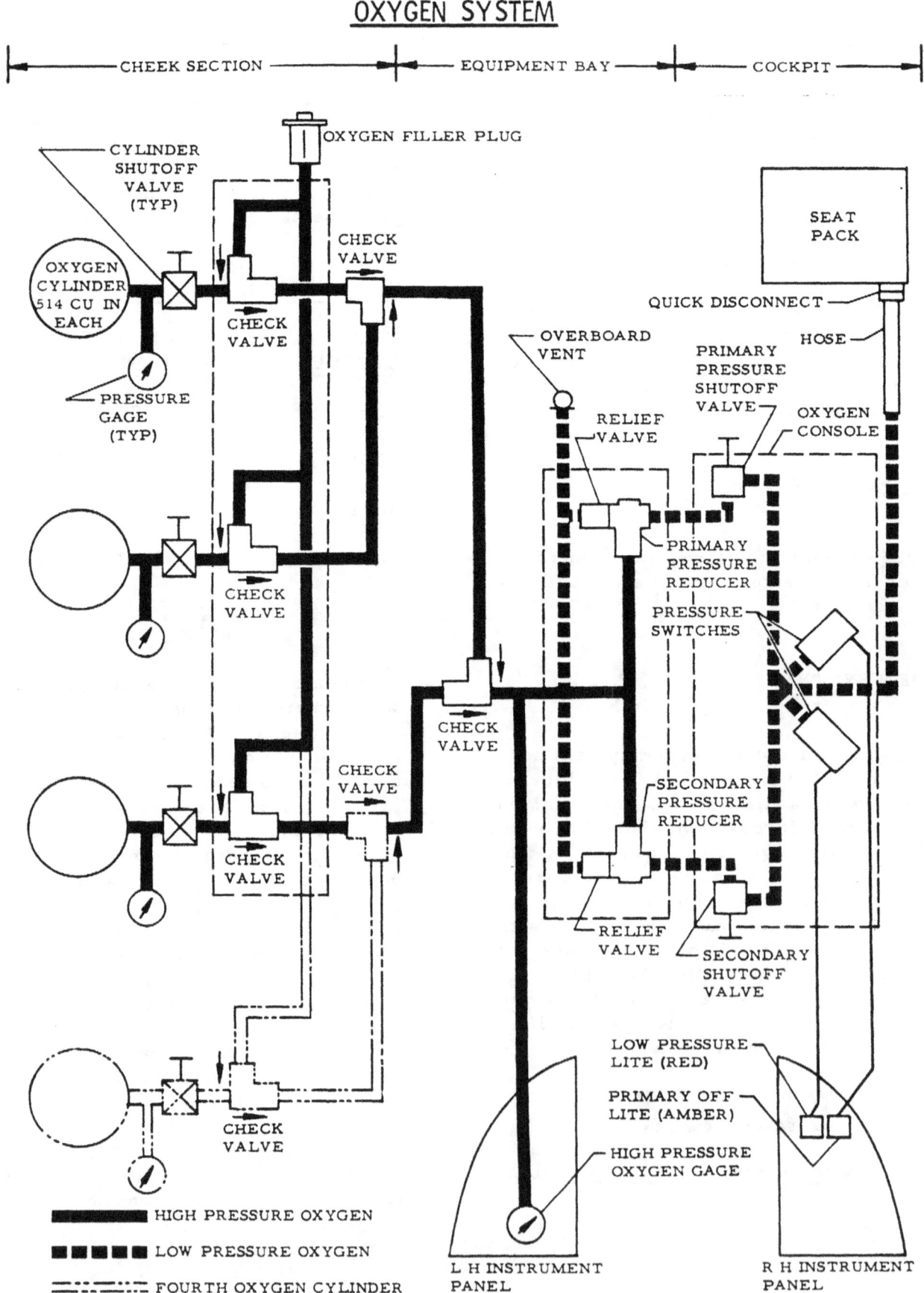

Figure 4-5

The oxygen control panel is on the left console. It includes two shutoff valves and two pressure switches. Two lights, actuated by the two pressure switches, are located on the annunciator panel: a red OXY WARN LOW PRESS light and an amber OXY WARN PRI OFF light. Either of these lights operates in conjunction with the Master Caution light.

The dual pressure reducer system is provided to prevent the blockage of one reducer from cutting off the pilot's oxygen supply. Icing is the most probable cause of blockage of a pressure reducer. Therefore, when blockage of the primary reducer does occur, it is assumed to be due to water in the oxygen and this will eventually result in blockage of the secondary reducer unless corrective measures are taken as detailed in EMERGENCY PROCEDURES. The approximate pressure settings of the reducers and the low pressure warning switches are:

Primary reducer	80 psi
Secondary reducer	60 psi
Primary off light (amber)	70 psi
Low pressure light (red)	50 psi

NOTE

The two low pressure warning lights should be used as the indication of proper operation of the primary and secondary pressure reducers. Pressure reducer and pressure switch settings require close tolerance for compatibility. They are checked by the ground crew with a precision gage supplied with the oxygen system checkout equipment as prescribed in the maintenance manual.

All aircraft are fitted with a quick disconnect that is stowed at the base of the control column and which mates with the seat pack portion. This disconnect system has a check valve to prevent oxygen flow when the pilots personal equipment is disconnected.

PERSONAL EQUIPMENT

The pilots specialized personal equipment includes a pressure suit, matching helmet and seat pack. The seat pack is attached to the parachute harness and stays with the pilot on ejection.

SEAT PACK

The following items are contained in the seat pack: survival gear, oxygen regulator, and emergency oxygen supply.

A press-to-test button is located on the seat pack. Holding this button down causes the oxygen regulator to increase both the breathing and suit capstan pressures. This is used periodically to check the regulator.

SEAT PACK DISCONNECT

The quick disconnect fitting connects the seat pack to the ships oxygen supply and also provides the electrical connections for face heat and communications.

Attachment is accomplished by aligning the two halves and pressing them together. When properly latched in, the release knob resets.

A cable connects the ships half of the disconnect to the floor of the aircraft.

During the initial travel of the seat when ejecting or if the pilot stands up in the cockpit the cable pulls the latching pip pin which disengages the two halves of the disconnect. Under these conditions the emergency oxygen supply is automatically activated.

The disconnect can be manually released by pulling up on the release knob on the upper half of the disconnect.

This motion releases the two halves of the disconnect but does not activate the emergency oxygen supply.

SECTION IV A AF (C)-1-1

EMERGENCY OXYGEN SUPPLY

The emergency oxygen supply is contained in a 57-cubic inch high pressure bottle fastened to the seat pack. It is provided to supply suit and breathing pressures in the event that the ships system becomes inoperable and/or during ejection. This bottle will furnish oxygen for a period of 15 to 20 minutes, which is sufficient for emergency descent to a safe altitude.

An emergency oxygen high pressure gage shows through a hole in the top of the pack. A bottle is charged to a minimum of 1800 psi. Actuation of the emergency oxygen supply is accomplished by extraction of a release pin from its engaged position. A flexible cable is connected to this pin and the other end terminated in a cable knob, commonly called the green apple. Seat packs for use in aircraft with the automatic disconnect have a second cable attached to the release pin and terminated in an automatic lever arm in the seat pack half of the disconnect. Separation of the disconnect halves during ejection automatically extracts the pin. Manual extraction is accomplished by pulling the green apple.

NOTE

Once the emergency oxygen supply has been activated, it cannot be deactivated until the bottle empties. Therefore, the bottle must be recharged on the ground whenever the release pin has been extracted.

When the emergency oxygen supply is actuated, a reducer drops the high pressure to 50 psi. A check valve at the input to the seat pack regulator prevents flow from the seat pack to the aircraft system. If the green apple has been pulled, the pilot must check the emergency oxygen supply gage for depletion at least every 10 minutes. If the emergency pressure drops, the pilot should make a fast descent to a cabin altitude of 10,000 feet or less.

A relief valve is provided to prevent an excess of pressure from being delivered to the seat pack regulator by a malfunctioning reducer. This valve is set to operate at 130 (± 10) psi.

SURVIVAL GEAR

The survival gear is carried in a fiberglas container within the seat pack. Contents of the container vary and are normally dictated by a specific mission. A nylon lanyard 25 feet long attaches the survival gear container to the seat pack. A 'Switlick' slidefastener quick release allows the container and a life raft, if carried, to be released from the seat pack during a parachute descent. The life raft is attached to the survival gear container lanyard 15 feet from the seat pack attaching point.

FACEPIECE

A molded facepiece has four embedded wire heat elements which prevent fogging or frosting. All elements are jumpered together for parallel operation.

There are two dc circuits to the facepiece elements; one for normal operation, the other for emergency operation. The normal operation circuit is protected by a FACE HEAT circuit breaker; the emergency operation circuit is protected by an EMERG FACE circuit breaker. A FACE HEAT rheostat, located below the landing gear handle, controls the heat level in normal operation. A preset resistor in the emergency heat circuit maintains a suitable, but nonselective temperature.

NORMAL FACE HEAT OPERATION

Power for normal operation is supplied directly from the battery, and when the FACE HEAT circuit breaker is closed, the circuit is completed to the facepiece elements. The EMERG FACE circuit breaker is left open during normal operation. Since power is not taken from the 28-volt dc bus, normal face heat is available so long as energy remains in the battery. In case of dc generator failure, the battery will heat the facepiece regardless of generator-battery switch position. Additional heat may be obtained if required by operating both normal and emergency face heat circuits simultaneously. About twice the heat will be available.

EMERGENCY FACE HEAT

The emergency face heat cord runs from the left-hand console to the seat and then to the left-hand side of the facepiece. The cord is normally connected; however, the emergency face heat circuit breaker remains out.

EMERGENCY FACE HEAT OPERATION

In the event of normal face heat failure, the connector plugs should be checked for proper seating. If condition persists, the normal cord is disconnected at the facepiece, the FACE HEAT circuit is pulled, and the EMER FACE HEAT circuit breaker is then pushed in for intervals as required. Power for emergency operation is taken from the 28-vold dc bus. The generator-battery switch must be in either the BAT or the BAT & GEN position for emergency heat.

Upon ejection, all power to the elements will be disconnected and the facepiece will probably fog over. This situation will exist until the facepiece is removed following automatic parachute deployment at a safe altitude.

In the event of ejection an emergency face heat quick disconnect is provided on the left side of the pilots seat for automatic separation from the aircraft. Another automatic disconnect is provided when the pilot separates from the seat.

PERSONAL EQUIPMENT SELECTION

On all flights above 45,000 feet, the partial pressure suit is mandatory.

A pressure reducing adapter must be used to permit the use of a MBU-5/P or A-13A oxygen mask in lieu of the MA-2 helmet for low altitude flight up to 45,000 feet.

WARNING

The MA-2 pressure helmet will not be worn without the suit as it requires tiedown provisions. Loss of cabin pressure can cause the helmet to rise dangerously high or it could be torn off during ejection causing loss of oxygen facilities.

NORMAL OPERATION

After the pilot enters the cockpit, the primary and secondary oxygen valves are turned on and the facepiece closed. The oxygen system preflight check is made at this time.

SECTION IV A AF (C)-1-1

EMERGENCY OPERATION

As long as the cabin pressure remains below approximately 39,000 feet the seat pack provides breathing pressure only. If the cabin altitude goes above 39,000 feet due to flameout or other cause, the seat pack will pressurize the suit as required. If the suit and breathing pressures do not respond fast enough or high enough as the cabin depressurizes, the press-to-test button should be used to increase the pressure.

If an ejection has been accomplished and a water landing is expected, the life raft should be released and inflated prior to landing. Pulling the red 'Switlik' release on the seat pack unlocks the slidefastener, inflates the life raft, and allows it and the survival gear container to fall free of the seat pack. A lanyard attaching the life raft, survival gear container, and seat pack together keeps them within reach of the pilot upon landing.

LIGHTING EQUIPMENT

LANDING AND TAXI LIGHTS

Two identical nonadjustable, lights are mounted on the main landing gear strut and serve the dual purpose of landing and taxi lighting. The right-hand light points straight ahead and the left-hand light points slightly to the left to illuminate the edge of the runway. These lights may not be operated independently.

Do not leave lights on after gear retraction or operate the lights on the ground any longer than necessary. Without proper cooling the lights will burn out in a very short time.

The switch for operating the landing and taxi lights is located on the left console aft of the throttle and is labeled LANDING LTS.

COCKPIT LIGHTING

Warning Lights

All warning lights are consolidated into a master CAUTION system except for the FIRE and O'HEAT lights, the landing gear warning light and Special Systems indicator lights. A test button is on the right side instrument panel to test the bulbs in the master caution light, annunciator panel, and all other indicator lights which are not checked by other means. The warning lights are dimmed by the initial movement of the instrument light rheostat from the OFF position.

There are two types of lights on the annunciator panel, Warning lights (left and center rows), and Advisory lights (right row). When a warning light on the panel is illuminated by a dangerous situation the master CAUTION light is also lighted. The master CAUTION light capsule is pressed. Any additional failures will then relight the master CAUTION light as before. The Advisory lights on the annunciator panel will not light the master CAUTION light. The only other light which will illuminate with the master CAUTION light is the autopilot auto trim failure light.

Instrument Lights

The primary engine and flight instruments are lighted by small post-type red lights. The instrument light rheostat, located on the upper right side console, controls the intensity of these lights. In the extreme counterclockwise or OFF position, the instrument lights are off and the warning lights are bright. The first clockwise motion of the rheostat dims all warning lights except fire warning, and the master caution light, as well as turning the instrument lights on dimly. Further clockwise motion rotation of the rheostat increases the instrument light intensity. Power for the instrument lights is from the CKPT LTS circuit breaker. The equipment panels on the left and right consoles contain integral lighting which is wired to the CKPT LTS circuit breaker. No intensity control is provided.

Panel Lights

All edge lighted panel instruments are connected to a rheostat next to the instrument light rheostat. It is operated in the same manner as the instrument lights but does not affect the warning lights.

Utility Spotlights

A utility spotlight is provided on each side of the cockpit for added illumination or map reading. Each of these lights contains an integral dimming rheostat and a spring coil type extension cord. Mounting sockets are underneath the left and right canopy sills. To avoid interference caused by these lights when not in use, they are stowed aft on the left and right console. Power is from the SPOT LTS circuit breaker.

NAVIGATION LIGHTS

There are two red rotating beacon lights, one located on the top of the fuselage and the other located below the fuselage.

The lights are not cut out automatically when the DC generator is not operating. Therefore, they must be turned off manually when it is desired to reduce the DC electrical load.

LIGHT SWITCH

A two-position switch (three-position in U-2F) on the lower instrument panel, controls the NAV light. The circuit breaker for this equipment is located in the equipment bay. Power is supplied by the generator, battery, or external source through the main DC bus. The lights can be operated with the DC power switch in either the BAT or GEN and BAT EMERG position.

DRIFTSIGHT SYSTEM

A driftsight system, figure 4-6 is in all aircraft. This equipment consists of an optical tracking system using a combination of mirrors and prisms to project a presentation of the local terrain on a scope in the cockpit. This information is used in pinpoint navigation. The system is composed of a scope or viewing screen, a hand control panel, a periscope and a junction box. All of the systems use the same viewing screen, basic periscope, and hand control.

The driftsight system, when operated by the MARK III hand control, requires only 28-volt DC power from the aircraft electrical system; when the MARK IIIA hand control is incorporated, inverter power is required as well, for operation of a servo system which replaces the manually driven flexible shafts used with the MARK III control.

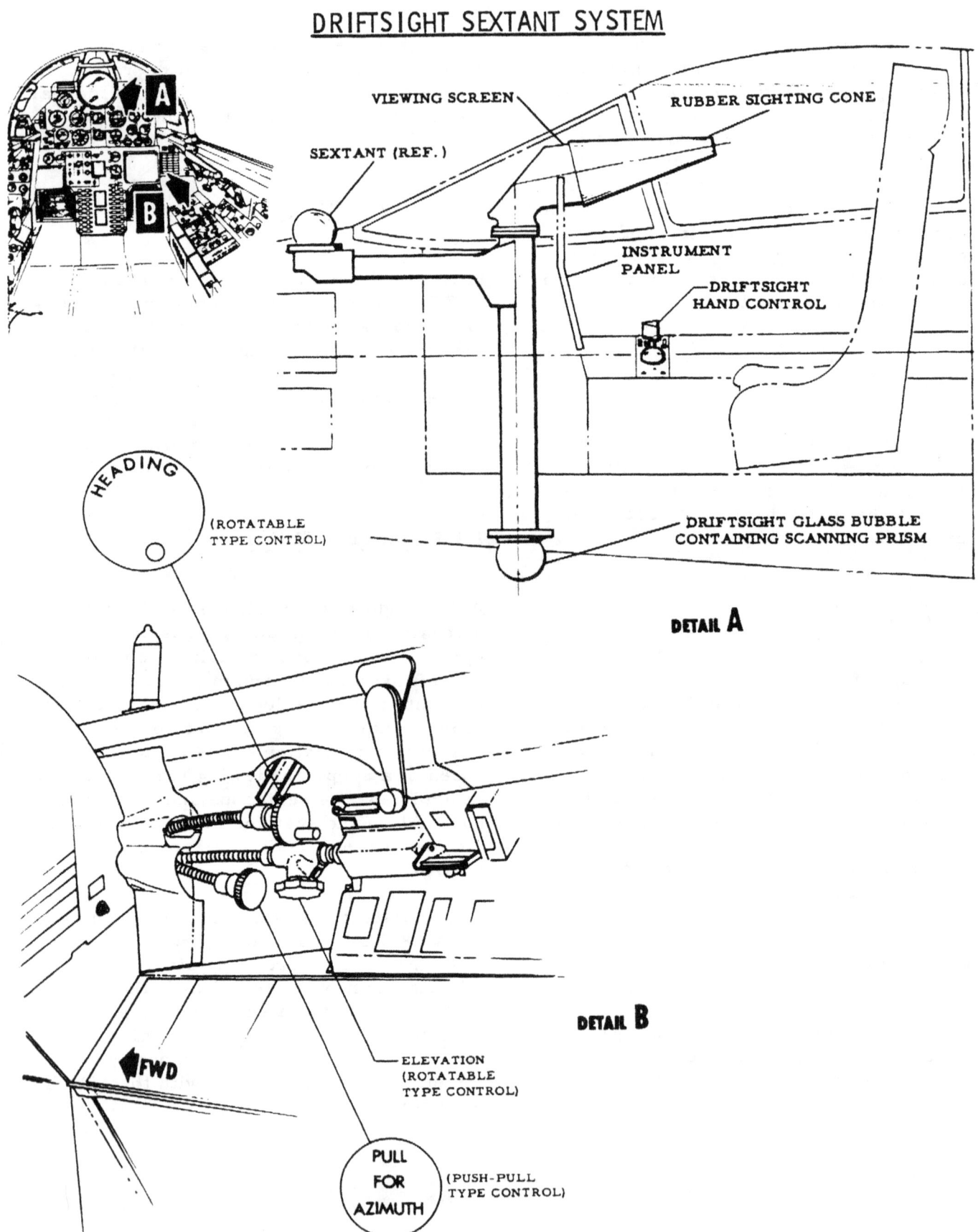

Figure 4-6

VIEWING SCREEN (SCOPE)

The viewing screen is six inches in diameter and is equipped with a drift grid to simplify drift detection. It is located directly above the instrument panel. This is the same screen used in conjunction with the sextant. Below and to the right of the screen is a push-pull knob labeled SEXTANT PULL. For driftsight use, this knob is pushed in.

PERISCOPE

The tracking periscope has a dual prism scanning head. Two magnifications or powers are available; .4X and 1X. These are mechanically varied by an electrically controlled power changer attached to the optical tube.

Scanning in azimuth and elevation is accomplished by controlling a scanning prism at the objective end of the optical system. The scanning prism is located in a nitrogen purged glass bubble underneath the aircraft and just forward of the cockpit.

JUNCTION BOX

The junction box provides mechanical and/or electrical connection points for the various units of the system.

MARK III HAND CONTROL PANEL

A MARK III hand control panel, figures 4-6 and 4-7, is mounted on the right-hand console sill.

MARK III HAND CONTROL PANEL

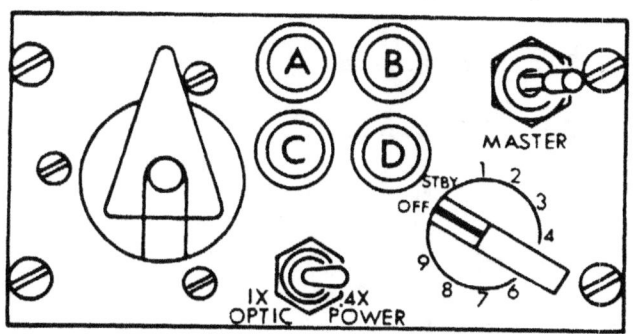

Figure 4-7

Controls

The control panel has a control handle, a master switch, a mode selector switch, and a power changer switch. The control functions are as follows:

Control Handle - The control handle can rotate the scanning head a full 360° in azimuth, and elevate it to an almost horizontal position. This provides complete coverage under the aircraft. When rotating the handle, either clockwise or counterclockwise, regardless of the fore and aft position, a slight detent may be felt. This is to aid the observer in determining when he is looking straight ahead of the aircraft. This aid is only effective when the aircraft is in straight and level flight. The control of the scanning prism is purely mechanical and no electrical power is needed for the scanning operation.

Master Switch - This two-position switch applies power to the tracking camera and provides control power at the Q-Bay military power receptacle, permitting control of special military equipment.

Changed 6 November 1967

4-21

SECTION IV A   AF (C)-1-1

The master switch controls the following systems and equipment:

1. Tracker camera.
2. Delta III camera.
3. "B" camera.
4. System 1.
5. System 1A.
6. System 3.
7. F-2 Foil.
8. A-1 camera.
9. A-2 camera.
10. "H" camera.
11. Military equipment disconnect plug.
12. Any special package requiring power and control through use of military equipment disconnect plug.

Mode Selector Switch - This is an eleven-position switch labeled OFF, STBY, and MODE 1, 2, 3, 4, 5, 6, 7, 8, 9. This switch is unrelated to the driftsight and periscopic sextant. It controls equipment in the equipment bay.

Power Changer Switch - The optical power changer switch changes the magnification of the driftsight. The switch positions are labeled .4X and 1X. Placing the switch in these respective positions causes the scope presentation to be in a .4 to 1 ratio or a 1 to 1 ratio with respect to its unaided field of vision. The field of view for each lens position can be estimated by observing the square reticle in the driftsight presentation which is approximately one nautical mile on a side when viewed from operational altitudes regardless of magnification.

Operation - The operation of the driftsight is relatively simple since there is no automatic tracking function. It is only necessary to select the desired magnification with the power changer switch and then scan using the control handle. It is possible to estimate drift by observing a fixed object on the terrain, but no ground speed estimate is possible.

MARK IIIA HAND CONTROL

The MARK IIIA hand control (see figure 4-8) is in the same location and performs the same functions as the standard MARK III control, but uses servo motors instead of cables to operate the driftsight.

MARK IIIA HAND CONTROL

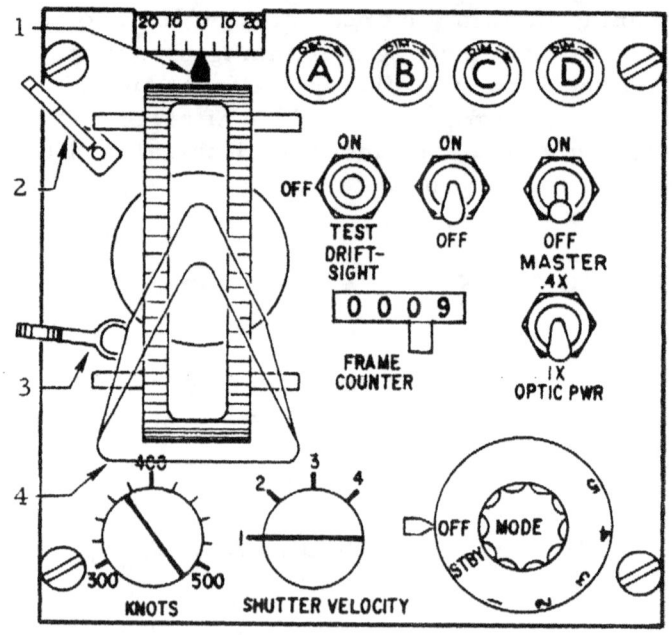

1 POINTING ANGLE INDICATOR
2 HORIZON VIEWING LEVER
3 YOKE CLAMP LEVER
4 CONTROL STICK

Figure 4-8

Controls

The MARK III A hand control includes the same functions as the MARK III. See Section IV B for controls and functions of the MARK III A.

SEXTANT SYSTEM

The sextant is a conventional bubble type sextant which is mounted rigidly to the aircraft, just forward of the windshield. The sextant optical system uses a portion of the driftsight optics, (see figure 4-6) and throws its measuring presentation on the face of driftsight display. The sextant can be adjusted from minus 4° to 90° in elevation, and 360° in heading to give complete coverage of the heavens, except for that part near 180° heading which is blacked out by the windshield and canopy. The optical system of the sextant is a unit-power telescope with a field of view of 15°. This wide field simplifies location and identifications of celestial bodies. The elevation prism is located inside a 5-inch diameter protective glass dome which projects through the skin of the aircraft. This prism is mounted on a horizontal rotary platform so as to provide completely separate adjustments for elevation and heading. This part of the sextant is located in an unpressurized portion of the aircraft, and is completely sealed with respect to ambient air. Information from the sextant is projected optically through the forward pressure bulkhead into the cockpit through a pressure-tight bulkhead coupling. The construction is such that in case of a broken glass dome or other break in the sealing of the sextant head, there will be no loss of cabin pressurization. The part of the sextant inside the cockpit between the pressure wall and the driftsight is open to the interior of the driftsight, but sealed from cabin air to prevent entry of dust and dirt.

A ball-and-disc type integrator is located on the right side of the cockpit near the other sextant controls and allows the operator to average a sight over an observation period of up to two minutes. The sextant presentation is introduced into the driftsight presentation by means of a mirror which can be rotated into the optical path of the driftsight. All motions and controls are purely mechanical, and there are no motors or electronic elements. The presentation is illuminated by aircraft type miniature light bulbs operated from 28 V DC.

CONTROLS

The controls for the sextant (see figure 4-6) consist of the following:

Sextant Pull Knob

This knob, located to the right of the scope, pulls out about 2 inches and rotates a 45° mirror into the driftsight optics, thereby throwing the sextant presentation into the driftsight. During night operation, this knob should be pulled out so as to prevent the landing lights from causing undesirable glare through the driftsight.

Exercise care in operating pull knob as abrupt movements are apt to damage mirror mounting.

Sextant Filter Pull Knob

This knob, located to the left of the scope, pulls out about 2 inches and inserts a sun filter into the sextant optics. In daytime operation this knob should be left pulled out. If the sextant is pointed at the sun and the operator looks through the sextant, he may be temporarily blinded by the glare without the sun filter.

Day-Night Switch

This two-position switch, located on the left side console, controls the lighting of the sextant presentation. The DAY position provides maximum sextant illumination for use with the filter. In the NIGHT position, a rheostat is incorporated in the lighting circuit to enable the pilot to control the sextant light intensity.

Sextant Dimming Rheostat Knob

This knob is located outboard of the throttle. It is used to vary the intensity of illumination of the sextant presentation when the DAY-NIGHT switch is in the NIGHT position. With the switch in the DAY position, the rheostat varies the intensity of the bubble illumination.

When not in use, the DAY-NIGHT switch should be left in the NIGHT position and the rheostat turned to the OFF position.

Heading Twist Knob

This knob, located below the right canopy sill, adjusts the heading of the sextant with respect to the aircraft. It is capable of continuous rotation in either direction and adjusts the heading approximately 10° per turn.

AZ-HD Push-Pull Knob

The AZ-HD push-pull knob is located below the right canopy sill. It has two positions and allows either the true azimuth of a celestial body or a desired heading to be set into the sextant. Azimuth information is set into the sextant by pulling the AZ-HD knob and then rotating the heading twist knob until the desired azimuth is obtained.

Heading information is set into the sextant by pushing the AZ-HD knob and then rotating the heading twist knob until the desired heading is obtained.

Elevation Twist Knob

This knob, located below the right canopy sill, adjusts the elevation setting of the sextant, and also drives the averager. It can be rotated in either direction, and one turn adjusts the elevation by 5°. There are mechanical stops built into the sextant which stop this knob at approximately minus 4° and plus 91°.

Averager Wind Lever

This lever, on the averager, pulls out and forward approximately 1 inch and simultaneously winds and resets the averager. It has a spring return so that upon release it returns to its original position. It winds the clock mechanism in the averager to approximately 3 seconds below zero, and after release of the lever the counter runs and stops automatically at zero.

Averager Start Lever

This lever, just below and forward of the elevation twist knob, swings from approximately 1/2 inch to start or stop the averager. It has an automatic spring return.

PRESENTATION

The presentation appears in the driftsight as shown in figure 4-9. The counter reads elevation angle directly in degrees and minutes. For angles below zero it reads 99°, 98°, and 97°, etc., for decreasing angles. There can be no confusion with high angles, as the lower stop is set at approximately 96° (minus 4°) and the upper stop at slightly over 90°. The heading is read by the setting of the bearing dial with respect to the lubber's line, and the azimuth is the setting of the arrow on the cursor with respect to the bearing dial. The cursor has the form of a cross with an open center. Movement of the elevation knob will move a target up or down parallel with the arrow on the cursor regardless of heading, and movement of the heading knob will move a target parallel to the cross bar of the cursor. The bubble appears as an illuminated ring near the center of the field of view. Since the bubble is projected optically into the field of view, it will not hide targets, and they can be seen clearly inside the bubble. The lighting intensity is controlled by the DAY-NIGHT switch and the Sextant Diming Rheostat knob as previously explained.

A rubber cone which slides onto the end of the driftsight viewing head is a supplementary item for daytime use of the scope. When removed from the scope head, it is normally stowed in the map case on the left side of the cockpit. The small opening at the end of the cone permits a full view of the presentation, and excludes a maximum of stray light. Proper eye position is about 3 inches behind the end of the cone piece.

AVERAGER

The averager is located above the right console directly behind the elevation twist knob. The averager contains a clock mechanism and a ball integrator which effects a continuous moving average over any observation period from 30 seconds to one minute. The averager is wound and reset by the wind lever. No other presetting of the sextant, timing mechanism or averager is required. Because it continuously integrates altitude against elapsed time, it may be stopped at any time after 30 seconds by depressing the start lever without disturbing its operation. However, it is desirable to use the full observation period if possible The average altitude angle is obtained at

SECTION IV A AF (C)-1-1

SEXTANT PRESENTATION

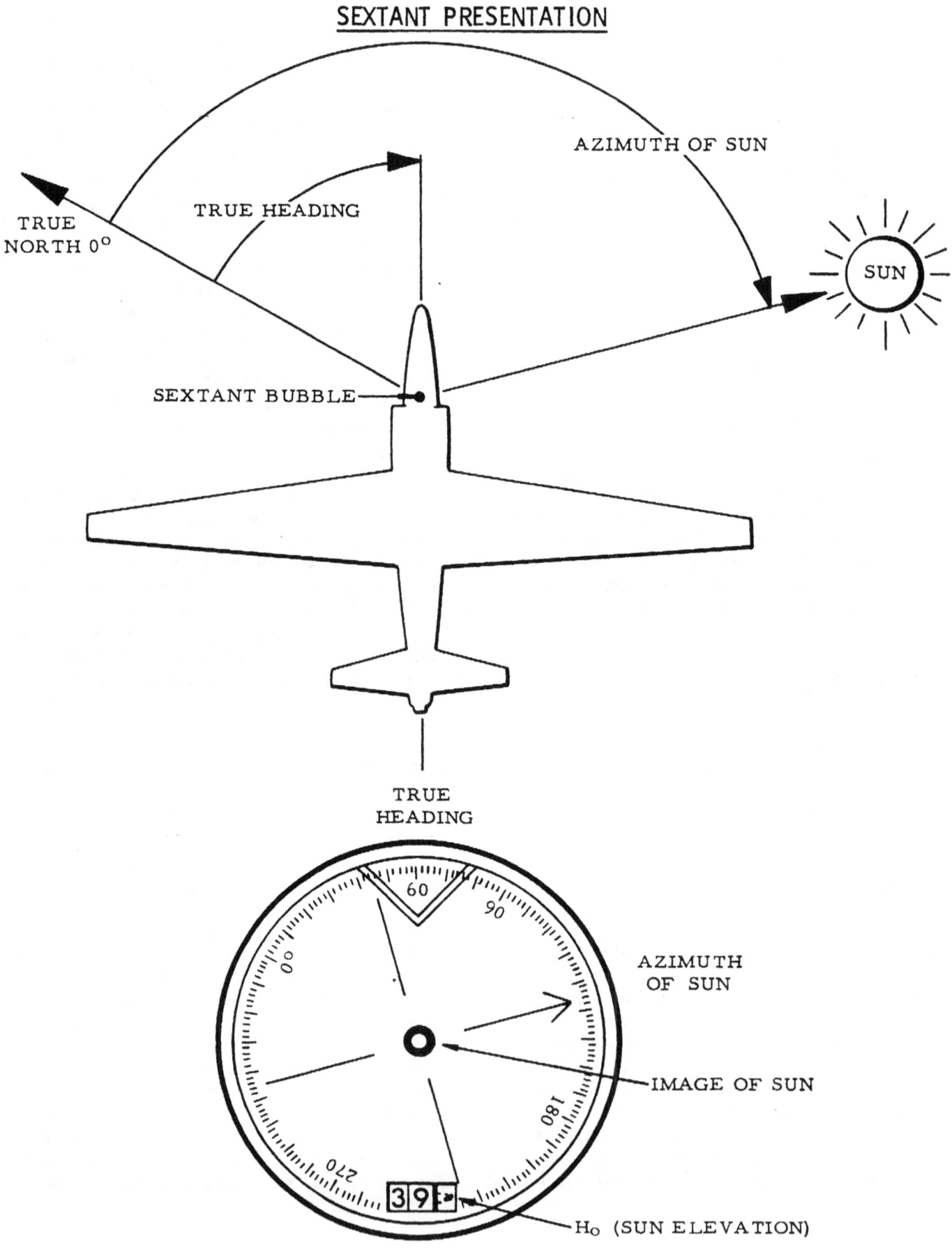

Figure 4-9

4-26

the end of an observation by recentering the average indice by means of the elevation twist knob. These indices can be observed by looking into the stainless steel mirror on the averager. When the indices are lined up, the counter in the presentation reads the average altitude directly. The time dial graduated in seconds indicates the half time of the observation, which indication may be added directly to the time of start to give the mean time of the average altitude. When an integration time of less than one minute is used, care should be taken in reading this time dial as it is viewed through a mirror and the numbers are mirror-image.

OPERATION

Locating a Body

In order to locate a celestial body, its approximate altitude and azimuth should be known. The first step is to pull out the sextant pull knob so as to introduce the sextant presentation into the driftsight optical system.

NOTE

For a daytime observation the sextant filter pull knob should first be pulled out so as to be sure that the pilot will not be blinded if the sun is in the field of view.

The elevation twist knob should now be turned so that the counter at the bottom of the field reads the proper precomputed angle of altitude (H_c). The AZ-HD knob should now be pulled out to lock the bearing dial in position so that the true azimuth of the celestial body may be set in.

True azimuth is set in by turning the heading knob until the cursor arrow points to the proper reading on the bearing dial. The AZ-HD knob should now be pushed in. This now locks the bearing dial to the cursor so that when the heading knob is turned again the dial and cursor will rotate together. The estimated true heading of the aircraft is now set in by turning the heading twist knob so that the proper angle on the bearing dial appears under the lubber's line at the top of the presentation. At the precomputed time, if all adjustments have been made correctly, the celestial body should appear well within the 15° field of view.

After the body has been located, it should be centered in the bubble. Rotation of the elevation twist knob will move the object along the direction of the arrow on the cursor, and rotation of the heading twist knob will move the object at right angles to this arrow. In the case of night use, the initial adjustments should be made with the DAY-NIGHT switch in the NIGHT position and the rheostat turned towards brighter. The final adjustments should be made with the rheostat adjusted for best star-bubble visibility.

Elevation Readings

After the sextant has been adjusted and the celestial object is in the field of view, it is necessary to center the object inside the bubble by using the elevation and heading twist knobs. It is not necessary for the bubble to be in the center of the field to take a reading, as the bubble and the image of the object will move together in the plane of the sextant elevation. However, if the bubble is far off it may indicate that the aircraft is not on a straight and level course, and to obtain accurate readings it is important that the aircraft be operated as straight and level as possible. Spot readings may be taken, but they will be subject to more error than those taken with the averager. The counter in the presentation reads the altitude directly and no corrections are required. The exact time of the shot must be noted as well as the altitude.

Use of Averager

In preparing to take a shot using the averager, first operate the wind lever on the averager and make sure the timing disc runs a few seconds and stops at zero. Then locate the celestial body as previously described. After the body is centered in the bubble, note and record the exact time and push the start lever on the averager. For the next minute adjust continuously both the elevation and heading twist knobs to keep the object centered in the bubble. After one minute has elapsed, turn the elevation twist knob so as to recenter the averager indices, and read the averaged altitude directly from the counter. If the full one minute period has been used, add 30 seconds to the starting time to obtain the correct time for the averaged reading. If the averager has been stopped before the end of the one minute period, add the number to the starting time. No error will be introduced if the object is tracked more than one minute, as the averager clock work has been shut off automatically and no averaging takes place after the one minute regardless of motions of the elevation twist knob.

Heading Readings

True heading readings may be obtained by setting in precomputed azimuth, heading, and altitude information as previously described. After the chosen celestial body has been located and centered in the bubble, at the precomputed time the true heading can be read directly from the lubber line and the bearing dial. The accuracy of this reading will depend upon the accuracy of the time measurement and the assumed position used in the precomputations. It will also depend on the altitude of the celestial object. Therefore, it is desirable to select celestial bodies at least 30° away from the zenith.

Daytime Checklist

1. Pull out the sextant filter pull knob which places a light filter in the optical system to permit looking directly at the sun.

2. Pull out the sextant pull knob. This lowers a mirror into the driftsight barrel and places the sextant presentation on the driftsight.

3. Place DAY-NIGHT switch in DAY position and rotate rheostat full clockwise.

4. Set in azimuth of the sun (Zn from pre-comp sheet) by pulling out azimuth knob and by turning heading knob until scale reads Zn against cursor arrow.

5. Set in approximately one true heading from compass (correcting to approximately true) by pushing in azimuth knob until scale shows approximately true heading up against lubber pointer.

6. Set in computed elevation of the sun (H_c from precomp sheet) by turning elevation knob until counter reads H_c.

7. The sun should now appear in field of view. Center sun in bubble by adjusting heading and elevation knobs. Field of view is about 15° across. Adjust the dimming switch if necessary.

8. Use averager as required.

9. Read true heading indicated on scale.

10. Read elevation of the sun H_o on counter.

11. $H_o - H_c$ = intercept.

Night Checklist

1. Push in sextant filter pull knob.

2. Pull out sextant pull knob.

3. Throw toggle switch to NIGHT and adjust rheostat for best visibility.

4. Set in precomputed data, search for star in field of view, proceed as with daytime operation.

5. Use averager as required.

6. Read true heading and H_o. (Adjust rheostat if necessary.)

AERIAL REFUELING SYSTEM (ARS)

The ARS is installed primarily to increase the operational capability of the U-2C. It allows the aircraft to receive as much as 925 gallons of fuel in the main tanks during flight. (See figures 4-10 and 4-11.)

Changes to the basic U-2C fuel system are:

1. The fuel transfer sequence is revised to feed from the main tanks first then from the auxiliary tanks. (Slipper tanks still feed into auxiliary tanks.)

2. A cross transfer pump is installed between the auxiliary tanks.

NOTE

To preclude pumping fuel overboard, do not use auxiliary cross transfer pump unless fuel load is below 300 gallons.

3. A float switch is installed in the sump tank to operate a solenoid valve which shuts off the air to the main tanks when they are empty.

4. A cockpit switch is provided to override the float switch which shuts off the air to the main tanks when they are empty.

NOTE

This switch is used to repressurize the main tanks to restore the sump tank fuel to normal operating level.

The ARS is composed of the following:

1. A retractable receptacle located in a fairing on the top of the fuselage aft of the equipment bay.

2. A pressure switch in the fuel manifold downstream of the receptacle.

3. A dual shutoff valve located at the inboard end of each main tank.

4. A dual float operated pilot valve located near the outboard end of each main tank.

5. Four solenoid operated fuel valves for checking the pilot valves and their control panel on the instrument panel.

6. A PART FULL float switch located inboard in each main tank.

7. A FULL float switch located outboard in each main tank.

8. A pressure switch in the main tank inboard vent line.

9. A motor operated bypass valve installed between the main tank outboard vent line and the overboard vent.

10. An amplifier system for automatic operation of the ARS.

11. An override system for manual operation of the ARS.

Changed 1 September 1968

SECTION IV A AF (C)-1-1

ARS RECEPTACLE

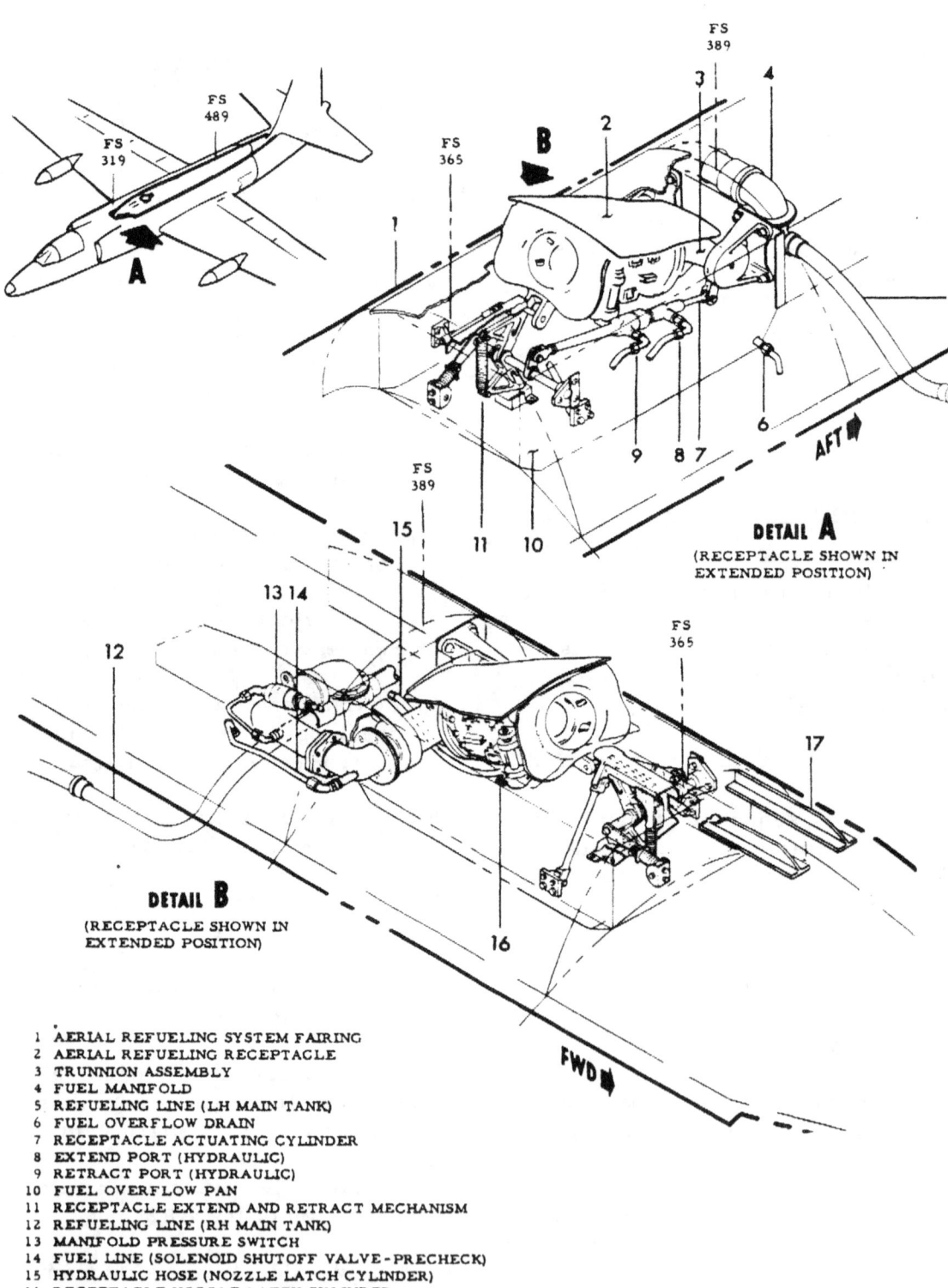

1. AERIAL REFUELING SYSTEM FAIRING
2. AERIAL REFUELING RECEPTACLE
3. TRUNNION ASSEMBLY
4. FUEL MANIFOLD
5. REFUELING LINE (LH MAIN TANK)
6. FUEL OVERFLOW DRAIN
7. RECEPTACLE ACTUATING CYLINDER
8. EXTEND PORT (HYDRAULIC)
9. RETRACT PORT (HYDRAULIC)
10. FUEL OVERFLOW PAN
11. RECEPTACLE EXTEND AND RETRACT MECHANISM
12. REFUELING LINE (RH MAIN TANK)
13. MANIFOLD PRESSURE SWITCH
14. FUEL LINE (SOLENOID SHUTOFF VALVE-PRECHECK)
15. HYDRAULIC HOSE (NOZZLE LATCH CYLINDER)
16. RECEPTACLE NOZZLE LATCH CYLINDER
17. REFUELING NOZZLE "T" GUIDE

Figure 4-10

AERIAL REFUELING TEST PANEL

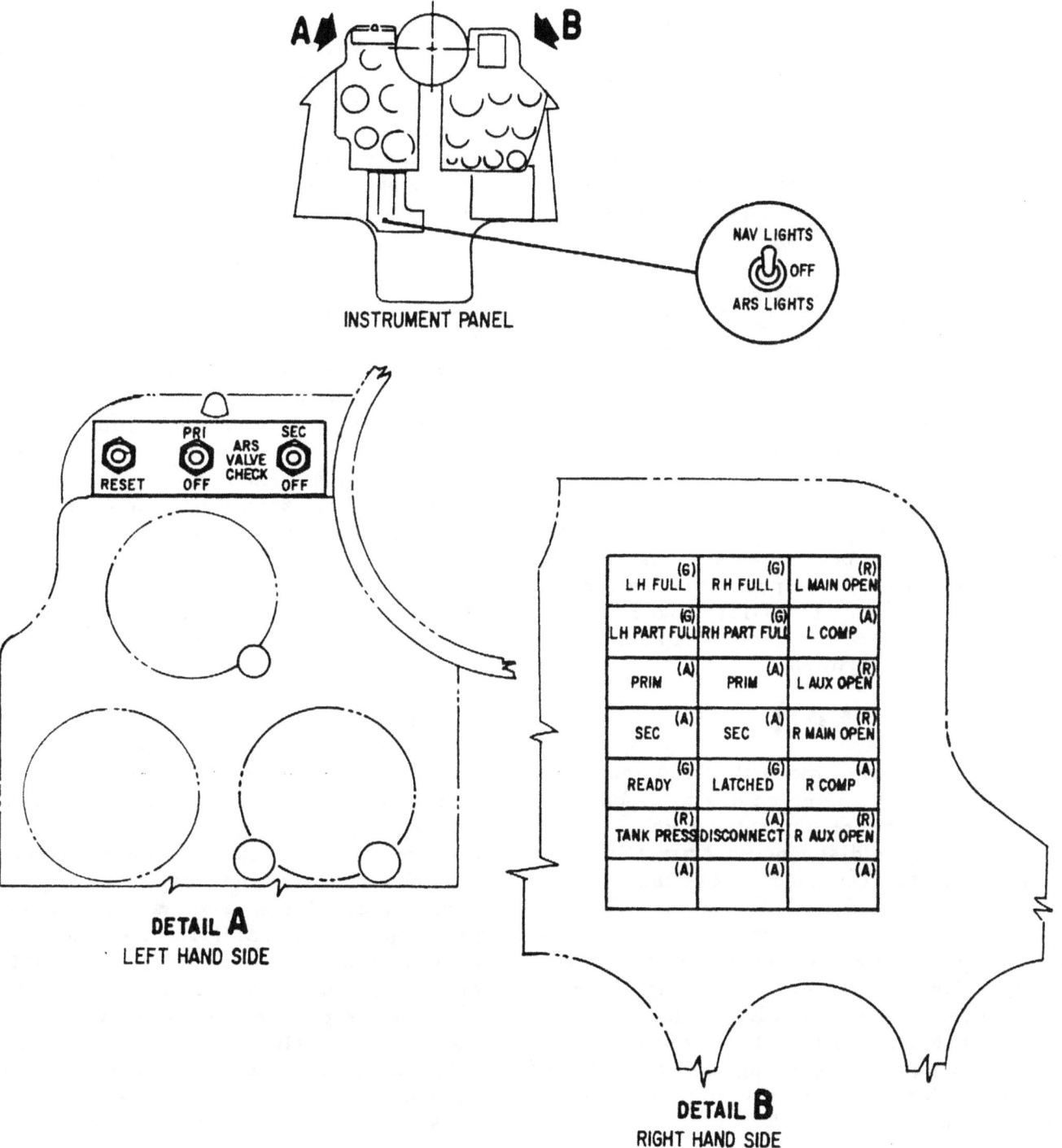

Figure 4-11

SECTION IV A   AF (C)-1-1

12. A control panel on the left console.

13. An indicator panel on the instrument panel.

14. Two electrically actuated doors to close the engine compartment cooling air inlets on the top of the fuselage during refueling.

15. Provisions to deactivate the autopilot during refueling and use of the autopilot disconnect switch as in ARS disconnect.

16. Provisions to deactivate the landing gear warning horn signal so that the horn will not sound with the throttle in idle during refueling.

An additional 514 cubic inch bottle has been added to the oxygen system, making a total of four.

The ARS system is designed to operate with tankers using the boom type system. Techniques have been developed for operation with both the KC97 and KC135 tankers. The U-2F is capable of receiving fuel at the rate of 250 gallons per minute.

NOTE

Refueling rate of more than 250 gpm should not be attempted due to probable overpressure and subsequent overpressure disconnects. Normally, a full load is received in less than five minutes.

Refueling pressure is approximately 45 psi at the refueling nozzle. To avoid structural damage to the aircraft during refueling several precautions have been taken. These are concerned with avoiding high tank pressures which could lead to structural failure of the wing.

As a means of rapidly discharging the air displaced by the incoming fuel, a bypass valve connects the main tank outboard vent line directly to the overboard vent line. During refueling flight tests, the tank pressures recorded were less than 2 psi. A pressure switch set for 3 psi is installed in the wing tank inboard vent line. This switch, when actuated, causes an immediate disconnect in either automatic or override operation. It also lights a red light on the panel to advise the reason for disconnect.

NOTE

Failure of the tank vent bypass valve to open would cause the tank pressure to exceed 3 psi and close the pressure switch. With this condition, it may be possible to receive fuel from the tanker at a reduced rate, by the tanker pump being turned on and off intermittently. This should only be attempted under extreme emergency conditions.

Dual fuel shutoff and pilot valves are installed to assure the fuel shutting off when the tanks are full. Excessive tank pressures would be experienced if the fuel failed to shut off. Therefore, a system for checking the valve operation has been installed.

The pilot valve float is encased in a perforated cup. During normal operation the fuel level in the tank rises slowly enough to gradually raise the float as it enters the cup. To check the valve, a separate line from the fueling manifold flows fuel directly into the cup. The volume of fuel is sufficient to raise the float and check its ability to close the pilot valve which in turn closes the refueling valve. The check can only be performed when fuel is flowing. It is accomplished by turning on a test switch in the cockpit. The switch opens a solenoid valve in the test line. When the fueling valve closes, it lights an indicator light in the cockpit. Each dual fueling valve has a primary and secondary poppet. There is a separate float, check switch, and indicator light for each poppet.

Electric actuators close doors over the engine compartment cooling air inlets on the top of the fuselage during refueling. This is to prevent spilled fuel entering the engine compartment.

The ARS control panel is on the left console. The control consists of two switches; an ON-OFF switch and an OVERRIDE switch. Normally the ON-OFF switch is the only one used. Turning this switch ON accomplishes the following:

1. The receptacle is extended hydraulically.

2. The latching system is armed.

3. Power is supplied to the amplifier system.

4. Power is supplied to the indicator panel.

5. The tank vent bypass valve is opened.

6. The engine compartment cooling air inlet doors close.

7. The landing gear warning horn signal cutout is energized.

8. The autopilot disconnect switch on the control wheel serves as the manual ARS disconnect switch during refueling.

NOTE

The autopilot cannot be turned on when the ARS is ON.

Turning the switch OFF reverses the above functions.

The amplifier system allows automatic sequencing of events during refueling. Through an induction coil in the receptacle, the receiver and tanker are connected electrically. With this system in operation the tanker gets a contact light and the operator knows when to start fuel flowing. Disconnects can also be initiated from the tanker. These disconnects may be initiated by the tanker pilot, boom operator, or limit switches on the boom when it exceeds the refueling envelope. The latter function may be cut out at the boom operator's discretion.

The override system cuts out the amplifier system. In override, the latches latch when contact is made but must be disconnected by the receiver pilot.

NOTE

The tank pressure warning system will still cause a disconnect in override.

Panels for the ARS are located in the upper left and upper right portions of the instrument panel as shown in figure 4-11, and consist of the following:

1. PRIMARY test switch.

2. SECONDARY test switch.

3. TANK PRESS WARNING light - red.

4. Four amber lights indicating valve closure - PRIM, SEC, (For LH) and PRIM, SEC, (For RH).

5. Four green lights indicating fuel level, LH FULL, LH PART FULL; RH FULL, RH PART FULL.

6. READY light - green.

7. DISCONNECT light - amber.

8. LATCHED light - green.

9. RESET switch.

FUEL CAPACITY AND TOTALIZER SETTING

When the airplane is fueled on the ground the total capacity with slipper tanks is 1520 gallons. 925 gallons of this fuel is carried in the main tanks.

During air refueling, however, the main tanks will receive only about 900 gallons. In addition, the sump tank fuel level is drawn down about 15 gallons during engine start and takeoff, and about 40 gallons of fuel is transferred from the auxiliary and slipper tanks during climb. Therefore, 80 gallons of fuel is consumed which cannot be replaced during air refueling. This fuel is accounted for by resetting the totalizer to 1440 gallons after refueling. However, this totalizer setting should be used only when it is known that main fuel was being used at the start of refueling, and the tanks were refueled to capacity.

NOTE

If the totalizer reading is below 540 gallons at the start of refueling, auxiliary and slipper tank fuel will be burned but will not be accounted for if the counter is reset to 1440.

If the main tanks are refueled completely with less than 540 gallons remaining. 900 should be added to the totalizer reading and the counter reset accordingly. If desired, this need not be done immediately after refueling.

The MAIN TANK PRESS switch must be moved from NORM position to PRESS position after refueling, for approximately 15 minutes. This switch is located forward of the lateral fuel transfer switches on the left side console. Placing the switch in the PRESS position overrides the float switch in the left-hand sump tank which in turn operates the air shutoff solenoid valve. This action would then allow the fuel to rise above the float valve level thereby restoring normal air pressurization for the main tanks. The switch must be returned to normal position sometime before the main tanks are again emptied in order to assure slipper tank pressurization and complete fuel transfer.

If the totalizer reading is below 540 gallons and the main tanks are not refueled completely, figure 4-12 should be used to determine the gallons received. This amount can then be added to the totalizer reading. The tanker will inform the receiver pilot how many pounds of fuel have been transferred and figure 4-12 can be used to convert to gallons. This chart can be used as a crosscheck on the quantity of fuel received at any time.

ARS DAY OPERATION

Refueling Position

Prior to reaching observation position, the gust control should be in GUST. The aircraft should remain in GUST throughout the refueling operation. Airspeed and altitude when refueling from the KC135 is between 200 and 220 knots at 35,000 feet. When refueling from the KC-97, the ideal airspeed is 170 knots at 20,000 feet.

With the U-2 stabilized as above, the tanker passes the U-2 on the right, at the same altitude at slightly higher speed and with a separation of approximately 1/4 mile.

GALLONS VS POUNDS

MIL-F-25524B FUEL

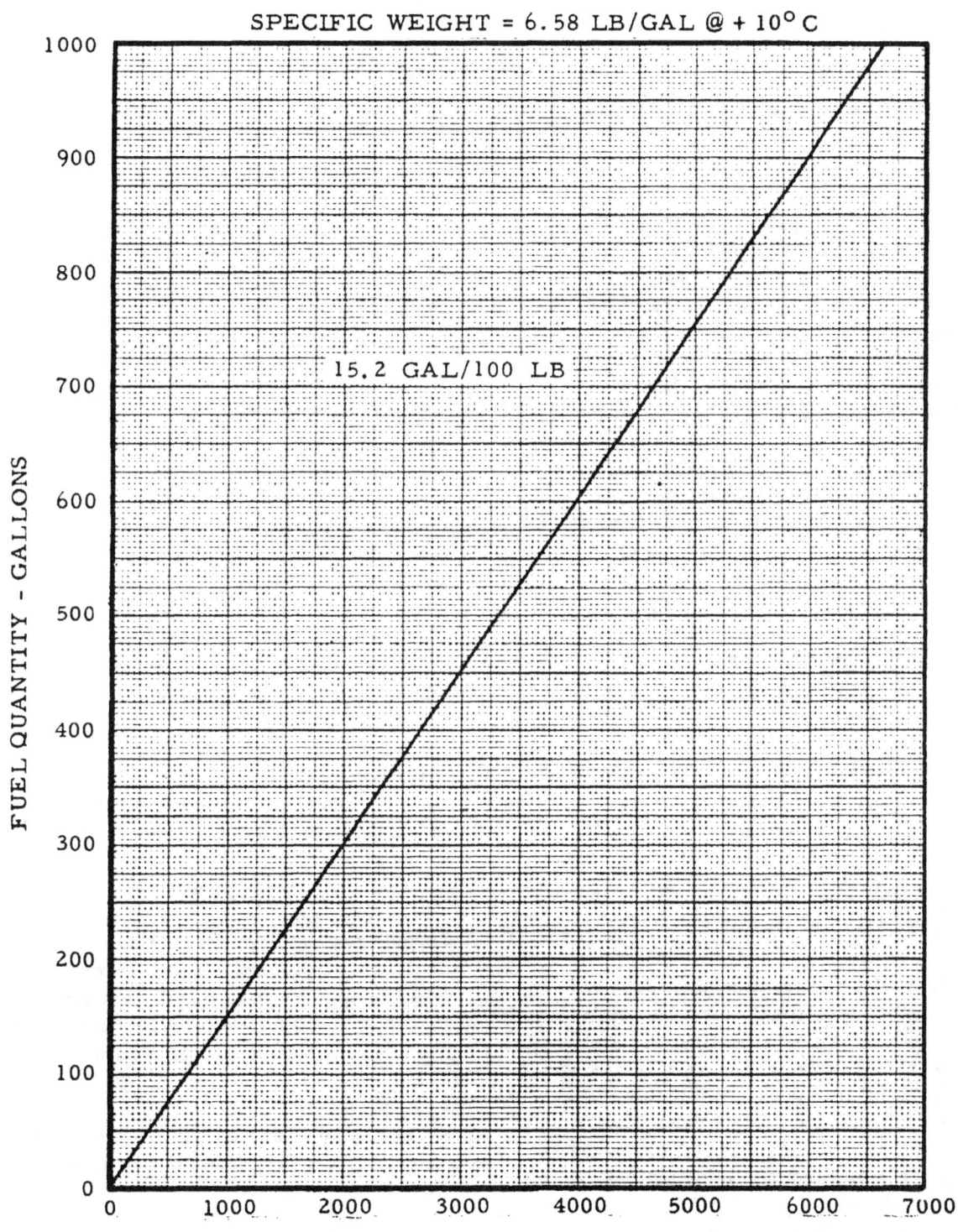

Figure 4-12

SECTION IV A   AF (C)-1-1

As the tanker passes the U-2, it reduces speed to match that of the U-2. At this point, the U-2 starts sliding in toward the tanker at the same time descending to an altitude approximately 500 feet below the tanker. While getting into this position, turn the ARS switch ON. A slight rudder buffet will be felt with receptacle extended. When the U-2 is below and in line with the tanker, it rises slowly to the refueling position.

There is very little turbulence when using this approach. Stay below the level of the boom operators window to avoid turbulence and to have the desired boom angle. Use power as required to comply with the boom operator's directions. As the boom is lowered a small amount of rudder buffet may be felt. This is caused by the boom rudevators. When the U-2 is in optimum refueling position, very little aerodynamic effort from the tanker is felt. If the U-2 is off center, a rolling effect is encountered as the wing tips enters the tanker downwash. Countering aileron easily corrects this effect.

WARNING

The areas directly aft of the fuselage, and aft and below the wings of the tanker should be avoided because of turbulence. Structural failure may result.

Normal Refueling

When the ARS switch is turned on, several lights in the ARS panel may come on, depending on fuel load and static positions of valves. These lights are not important at this time. If the READY light is not on, push the RESET switch to illuminate it. If the READY light is still not on recycle the ARS ON-OFF switch. After the boom has been fully engaged in the receptacle, the LATCHED light will come on and the READY light will go out.

When fuel starts flowing all PRIMARY and SECONDARY lights should go out. If the main tank fuel level is below approximately 100 gallons, the PART FULL and FULL lights will be out.

After the tanker starts transferring fuel the shutoff valves should be tested individually in the following sequence:

1. Left Primary.

2. Right Primary.

3. Left Secondary.

4. Right Secondary.

NOTE

The pressure regulator in the tanker requires time to settle out due to the low fueling rate of the U-2. Therefore, if check is initiated too soon after tanker starts fuel flowing, a pressure disconnect may be experienced. This requires pushing the RESET button to get the READY light and making another contact.

As each valve closes, its light will come on. Operating time for the valve is normally less than 10 seconds.

NOTE

Do not try to check a second valve until the light for the first one has gone out. Do not check Left and Right simultaneously or a pressure disconnect will be experienced.

After the valve test has been satisfactorily accomplished it is safe to continue filling the tanks to capacity.

NOTE

It is safe to complete fueling as long as one valve on each side passes the test. However, any valve malfunction should be reported to Maintenance.

WARNING

If both valves on one side fail to pass the test during fueling, a DISCONNECT should be made at once. During fueling, failure of a valve in both primary and secondary can cause structural failure of the wing when the tank fills.

Shortly after the FULL lights come ON, the valves will close automatically and the boom will be disconnected automatically.

Place the ARS switch in OFF when the boom is clear. Blinking of the DISCONNECT light after turning the ARS switch off is normal.

NOTE

If the DISCONNECT light remains ON after turning the ARS switch off, it indicates the bypass valve in the vent system has not closed. This is not a serious situation since the main tanks will gravity feed but the condition should be reported to Maintenance.

CAUTION

During refueling operation, engine compartment cooling air doors are closed. There are no indications if the cooling air doors do not reopen. However, the mission may be completed. Continued use without the doors opening will cause overheating and weakening of the structure.

There is no indicator for the receptacle. However, it may be checked with the rear view mirror. If it fails to retract return to base.

Pull FUEL COUNTER circuit breaker, reset FUEL COUNTER, push in circuit breaker.

Override Refueling

If attempting a normal refueling and the READY light will not come on or premature disconnects are experienced, place the OVERRIDE switch in the ON position and proceed as described in Normal Refueling.

WARNING

Failure to obtain a ready light may indicate that the fuselage cooling air doors are not closed. If boom operator advises that they are open, fueling should not be attempted. Fuel spillage may enter the engine compartment and cause an explosion.

NOTE

If failure to obtain a READY light is caused by the vent bypass valve not opening, a TANK PRESS WARNING light and a DISCONNECT will probably be experienced as fuel starts to flow. If this happens return to base. If it is absolutely necessary to receive fuel, proceed and have the tanker reduce the flow rate by turning his pump switch off and on or accept gravity flow from the tanker.

When the tanks are full an automatic disconnect will not be obtained and the receiver pilot must press the DISCONNECT switch.

NOTE

In OVERRIDE, both READY and LATCHED lights remain on.

If maximum amount of fuel is desired, do not disconnect until tanker advises fueling has stopped.

Upon completion of refueling, the U-2 will decrease power and descend 100 feet or more below the tanker. The U-2 will then slide away from the tanker, left or right. When approximately 1/4 mile abeam of the tanker, the U-2 will start the climb on a parallel course with the tanker until well clear and above the tanker, in order to avoid any possible contact with turbulence from the tanker.

EMERGENCY REFUELING

If latching cannot be obtained in Normal or Override, fuel may be transferred under emergency conditions. This requires the receiver pilot using sufficient power to hold the nozzle and receptacle together. Fuel will probably be spilled during this operation and it is not recommended except in an extreme emergency.

ARS NIGHT OPERATION

The procedures for joinup and refueling position are the same for night operation as those described for day operation. The APN-135 rendezvous beacon is used to rendezvous with the tanker under radio silence conditions. Its use and operation are described under separate heading in this section.

Both the KC-97 and KC-135 tanker airplanes have refueling light systems which can be used, depending upon mission requirements, as an aid in maintaining proper position. Current information on these tanker light systems should be obtained from their respective Flight Manuals.

On the U-2, ARS refueling lights are installed to facilitate joinup and boom operator's contact. Two white lights are located in aerodynamic shrouds to the left and right and slightly forward of the fueling receptacle to light the fueling receptacle. A third light also located in a shroud is installed on the aft portion of the ARS fairing to light the aft fuselage and the tail surfaces.

The lights are controlled by a three-position switch which replaces the present navigation lights switch.

In the up position, anticollision lights are available when mission requirements allow their use. With the switch in the down position, ARS refueling lights are selected. The center position is the OFF position for both sets of lights. ARS and anticollision cannot be selected at the same time.

To provide added reliability, the left-hand and aft ARS lights are on a separate circuit breaker and the right-hand ARS light is on the same circuit breaker as the anticollision lights. Both circuit breakers are located in the Q-Bay. The lights are approximately 32-candle power each.

For radio silence conditions the boom operator has an illuminated chalk board to communicate gallons received, latitude and longitude information to the receiver pilot.

COMMUNICATIONS AND ASSOCIATED ELECTRONIC EQUIPMENT

MICROPHONE AND HEADSET CONNECTORS

A quick disconnect fitting, located on the left-hand side of the seat near the floor, provides a connection for the pilot's microphone and headset. The connection will separate in the event of ejection.

A cable which bypasses the quick disconnect is stowed behind the seat. This cable can be used to connect a microphone and headset directly to the communications system. This cable is normally used by the pilot's assistant during the preflight aircraft and equipment checkouts. The pilot's use of this cable will often improve the quality of communications. However, automatic disconnection of the cable is not provided for in the event that ejection is necessary. It must be disconnected prior to ejection.

MICROPHONE SWITCHES

Two microphones thumb-switches are provided. They are both of the momentary button type. One is located on the throttle grip and the other is on the left control wheel grip.

C-823A/AIC-10 INTERPHONE CONTROL

The Interphone Control is used in this aircraft primarily to improve intelligibility of audio communication. The unit consists basically of an audio frequency amplifier which preamplifies the pilots transmissions and boost-amplifies radio receiver outputs.

COMM SELECTOR PANEL

The COMM selector panel, figures 4-13 and 4-14, is used to select the communication system. This panel is located in the No. 2 position in the right console step panel.

COMM SELECTOR PANEL - MODEL U-2C

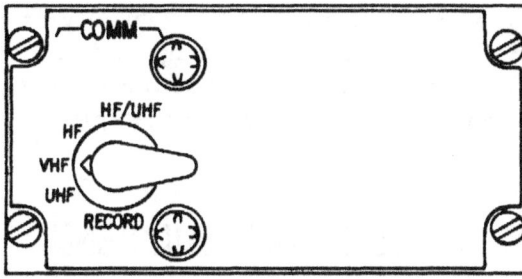

Figure 4-13

COMM SELECTOR PANEL - MODEL U-2F

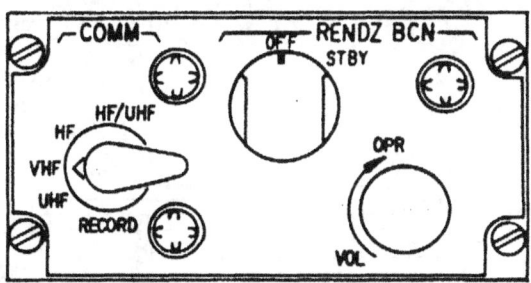

Figure 4-14

The COMM selector has RECORD, UHF, VHF, HF, and HF/UHF positions which determine whether voice is recorded or transmitted over the radio(s) selected.

Receiving is not affected by this switch, except that in the RECORD position, no receiving is possible. Receiving is controlled by the individual receiver switches and their respective volume controls.

4-39

AN/ARC-34 UHF TRANSCEIVER

A standard AN/ARC-34 UHF receiver-transmitter having a 10 watt output is provided for communications. This unit is located in the aft end of the equipment bay. The control unit is located on the left-hand console. (See figure 4-15.) This radio provides for voice communications in the frequency band range of 225 to 399.9 megacycles. In this range there are 1750 separate frequencies available. Twenty of these can be preset and selected by number with the selector switch. Any of the others can be manually selected in the cockpit.

AN/ARC-34 CONTROL PANEL

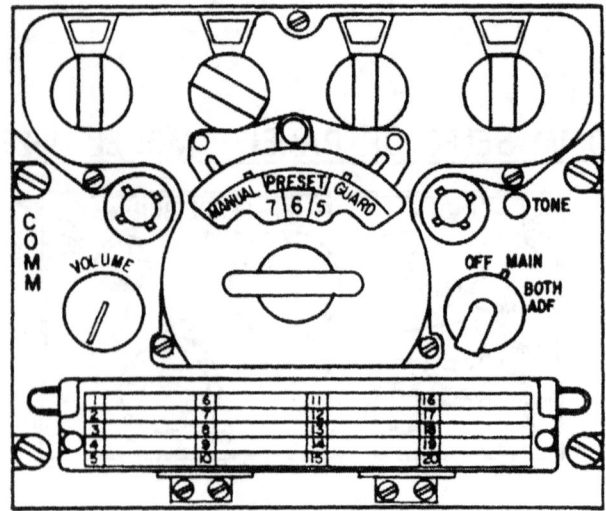

Figure 4-15

Operation

AN/ARC-34 circuit breaker in equipment bay must be closed.

1. Check the channel preset frequencies as indicated on the plastic write-in card. Change preset frequencies as required for the mission.

2. Close the interphone circuit breaker.

3. Rotate the function switch to BOTH if simultaneous monitoring of the preset channel and the guard channel is desired.

4. Set the mode switch so that PRESET is visible through the clear window.

5. Select the preset channel using the channel selector so that the channel number appears in the clear window.

6. Select UHF on COM selector switch.

7. Before transmission of a message, check for operation and warm up of the transmitter by using either the microphone button or tone button and listening for side tone.

8. If it is desired to transmit and receive on a frequency not preset on the channel selector, place the mode switch in the manual position and set up the new frequency with the manual frequency selector knobs.

9. Turn the function switch to OFF to deenergize the set.

618T-3 HF TRANSCEIVER

The Collins 618T is an airborne HF transceiver capable of operating in single band or AM modes in the frequency band of 2.5 to 24 megacycles. The major components and their locations are as follows:

1. Transceiver - Pressurized box in nose.

2. Antenna Tuner - Pressurized box in the top fairing.

3. Antenna - External wire between the antenna tuner and the vertical fin.

4. Control Panel - Cockpit on the LH console.

5. HF XMTR DISABLED light - Amber light on annunciator panel.

The control panel has a combination ON-OFF mode selector switch, tuning knobs, selected frequency window, tuning light, and volume control. (See figure 4-16.)

The HF transmitter is disabled if the cabin altitude (which is the pressure in the transceiver box) exceeds 39,000 feet. Operation will be restored when the cabin altitude drops to 34,000 to 35,000 feet. An amber light on the annunciator panel illuminates whenever the transmitter is disabled due to low pressure. The receiver is not disabled due to low pressure. A switch is provided on the right side upper console to override this pressure switch feature when operational requirements so dictate.

618T-3 CONTROL PANEL

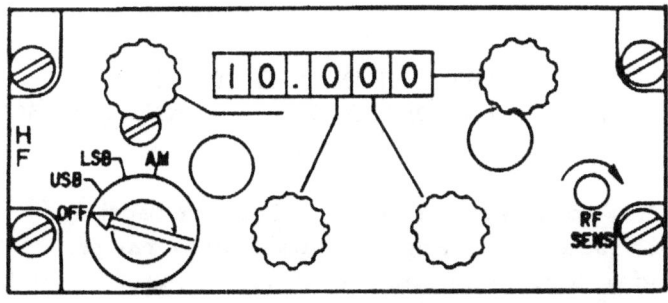

Figure 4-16

Operation

No. 1 or No. 2 main inverter power and DC power are required to operate the set. The HF circuit breaker in the equipment bay and the AIC-10 interphone circuit breaker in the cockpit must be closed.

The following procedures should be used to put the set in operation initially, or to put the set back in operation whenever the inverter is recycled for any reason.

1. Turn inverter switch to No. 1 or No. 2. (Set will not operate on emergency inverter.)

2. Move mode selector to the AM position. Allow several minutes for warmup.

3. Move COMM panel switch to HF or BOTH position.

4. Select desired frequency. (If desired frequency was already set, rotate 10 kc selector knob one digit off frequency and then back to desired frequency.) The transceiver will mute, indicating that it is setting up the new frequency.

5. When the 618T-3 is no longer muted, depress microphone button momentarily and wait for the 618T-3 to tune. A 1000-cycle tone will be heard in the headphones and the red light (if incorporated) on the control panel will glow until tuning is completed.

6. When tuning is completed, select desired mode position.

7. Set RF SENS (or VOL) control so that noise in headphones is barely audible.

8. Press microphone button to transmit.

NOTE

The 618T-3 radio should be on during descent and landing in order to prevent condensed moisture from forming on the transceiver box.

WARNING

Potentials dangerous to life are on the antenna during 618T operation.

AN/ARN-59 RADIO COMPASS (ADF)

The AN/ARN-59 set is installed in all U-2C and U-2F airplanes with the ARC-15F navigation set. A Bearing-Distance-Heading Indicator (BDHI) is used in conjunction with the AN/ARN-59 set to provide a display of both VOR & ADF information. The No. 1 pointer displays ADF and the No. 2 pointer displays VOR. The AN/ARN-59 has a frequency range of .19 to 1.75 megacycles, which is divided in three bands. The control panel used with this set (figure 4-17) is on the right console and includes a tuning meter. When the switch labeled BFO is on, it enables reception of continuous wave signals. This set requires DC power. Normal or emergency inverter power is required for the ADF needle.

AN/ARC-59 CONTROL PANEL

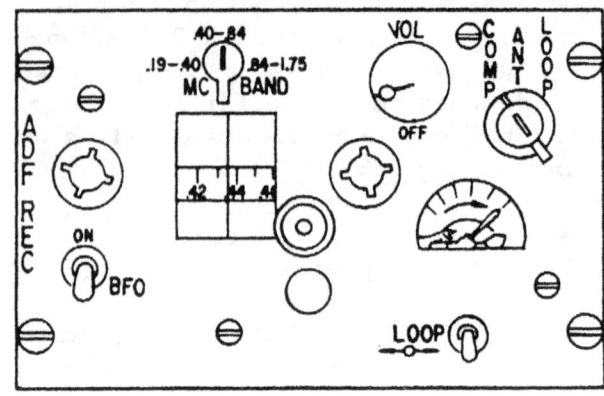

Figure 4-17

Operation

1. Turn VOL control clockwise to turn equipment on and allow approximately 30 seconds for equipment to warm up.

2. Turn COMP-ANT-LOOP switch to ANT.

3. Turn MC BAND switch to desired frequency band.

4. Adjust VOL control until background noise (or station if already tuned in) is heard.

5. Rotate tuning crank until station frequency is aligned with frequency dial hairline, then tune for maximum tuning meter deflection by slowing rotating tuning crank in the vicinity of desired frequency.

6. Readjust VOL control to desired audio level. Identify station.

7. Turn COMP-ANT-LOOP switch to position required for the desired function of system:

    a. To use as an automatic direction finder, set switch to COMP.

    b. To use as a low-frequency radio range receiver, set switch to ANT.

    c. To use for aural null procedures or during conditions of poor reception, set switch to LOOP.

8. Turn set off by turning VOL control counterclockwise to OFF.

ARC TYPE 15F VHF NAVIGATION EQUIPMENT (VOR) (MIL TYPE AN/ARN-30)

This equipment is installed for reception of visual omnirange, visual-aural range, localizer, and communication signals in the 108.0 to 126.9 megacycle frequency range. This equipment requires DC power. The control panel for the ARC-15F (see figure 4-18) is located on the right-hand console aft of the ARN-59 panel on most airplanes. This panel consists of the following controls: a combination power switch and volume control, a megacycle channel selector switch, a tractional megacycle channel selector switch, and a squelch disabling switch. The fractional megacycle channel selector switch selects frequencies in 0.1 megacycle steps and automatically selects the VOR mode in even-tenth positions from 108.0 to 111.8 mc, and in all positions from 112.0 to 117.9 mc. In odd-tenth positions from 108.1 to 111.9 mc, the VOR/LOC mode of operation is selected. Frequencies from 118.0 to 126.9 mc are available for VHF communication. VOR station information is displayed by the No. 2 pointer on the BDHI.

ARC TYPE 15F CONTROL PANEL

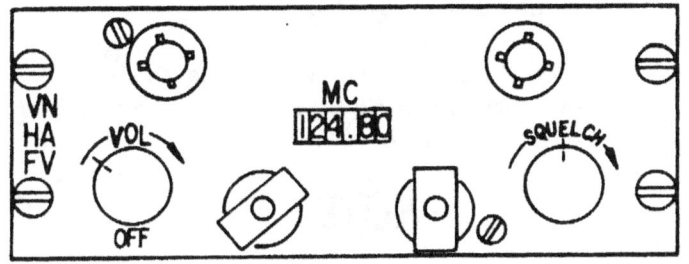

Figure 4-18

Course Indicator

This is an ID453 indicator located on the right side of the upper instrument panel. A course selector knob is located on the lower left-hand corner of the instrument for setting a desired track to or from a desired VOR station. The track and the reciprocal track are indicated by the positions of a triangle shaped pointer and a ball, respectively, on the fixed 360° dial. A TO-FROM indication on the face of the instrument shows whether the selected track is to or from the VOR station. A vertical bar pivoted at the top of the instrument moves to the left or right to supply visual indication of the lateral position of the aircraft with respect to the on-course signal of the VOR, VAR, or localizer signal. Fluorescent blue and yellow sectors are included for flying the visual courses of VAR and localizers. A red flag alarm at the bottom operates any time a signal is unreliable, weak, or nonexistant. The instrument also has a horizontal bar which is ordinarily used for a glide slope indication. This bar does not function, however, since no glide slope input is provided and the red flag associated with this position of the instrument will always indicate OFF. Either normal or emergency inverter power is required for the VOR indicator.

Operation

1. Turn set on by turning volume knob clockwise from OFF.

2. Set in desired frequency with the two channel selector knobs.

3. Adjust volume knob to desired audio level (VHF communication and VAR only).

4. Set in desired track with course selector knob on the indicator (VOR only).

5. Turn set off by turning volume knob counterclockwise to OFF.

ATC (AIR TRAFFIC CONTROL) SYSTEM

The ATC system, in conjunction with the Secondary Surveillance Radar, provides a means whereby ground control can identify the airplane.

The system remains in a quiescent state until triggered by an interrogation from ground control, at which time it automatically replies with a code which has been previously programed into the system. The pilot is made aware of the interrogation by the illumination of a yellow light on the control panel.

SECTION IV A   AF (C)-1-1

An additional identification feature is provided. When requested, the pilot, by momentarily pushing the IDENT button on the control panel, programs an additional identification pulse into the system for a period of 20 seconds. If interrogated during this period, the system automatically replies with the existing code plus the additional identification pulse.

The ATC system consists of a transponder mounted in the Q-Bay high on the 252 bulkhead, a functional tester mounted below the transponder, an antenna mounted under the nose of the airplane, and a control panel mounted in the LH console aft of the AN/ARC-34 control panel. Circuit protection is provided by a 5-ampere circuit breaker mounted in the Q-Bay power distribution box.

A pressure switch is incorporated in the power circuit to remove power as the Q-Bay altitude exceeds 34,000 feet. Power is restored descending through 31,000 feet. The pilot should not operate the ATC at Q-Bay altitudes above 30,000 feet unless absolutely necessary.

CONTROLS

The control panel (see figure 4-19) is located on the left console in the cockpit aft of the AN/ARC-34 controls.

ATC CONTROL PANEL

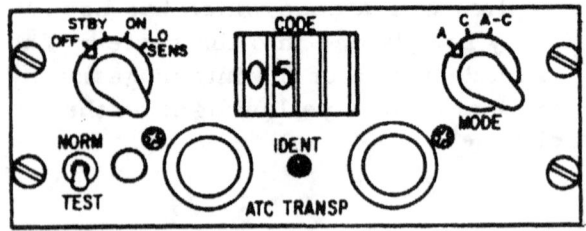

Figure 4-19

The panel incorporates the following controls:

1. POWER SWITCH - Has four positions with the following functions:

   a. OFF - In this position, the switch turns off the system.

   b. STBY - In this position, following a 35-second warmup period, the set is left warm and ready to be put into immediate operation.

   c. ON - The ON position turns the system ON, making it ready to transmit if interrogated.

   NOTE

   Direct switching from OFF to ON requires a 35-second warmup period.

   d. LO SENS - This position is used when requested by the ground station to descrease the system sensitivity.

2. CODE INDICATOR - Shows the code selected to be transmitted. Only the first and second digits on the left are used at present; the other two are locked out for future use. At present, 64 possible codes can be set up, and 4096 can be set up when the other two digits are put into use.

3. MODE - This mode switch has three positions:

   a. A - In the A position, the identification mode is selected. (This is the same as Military Mode 3.) When interrogated, the code shown on the code indicator (2 above) is automatically transmitted in reply.

   b. C - In the C position, the altitude reporting mode is selected. This position is inactive at present.

4-44

c. A-C - In the A-C position, both identification and altitude reporting modes are selected. At present, however, only the identification mode is operative (see 3. a. above).

4. NORM-TEST SWITCH - This switch has three positions:

   a. NORM - In this position, the system is set to indicate when it is being interrogated by lighting a yellow light at the right of the switch while the code is being transmitted. A 1000 cps buzz is also sounded in the pilots audio during the transmission.

   b. Mid Position - In the mid position of the switch, the indicator light is made inoperative. (Otherwise the system functions normally.)

   c. TEST - This position provides a means of checking the operational status of the system.

   Holding the switch in the test position interrogates the system. Illumination of the yellow light indicates satisfactory operation.

5. Indicating Light - The yellow indicator light located on the right of the NORM-TEST switch illuminates whenever the system is replying to an interrogation, remaining on during the cycle of transmitting the code. It also lights up when the system is replying on the IDENT if the system is interrogated during the 20-second interval. It also lights up when the NORM-TEST switch is held at the TEST position.

6. CODE SELECTOR CONTROLS - There are two dual coaxial code selectors. The one on the left sets the 2 left digits and the one on the right the 2 right digits. The latter is locked out at present. By means of these selectors it is possible to select 4096 different codes. However, only 64 codes are currently being used, these being selected by the left-hand coaxial knobs. The large knob selects the left-hand digit, the small knob the right-hand digit.

7. IDENT - The IDENT switch button, when pressed in momentarily, selects the code shown in the indicator and inserts an additional signal. If the set is interrogated within 20 seconds it will reply transmitting the code and additional signal. The IDENT automatically turns off after 20 seconds.

OPERATION

To operate the ATC, the 5-ampere circuit breaker in the Q-Bay DC power distribution box should be engaged, and the following action taken.

1. MODE SELECTOR - A position.

2. SELECT CODE - Left 2 digits (for present).

3. POWER SWITCH - ON (Allow 35 seconds for warmup).

4. NORM-TEST SWITCH - Hold to test momentarily to check system.

5. NORM-TEST SWITCH - To NORM position.

6. POWER SWITCH - Select LO SENS position when requested by Air Traffic Control.

7. IDENT - Press momentarily when requested by Air Traffic Control.

AN/APN-135 RENDEZVOUS BEACON

An AN/APN-135 (modified) rendezvous beacon is installed in U-2F aircraft. This is a beacon designed to serve as a navigational aid. The beacon operates in conjunction with X-band radar system equipped with appropriate beacon interrogating facilities. In response to an appropriate interrogating signal from a radar system, the beacon transmits a coded reply which results in a presentation on the radar display indicating the range, bearing, and identity of the beacon equipped aircraft.

The control panel, figure 4-20 is located on the right console Comm Selector panel. The panel contains a rotary switch with OFF, STBY, and OPR positions used to select the mode of operation, and a VOL control knob. When the power switch is in the standby position and the beacon is being interrogated, a signal is applied to the headset. The beacon does not respond to interrogation in this mode, but it does acknowledge the interrogation by an aural signal in the headset. When the power switch is in the OPR position, the beacon will respond to interrogation with a coded reply. Also, an aural signal is applied to the headset whenever the beacon replies to an interrogation.

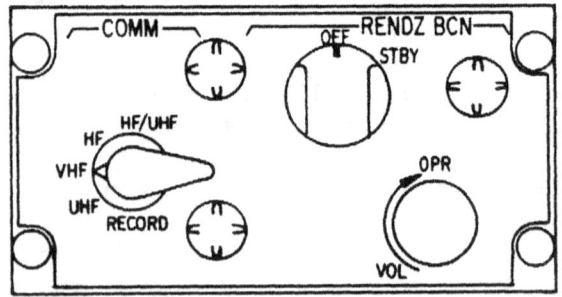

Figure 4-20
(USED ON U-2F AIRPLANES ONLY)

Operation

1. Turn inverter switch to No. 1 and battery switch to BAT & GEN. Turn beacon on by turning rotary switch to STBY. Approximately one minute warmup time is required.

NOTE

In the event the AC generator fails, power for continuous ignition is supplied by the inverter. In order to prevent inverter overload under this condition, the beacon and continuous ignition should not be operated simultaneously. Either the boost pump and continuous ignition or the beacon should be turned off as operating conditions dictate.

2. Turn volume control full clockwise. Adjust to desired level when signal is received.

3. When aural signal is received and a beacon reply is required, turn switch to OPR. An aural signal will be heard whenever a beacon reply is made.

4. To turn set off, return switch to OFF position.

WARNING

After visual contact is made, turn beacon off before joinup for refueling.

AIRCRAFT RADIO CORPORATION, TYPE 12 VHF TRANSCEIVER

Certain airplanes are equipped with an Aircraft Radio Corporation, Type 12 VHF transceiver in addition to the AN/ARC-34. This set consists of a very-high-frequency range of 118 to 148 megacycles, a transmitter which covers the frequency range of 116 to 132 megacycles, a control panel, a power supply and an oscillator-relay unit. The control panel is shown in figure 4-21.

ARC TYPE 12 CONTROL PANEL

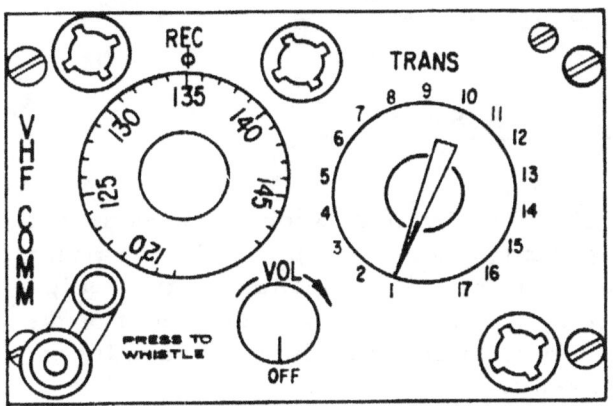

Figure 4-21

When the set is turned on, transmission is controlled by the microphone buttons on the throttle and on the control wheel. Five channels may be preset into the transmitter. No manual transmitter tuning is provided. The receiver must be manually tuned. A whistle-through switch is provided which gives a high-pitched side tone to aid in tuning. This side tone is adjusted to a maximum to get on-station.

Operation

ARC-12 circuit breaker in the equipment bay must be closed.

1. Close interphone circuit breaker.

2. Turn set on by rotating volume knob clockwise from OFF position and allow one minute warmup.

3. Select VHF on COMM selector panel.

4. Set transmitter switch to channel number corresponding to desired transmitting frequency. Refer to VHF frequency card on instrument panel.

5. Tune receiver to frequency corresponding to desired receiving frequency. After coarse tuning by reference to numbered dial, push in on tuning crank to obtain high-pitched side tone for fine tuning, tune side tone to maximum and release crank.

6. Turn volume switch counterclockwise to OFF to deenergize the set.

AN/ARC-3 VHF RADIO SET

The AN/ARC-3 radio set replaces the Type 12 radio set on some airplanes. The ARC-3, when used, is installed on the lower equipment bay hatch.

This radio set operates in the VHF frequency range of 100-156 mc. The set consists of a control panel, in the right side console panel; a transmitter; a receiver and a power junction box. The control panel is shown in figure 4-22.

AN/ARC-3 VHF CONTROL PANEL

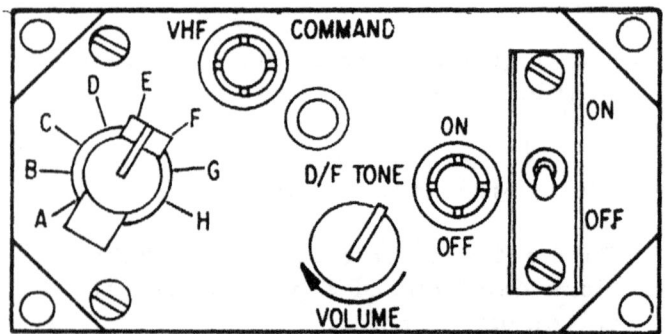

Figure 4-22

Transmission is controlled by the microphone keying switches on the throttle and the control wheel. Eight preset channels are selectable at the control panel. No manual tuning of the transmitter or the receiver is provided.

Operation

ARC-3 circuit breaker in the equipment bay must be closed.

1. Close INTERPHONE circuit breaker.

2. Start set by placing ON-OFF switch in ON position. Allow one minute warmup.

3. Select VHF on COMM selector panel.

4. Set channel selector to desired channel. Refer to VHF frequency card on instrument panel.

5. Adjust volume at VOLUME control.

6. A direction finder tone may be transmitted by the pilot by depressing the D/F TONE button.

7. Move ON-OFF switch to OFF to de-energize the set.

AUTOMATIC OBSERVER

The automatic observer is provided to record pertinent information regarding the sample and the environment conditions from which it was collected. It is composed of a small instrument panel automatically photographed by a 16-mm camera. It is installed in the equipment bay. The camera operates intermittently during the time that a filter is being exposed. The following items are provided on the automatic observer panel:

A Rate Rate meter, an altimeter, a clock, a free air temperature indicator, six filter-in-position lights for the F-2 foil, and six indicator lights for use with the P-3 Platform system when installed.

F-2 FOIL SYSTEM

The F-2 Foil System hatch contains a particulate atmospheric sampler. This hatch is mounted to the lower opening of the equipment bay. When it is carried, most of the other major items which may normally be installed in the equipment bay cannot be used due to lack of space. The one exception is the P-3 Platform System which is compatible with the F-2 hatch.

The samples are collected on a series of six 16-inch diameter filter papers. These filter papers are rotated into and out of sampling position by an electric actuator. When a filter is in place, its exposure is controlled by a door in the air inlet duct. The door is opened by an electric actuator allowing outside air to be ducted in, passed through the filter, and exhausted back outside.

CONTROLS

F-2 Foil Control Panel

Power to the system is controlled by master switch on MARK III hand control.

The control panel, figure 4-23, is located in the left systems panel.

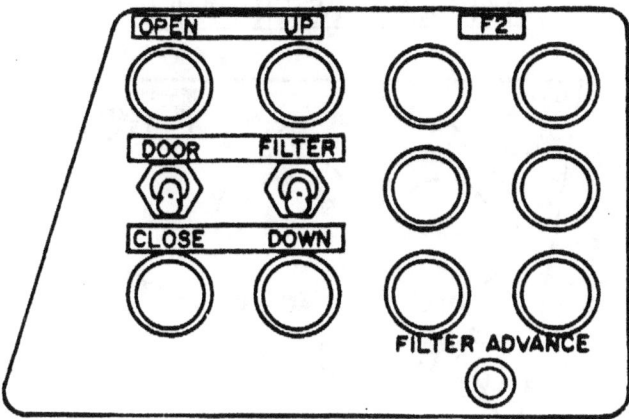

Figure 4-23

The layout of switches and lights is as follows:

1. Door switch - This switch opens and closes the air duct closure door. When moved to the CLOSE or OPEN position the door travels to the selected position. A corresponding amber light glows when the door is fully open or fully closed.

   The automatic observer camera circuits are energized when the door is in the open position.

2. Filter switch - This switch extends the filter for exposure when placed in the DOWN position and retracts the filter when placed in the UP position. A corresponding amber light glows when the filter is in the fully up or fully down position.

NOTE

The UP indicator light will go out if the door is open while the filter is in the UP position. This would normally occur only during duct flushing operation.

3. Filter advance switch - This switch advances the next filter into position for exposure. It becomes armed during the normal sequence of placing the filter in the ducts.

NOTE

Should the filter fail to advance the switch should be actuated repeatedly until the filter is free.

4. Filter position lights - These green lights show which of the six filters is in position to be exposed.

Normal Operating Sequence

The normal sequence of events is as follows:

Before starting:

Door switch - CLOSE.
Amber CLOSE light - ON.
FILTER switch - UP.
Amber UP light - ON.
Green #1 FILTER POSITION light - ON.
Rate Meter - Operating and adjusted.

To expose filter:

FILTER switch - DOWN.
Amber DOWN light - ON.
DOOR switch - OPEN.
Amber OPEN light - ON.
Green #1 FILTER POSITION light -
Remains ON.

To terminate sample after the required exposure time:

DOOR switch - CLOSE.
Amber CLOSE light - ON.
FILTER switch - UP.
Amber UP light - ON.
Green #1 FILTER POSITION light -
Remains ON.

This completes one cycle of operation. The #2 filter may now be advanced for exposure at any time by pressing the FILTER ADVANCE switch. Green #1 light goes out, #2 comes on.

NOTE

In the event that the door actuator or filter actuator jams in mid-position, place the DOOR switch in center OFF position. This will remove the electrical power from both actuators and prevent possible overheat damage.

Duct Flushing

When the sixth sample has been terminated all of the filter-position lights will be out, except position #6. The door-switch and filter switch lights remain on. The duct will be unobstructed at this time and the duct may be flushed by placing the door switch in the OPEN position for as long a period as desired. Actuating the FILTER ADVANCE switch one more time will snuff the #6 position light, if desired.

B/400 RATE RATE METER SYSTEM

The B/400 rate rate meter system is provided to monitor the radiation characteristics of the airflow in the F-2 Foil.

CONTROLS

The control panel for this system (see figure 4-24) is located in the cockpit in the right-hand system panel.

RATE RATE METER PANEL

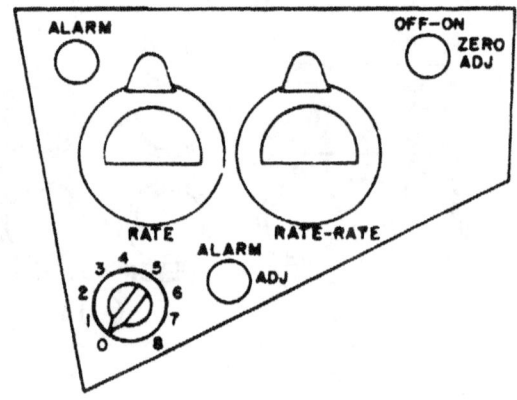

Figure 4-24

The panel incorporates the following:

1. ALARM ADJ Control.

2. RANGE Selector.

3. OFF-ON, and ZERO ADJ Control.

4. ALARM Light (Dimming, Push-to-Test).

5. RATE Indicator.

6. RATE-RATE Indicator.

The function of the panel mounted equipment is as follows:

1. ALARM ADJ - This control is preset so ALARM light comes on at different values of radiation depending on which range control position is selected. This control can be used to turn off the alarm light when it comes on, if it becomes annoying.

2. RANGE Selector - The Range Selector has 9 positions as follows:

POSITION	CPM
0	Meter setting position
1	0 to 1,000
2	0 to 2,000
3	0 to 5,000
4	0 to 10,000
5	0 to 20,000
6	0 to 50,000
7	0 to 100,000
8	0 to 200,000

3. OFF-ON ZERO ADJ Switch - This rotary switch is turned clockwise to activate the system. When the RANGE selector is in the 0 position, the OFF-ON ZERO ADJ switch is rotated as necessary to set the meter to zero. (Is also used in presetting the alarm light operating point, see OPERATION.)

4. ALARM Light - The alarm light is a push-to-test type and can be dimmed by turning the light shield in the clockwise direction. (The light is preset to come on at different values of radiation, depending upon which range is selected, by adjustment of the ALARM ADJ switch.)

5. RATE Indicator - This meter indicates the condition (CPM) of the air being sampled in the F-2 Foil air duct.

6. RATE RATE Indicator - This meter shows the rate of change (increase or decrease) of CPM value.

OPERATION

NOTE

The rate meter should be set at zero before reaching the sampling area.

1. AC Generator Switch - AC GENERATOR position.

2. RANGE Selector - 0.

3. OFF-ON, ZERO ADJ Switch - ON for 3-minute warmup, then clockwise until RATE Indicator is at point where ALARM light operation is desired in flight.

4. ALARM ADJ Switch - Counterclockwise to stop, then slowly clockwise just to point where light comes on.

5. OFF-ON, ZERO SET Switch - Rotate as necessary to return RATE Indicator to zero.

6. RANGE Selector - Set to CPM level required.

P-3 PLATFORM SYSTEM

This system is for the purpose of obtaining and storing air samples. It may be installed in the equipment bay and is not compatible with any other major equipment bay installation except the F-2 Foil System hatch.

The air samples are taken from the engine compressor, compressed to 3000 psi, and stored in six spherical shatterproof bottles. Each bottle is 13 inches in diameter and has a volume of 944 cubic inches. The air is compressed by a 4-stage compressor. The compressor is driven by a 2 3/4-horsepower 28-volt DC motor.

Operation of the system is automatic and no pilot attention is required to switch from a full bottle to an empty one. The same automatic observer installed for the F-2 Foil System hatch is used.

CONTROLS

The control panel is located on the lower left console as shown in figure 4-25.

P-3 PLATFORM CONTROL PANEL

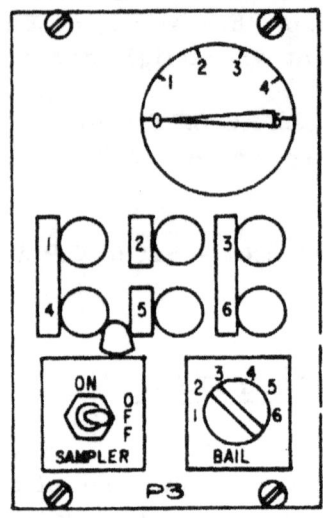

Figure 4-25

The following controls and indications are provided:

1. SAMPLER switch - This switch turns on the sampler.

2. Bottle Full indicator lights - Six green lights are provided to show when each successive bottle is filled.

NOTE

After a bottle has been filled and its bottle full indicator light is ON, it is possible for a leak or cooling of the pressure to decrease, causing the bottle full indicator light to go OUT again. It is not possible to go back and refill this bottle.

3. Pressure gage - Indicates the pressure in the bottle being filled.

4. Six position SELECTOR switch - This selector is provided to give manual change over or selection of bottle to be filled if it is required to override the automatic system. It is possible to select successive bottles with this selector switch, but it is not possible to return to a bottle already passed over.

NOTE

When entering cockpit on a mission using this equipment, the SELECTOR switch should be in the #1 position. If the switch is in any other position, check with the ground crew for equipment status.

OPERATION

Normal operation is as follows:

1. Sampler switch - SAMPLER when a sample is desired.

2. Sampler switch - OFF when bottles are full or when it is desired to save empty bottles for later use.

NOTE

At altitudes between 40,000 and 50,000 feet, the cockpit defroster valve should be placed in a partially open position. The pressure should thereafter build up from zero to 3000 psi in 30 minutes. This is at the rate of 100 psi increase per minute. Operation at altitudes above 50,000 feet will reduce pumping rate; this can be improved by completely closing the defroster valve to make more air available. At altitudes above 50,000 feet the pumping rate will drop as altitude is increased and may take as much as one hour to fill a ball from zero to 3000 psi. If it appears that the pumping rate will not meet these minimum requirements, the next ball should be selected. If inadequate rate is still observed, shut down equipment.

Do not operate system at altitudes lower than 40,000 feet as overloading of DC motor could result.

MISCELLANEOUS EQUIPMENT

SUNSHADE

The forward section of the canopy is provided with a movable sunshade for pilot comfort.

DEFROSTER FAN

A small, rubber-bladed fan is provided for defrosting the canopy. It is mounted on the canopy on the left-hand side. The fan is powered by 28 V DC. A HI-OFF-LO control switch is on the left console. The fan blades are guarded by a circular shield

MAP CASE

The map case has no cover and is located to the rear of the left console.

RELIEF BOTTLE

A relief bottle is stored in a bracket between the left console and the seat. This bottle has a funnel top having a spring-valve. After use, the bottle should be replaced in the bracket.

INTERNAL REAR VIEW MIRROR

There are two swivel mounted rear view mirrors located on each side of the windshield frame inside the cockpit.

WINDSHIELD SWAB

A windshield swab is provided with the aircraft. It is usually stored in the forward canopy to the right or left of the scope. This swab consists of absorbent cloth fixed to an 18 inch stick, and is used for cleaning the windshield side panels in flight.

Changed 6 November 1967

The other end of the stick has a hook that can be used for returning fold-over rudder pedals to normal position.

RUBBER CONE

A rubber cone is provided to attach to the driftsight and sextant display scope. The cone serves two purposes. First, it keeps out stray light when using the sextant. Second, its length places the pilot's eyes at the correct distance from the scope when he uses the optical system.

PINGER

Two small lightweight beacons are mounted in the airplane. One is on the autopilot access door on the right side of the nose section and the other is on the bottom center of the aft fuselage aft of the dive flaps.

These pingers provide detection and recovery capability for the aircraft if it becomes submerged. They each employ a mercury battery which has a life of approximately 21 days in continuous service, and a shelf life of one year.

Operation is immediate and automatic when water immersion closes the circuit between the two ends.

The pinger is compatible with AN/PQS-1 diver held sonar.

DEAD RECKONING NAVIGATION SYSTEM

The dead reckoning navigation system combines a modified AN/APN-153(V) doppler system with a modified AN/ASN-66 dead reckoning navigation computer system. The Doppler system measures drift angle and ground speed and feeds this information to the navigation computer as well as displaying it on the Doppler control panel. The navigation computer uses the drift angle, ground speed and ships compass information to compute and display continuous present position latitude and longitude. The computer also displays the course to follow and distance-to-go from present position to a selected target or destination. The coordinates of three different targets can be set into the navigation computer at any given time: target 1, target 2, and BASE. Targets 1 and 2 are used alternately on successive legs of a mission. In this manner, the system can be used to navigate to an unlimited number of targets. The computer also uses Doppler information to compute and display wind true direction and velocity.

Steering information and distance-to-go information is displayed on the BDHI. (See figure 4-26.) The number 1 pointer shows actual ground track and the number 2 pointer shows relative bearing to the selected target. Distance-to-go to the selected target is shown in digital form up to a maximum of 1999 nautical miles.

BDHI INDICATOR

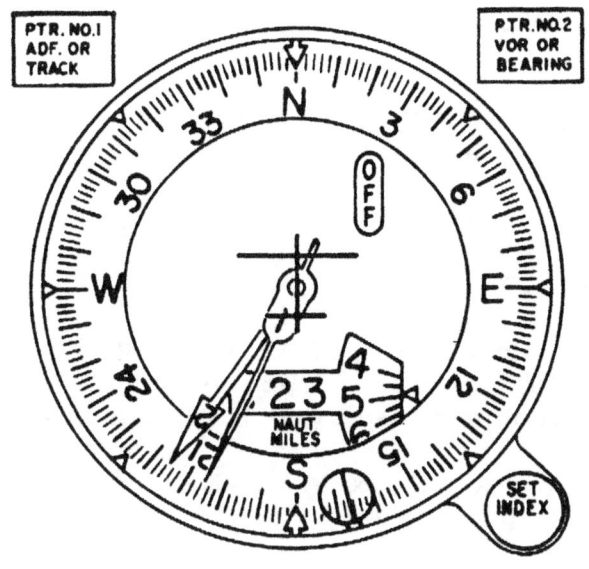

Figure 4-26

When the No. 1 and No. 2 pointers are kept together, the path of the airplane follows a great circle course to a point 200 nautical miles from the target and a rhumb line course from that point on. (At latitudes greater than 72 degrees, the great circle course continues to the target.) If there is no crosswind, the two pointers will both be on the index mark at the top of the BDHI (aircraft magnetic heading). If there is a crosswind from the right, and the number 1 pointer is "flown" to the number 2 pointer, both pointers will be to the left of the index mark. Conversely, if there is a crosswind from the left, both pointers will be to the right of the index mark. In any case, if the two pointers are kept together, the ground track will follow the great circle/rhumb line course described above. The number of degrees between the BDHI index mark and the two pointers is the drift angle due to crosswind.

Since the two pointers on the BDHI can be used either for ADF/VOR or for the navigation system functions, a transfer switch is provided on the right-hand console trim panel to enable the pilot to select either ADF/VOR or ASN-66 information to be fed to the pointers. A NAV light located near the lower left-hand corner of the BDHI illuminates when the switch is in the ASN-66 position. The navigation system continues to function regardless of the position of this transfer switch. The navigation system steering information, of course, is only available as long as the switch is in the ASN-66 position. Neither the distance-to-go, nor the VOR course indicator (ID 249) functions are effected by the position of the ADF/VOR-ASN-66 selector switch.

The navigation system can be used for navigation by grid maps. The system is set up and used in exactly the same manner, except that grid coordinates rather than standard coordinates are used, and the compass system is operated in the D-G mode.

The Doppler system may lose its signal return under certain conditions such as during turns at bank angles exceeding 12 degrees, or over smooth water. During these periods of return signal loss, the memory light located near the lower right-hand corner of the BDHI will illuminate. As long as the memory light is illuminated, the wind direction and velocity information showing on the navigation control panel will be fed into the navigation computer, and the Doppler ground speed information is replaced with either of two "canned" true airspeeds: 394 knots or a speed equivalent to Mach 0.72 for the specific mission day forecasts. (This speed would be 414 knots on a standard day.) A NAV TAS switch located just above the navigation control panel allows the selection of either STD (394 knots) or .72 M (414 knots on a standard day). When the Doppler return signal is picked up again, the memory lights will extinguish and the wind and speed information functions will automatically revert to the Doppler system. If the Doppler system has failed for an extended period of time as evidenced by the memory lights, wind information from forecasts, if known, can be inserted manually on the navigation panel.

The navigation power switch, energizes the number 2 inverter when the normal inverter selector switch is in the INVERTER NO. 1 position. The power system has been designed with a priority circuit such that the navigation system will be shut off automatically if:

1. The normal inverter selector switch is in the INVERTER NO. 2 position or,

2. During normal operation, the AC generator fails causing transfer of System 13 to the number 2 inverter. If System 13 is not turned on, however, the navigation system will continue to operate.

NOTE

The Doppler control panel should be monitored occasionally to verify operation. Under stabilized conditions, very little activity, if any, may be detected; and no activity if the Doppler system has lost power. If loss of power is suspected, press memory push-to-test lamp. No illumination verifies loss of system power.

In the event there is excessive operation of the memory light, the TEST position should be selected for approximately 2 minutes. If the light is still illuminated, the OFF position should be selected for a few seconds, then STBY for 1 minute, then to the applicable operating condition to allow recycling of the equipment.

CONTROLS

NAVIGATION CONTROL SWITCHES
RIGHT CONSOLE TRIM PANEL

Three two-position switches are on a panel on the right console trim panel. (See figure 4-27.)

The foremost switch, NAV POINTER TRANSFER, selects ASN-66 or ADF/VOR information to be fed to the BDHI pointers.

The center switch, NAV POWER, has a No. 2 INV position and an OFF position.

The aft switch, COMPASS POWER, selects aircraft power or external power.

NAVIGATION CONTROL SWITCHES

Figure 4-27

APN-153(V)

The APN-153 (V) Doppler control panel, (figure 4-28) contains: a rotary function switch with five positions - OFF, STBY, LAND, SEA, and TEST; a drift angle slewing knob; a ground speed slewing knob; a memory light; and an indicator which displays a drift angle and ground speed.

In the TEST position of the rotary switch, the system automatically performs a self-test. Proper operation, after a 5-minute warmup period, is indicated when in the TEST position, the drift angle reads 0° (± 2°) and the ground speed reads 121 (± 5) knots. LAND or SEA position is selected prior to takeoff. LAND position is used when flying over land, and SEA position is used when flying over water. The memory light glows when the system is not providing reliable data to navigation system. The drift angle and ground speed slewing knobs will not be used during flight because of other provisions contained in the ASN-66.

AN/APN-153(V) CONTROL PANEL

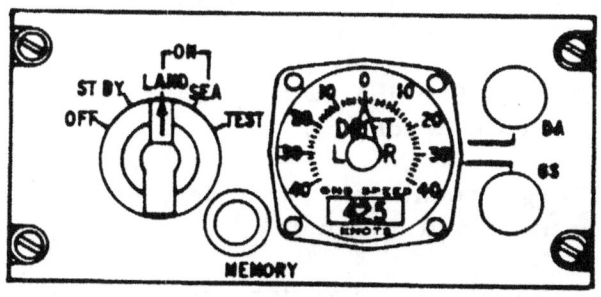

Figure 4-28

ASN-66

The ASN-66 navigation control panel, (figure 4-29) contains: a six position rotary function switch - TEST, OFF, STBY, TGT 1, TGT 2, and BASE (this rotary switch has a press-to-rotate bar on the knob which must be pressed in order to select BASE position); a destination (DEST) switch to select either TGT 1 or TGT 2; a variation counter with slewing knob; target latitude and longitude counters with slewing knobs; present position latitude and longitude counters with slewing knobs; a wait light; a wind from degrees true counter with slewing knob; and a wind knots counter with slewing knob.

NAVIGATION CONTROL PANEL ASN-66

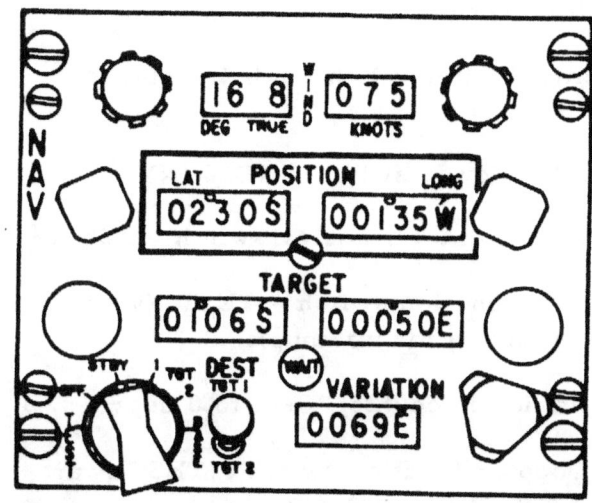

Figure 4-29

In the TEST position of the rotary switch, the system automatically performs a self-test. Proper operation is indicated when, in the TEST position, the wind counters read 240 (± 1.5) degrees and 167 (± 5) knots. The variation counter must be set to 0 degrees during self-test.

The TGT 1, TGT 2, and BASE positions of the rotary switch are selected while slewing in respective target and BASE coordinates on the target counters. After selecting a TGT or BASE position, slew the proper coordinates into target counters and wait until the WAIT light extinguishes and verify that the proper coordinates remain on the target counters before selecting another rotary switch position. The position of the rotary switch in either TGT 1 or TGT 2 has no effect on the navigation steering computation or operation, nor does it change the target coordinates "memorized". The position of the DEST switch determines which target steering information is being displayed. In the BASE position, however, all other functions are overriden and the computer navigates to and displays steering information to the BASE site coordinates. If desired, the selector can be returned to TGT 1 or TGT 2 and the normal techniques resumed.

OPERATION

Accomplish items 1 through 23 before takeoff.

1. NAV Power switch - NO. 2 INV.

2. NAV pointer transfer switch - ASN-66.

3. NAV function switch - TEST (wait 2 minutes for warmup).

4. Variation counter - set to 0 degrees.

5. Doppler function switch - TEST (wait 5 minutes for warmup).
   Wind direction counters should read 238.5 to 241.5 degrees.
   Wind-knots counters should read 162 to 172 knots.
   Ground speed counter should read 116 to 126 knots.
   Drift indicator should indicate 0 (± 2) degrees.
   Memory light should be off.
   Present position counters should move Northerly and Easterly.

6. NAV-TAS switch - set as briefed.

7. ASN-66 function switch - STBY.

8. Doppler function switch - STBY.

9. Variation counter - set to correct magnetic variation for first leg of mission.

10. Wind Direction and Velocity counters - set to approximate values.

11. ASN-66 function switch - set to BASE and pull switch out.

12. TARGET counters - set in BASE latitude and longitude.

13. ASN-66 function switch - push in and allow WAIT light to extinguish and confirm BASE readings.

14. ASN-66 function switch - set TGT 2 and pull switch out.

15. Target counters - set in TGT 2 latitude and longitude.

16. ASN-66 function switch - push in and allow WAIT light to extinguish and confirm TARGET readings.

17. ASN-66 function switch - set to TGT 1 and pull switch out.

18. Target counters - set in desired TGT 1 latitude and longitude.

19. ASN-66 function switch - push in and allow WAIT light to extinguish and confirm TARGET readings.

20. ASN-66 function switch - STBY.

21. POSITION counters - set to desired starting point latitude and longitude.

22. DEST switch - TGT 1.

NOTE

Destination (DEST) switch position determines selection of target steering and distance-to-go information delivered to the BDHI.

With the function switch pushed in (normal), stored target coordinates are displayed on the target counters. The switch is pulled out to insert new target coordinates.

The function switch incorporates a detent feature to prevent accidental switching to OFF or TEST and has a press-to-rotate bar to prevent accidental selection of BASE.

Immediately prior to takeoff roll:

23. Doppler function switch - LAND.

24. ASN-66 function switch - TGT 1 as briefed. This displays TGT 1 coordinates and starts navigation system.
(Navigation system does not start to compute until a target is selected.)

25. Steer to TGT 1 by flying the number 1 pointer to align with the number 2 pointer.

26. When over TGT 1 - DEST switch to TGT 2 and steer to TGT 2.

NOTE

When over TGT 1, the TARGET counters and POSITION counters should read the same, and the number 2 pointer will swing through 180 degrees, and the distance-to-go counters will start reversing.

The POSITION counters can be adjusted, if necessary, BEFORE the DEST switch is switched to the TGT 2 position.

27. VARIATION - set, as necessary, for the second leg of the mission.

28. ASN-66 function switch - TGT 1 and pull switch out.

29. TARGET counters - set in target 3 latitude and longitude.

30. ASN-66 function switch - push in and allow WAIT light to extinguish and confirm TARGET readings.

31. ASN-66 function switch - TGT 2 to display TGT 2 coordinates. For additional targets repeat steps 26 through 30 using TGT 1 and TGT 2 alternately on successive legs of the mission.

The POSITION counters can be corrected at any time by moving the ASN-66 function switch to STBY and setting the POSITION counters to the proper coordinates for the known check point. Move the ASN-66 function switch back to the appropriate TGT position when directly over the known checkpoint. Obviously, the procedure should not be used if there is any doubt about the identification of the checkpoint or its coordinates.

AF (C)-1-1

OPERATING LIMITATIONS

TABLE OF CONTENTS

	PAGE
OPERATING LIMITATIONS	5-1
J-75 ENGINE LIMITATIONS	5-1
PROHIBITED MANEUVERS	5-2
AIRSPEED LIMITATIONS	5-4
GUST CONTROL LIMITATIONS	5-4
INSTRUMENT MARKINGS	5-4
ACCELERATION LIMITATIONS	5-4
JP-4 AND JP-5 FUEL LIMITATIONS	5-4
LIMITATIONS WITH SLIPPER OR DROP TANKS INSTALLED	5-6
WEIGHT AND BALANCE LIMITATIONS Ⓒ	5-6
WEIGHT AND BALANCE LIMITATIONS Ⓕ	5-6

OPERATING LIMITATIONS

This airplane was designed for a specific mission and has performance capabilities in certain categories exceeding anything previously available. In order to meet these requirements, it is necessary to impose very definite structural and aerodynamic limitations. Therefore, the airplane should only be flown in a manner required to accomplish the basic mission. If this fact is remembered by the pilot, he should have no difficulty staying within the placard limitations.

The airplane has received extensive static load tests, flight tests and operational experience. Safe flight limitations were established and demonstrated in flight.

J-75 ENGINE LIMITATIONS

Above 60,000 feet, operate the engine to the maximum EPR limits shown on figure 5-1 to provide standard airplane performance and maneuvering characteristics with respect to wing buffet. The maximum EGT limit for the engine is 665°C; however, operation in excess of maximum EPR limits will decrease wing buffet margin or result in wing buffet and decrease the maneuvering capability.

SPEED

Gate power for normal takeoffs (approximately 94%).

100% RPM for full power (emergency only).

Avoid use of maximum power for takeoff except in emergencies, due to excess thrust of the 13B engine.

OVERSPEED

103.0% Maximum.

IDLE

45% ± 1% RPM.

EXHAUST GAS TEMPERATURE AND ENGINE PRESSURE RATIO

Normal Operation - Climb and Descent

630° C Maximum, Takeoff to 40,000 feet.

485°C Maximum, 40,000 feet to 60,000 feet.

665° C Maximum, 60,000 feet and above.

Changed 6 November 1967

SECTION V   AF (C)-1-1

During climb, the EGT and EPR limits as shown in figure 5-1 should be observed to avoid engine roughness and compressor stall. During descent, the minimum EPR values listed in figure 5-1 should be observed.

Starting

400°C Maximum.

Engine Overtemperature Limits
Acceleration and Steady State

630°C Maximum, Takeoff to 60,000 feet.

665°C Maximum, 60,000 feet and up.

Overspeed and overtemperature conditions should be corrected immediately and recorded in DD Form 781 giving temperature and duration. If condition cannot be corrected, a landing should be made as soon as practicable.

OIL PRESSURE

Normal oil pressures are 40 to 55 psi. Oil pressure of 35 to 40 psi or 55 to 60 psi are cause for discontinuing the mission and should be reported on Form DD 781. Oil pressures of less than 35 psi or more than 60 psi are unsafe. The engine should be operated at the lowest RPM required to maintain flight, and a landing should be made as soon as practicable. Minimum oil pressure at idle RPM during ground operation is 35 psi. A fluctuation within a limit of plus or minus 2 psi is acceptable.

OIL TEMPERATURE

121°C Maximum.

15 - 121°C Normal.

FUEL PRESSURE

25 psi Maximum.

14-25 psi Normal.

NOTE

At full throttle takeoff power fuel pressure may drop to zero.

POWER REGULATION AT ALTITUDE

The primary engine operating variables are engine pressure ratio (EPR) and exhaust gas temperature. Above 60,000 feet, throttle position is varied in order to keep the exhaust gas temperature at or near the 665°C limit without exceeding maximum engine pressure ratio as tabulated in figure 5-1.

Normal descents may be made by opening the bleed valves and then retarding the throttle.

To avoid any possibility of flameout during a descent, the minimum EPR schedule should be closely observed.

PROHIBITED MANEUVERS

All acrobatic maneuvers are prohibited.

Changed 1 September 1968

AF (C)-1-1 SECTION V

EGT AND EPR LIMITATIONS

ALT (FT x 10³)	EGT °C		EPR MAXIMUM	EPR - MINIMUM	
	MAX.	MIN DESCENT		BLEED VALVES OPEN	BLEED VALVES CLOSED
S.L. (After T.O.)	*540				
15	*540				
25	*540	130			1.35
35	*540	130			1.35
36	*540	135			1.35
38	*540	145			1.35
40	485	155			1.35
42	485	165			1.35
44	485	180			1.35
45	485	185			1.35
50	485			1.27	1.44
55	485			1.42	1.61
60	485		3.30	1.65	1.82
62	665		3.29	1.74	1.93
65	665		3.27	1.90	2.12
67	665		3.24	2.02	2.25
68	665		3.23	2.08	2.31
69	665		3.20	2.15	2.38
70	665		3.18	2.24	2.45
71	665		3.16	2.32	2.54
72	665		3.14	2.40	2.61
73	665		3.11	2.49	
74	665		3.09	2.59	

*RECOMMENDED

Figure 5-1

SECTION V   AF (C)-1-1

AIRSPEED LIMITATIONS

The airspeed limitations are shown in the following table. The airspeed indicator maximum allowable needle is set for 0.80 Mach number or 240 knots, whichever is less. The following limitations must be observed.

CONFIGURATION	MAX. ALLOWABLE AIRSPEED
SMOOTH AIR	
Faired	220 Knots
Gust Control On	240 Knots
ROUGH AIR	
Faired	150 Knots
Gust Control On	170 Knots
GENERAL	
Landing Gear Extended	240 Knots
Wing Flaps Extended	130 Knots
Speed Brakes Extended	240 Knots

WARNING

Never extend the wing flaps above 45,000 feet since dangerous pitching moment effects are produced.

GUST CONTROL LIMITATIONS

When actuating the gust control from the GUST position to the FAIRED position, there is a remote possibility that the flaps might continue down. Since this type of malfunction could be hazardous, the gust control should not be actuated from GUST to FAIRED until the aircraft is slowed to 150 knots or slower.

The gust control should not be actuated from FAIRED to GUST at airspeeds above 220 knots. Since this is the placard speed for the FAIRED configuration, this condition will not normally arise. However, should the aircraft reach a speed in excess of 220 knots in the FAIRED configuration, the speed should be dissipated as carefully as possible.

INSTRUMENT MARKINGS

Cognizance must be taken of the instrument markings as shown on figure 5-2, since they represent limitations that are not necessarily repeated elsewhere.

ACCELERATION LIMITATIONS

The load factors are indicated in the flight strength diagrams, figures 5-3 through 5-6.

JP-4 AND JP-5 FUEL LIMITATIONS

JP-4 grade fuel can be used if the primary fuel is not available and maximum range is not required. Frequent or continued operation with JP-4 fuel, however, may result in coking of the fuel nozzles; therefore, the use of JP-4 fuel should be kept to a minimum. Due to the high vapor pressures of JP-4 fuel, the rate of climb must be restricted to a maximum of 3,000 feet per minute to avoid dangerously high pressure in the wing tanks and loss of fuel overboard through the vent system. The maximum allowable altitude with JP-4 fuel is 70,000 feet. To allow for possible variations in JP-4 fuel weight the range figures in the performance section should be reduced 3% for planning purposes.

INSTRUMENT MARKINGS

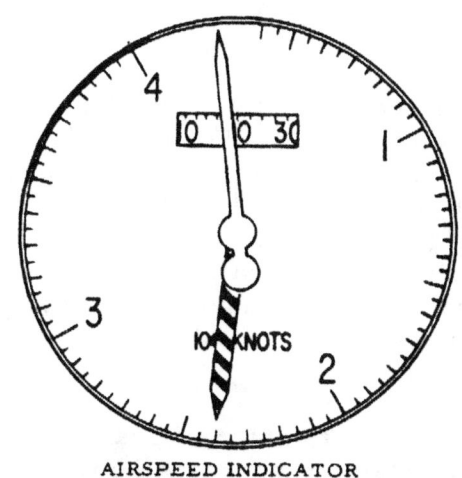

AIRSPEED INDICATOR

THE INSTRUMENT SETTING IS SUCH THAT THE MAXIMUM ALLOWABLE AIRSPEED POINTER WILL MOVE TO INDICATE THE LIMITING STRUCTURAL AIRSPEED OF 240 KNOTS, OR THE AIRSPEED REPRESENTING THE LIMITING MACH NUMBER OF 0.8, WHICHEVER IS LESS.

COLOR CODE
- RED
- GREEN

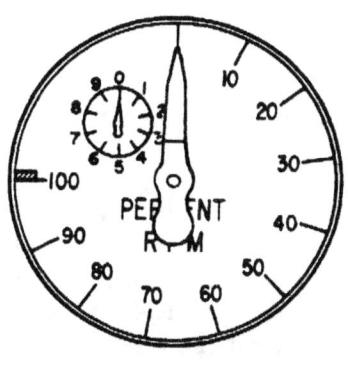

TACHOMETER
103% MAXIMUM OVERSPEED

EXHAUST GAS TEMPERATURE
P-13B ENGINE - 665°C MAXIMUM

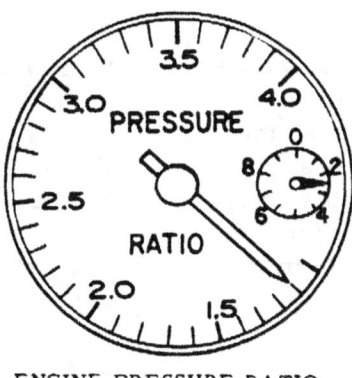

ENGINE PRESSURE RATIO
FLUCTUATION LIMITS ± 0.02

FUEL PRESSURE
25 PSI MAXIMUM
14-25 PSI NORMAL
0 PSI DURING FULL POWER TAKEOFF.

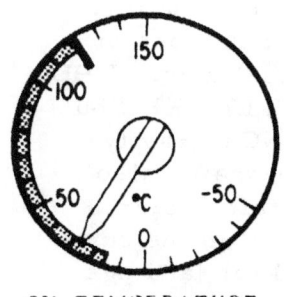

OIL TEMPERATURE
15-121°C NORMAL
121°C MAXIMUM

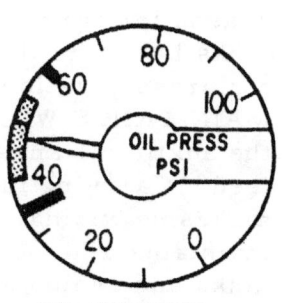

OIL PRESSURE
40-55 PSI NORMAL
35 PSI MINIMUM
60 PSI MAXIMUM

FLUCTUATION LIMITS ± 2 PSI

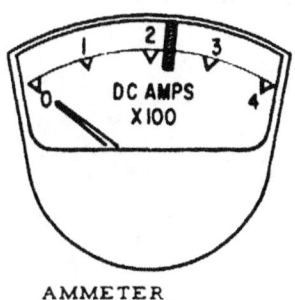

AMMETER
225 AMPS MAXIMUM

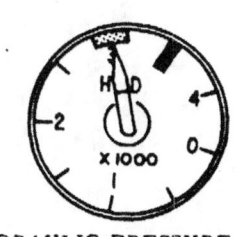

HYDRAULIC PRESSURE
2850-3150 PSI NORMAL
3500 PSI MAXIMUM

Figure 5-2

SECTION V

JP-5 grade fuel, although not tested, may be used as an emergency fuel under the following condition:

For occasional usage, such as for ferry purposes - no inspections required.

NOTE

JP-5, like MIL-F-25524, is a low vapor pressure fuel. Therefore, no rate of climb restrictions are imposed.

No fuel other than those mentioned may be used.

LIMITATIONS WITH SLIPPER OR DROP TANKS INSTALLED

The addition of slipper tanks or drop tanks does not change the aircraft limitations. However, drop tanks should be jettisonned when empty. (If retained, residual air pressure transferred from the empty drop tanks to the sump tank could cause fuel slugging. Further, the drag of the empty tanks offsets the range advantage gained in utilizing drop tank fuel.)

CAUTION

Drop tanks and slipper tanks should not be carried at the same time.

WEIGHT AND BALANCE LIMITATIONS (C)

The zero fuel weight center of gravity limits for the U-2C are 26.5% and 28% MAC, gear down. Since variations occur between aircraft it is necessary to determine the zero fuel weight center of gravity position for each individual aircraft. With certain loadings, a small amount of tail ballast may be required to stay within the forward center of gravity limit.

The variation of center of gravity position with fuel usage is shown in figure 5-7.

Reference should be made to the Weight and Balance Data Handbook for the particular airplane if detailed weight information is desired.

WEIGHT AND BALANCE LIMITATIONS

The CG curves, figures 5-8 through 5-10 reflect a representative airplane configuration and the Weight and Balance Handbook should be consulted for more detailed data.

The zero fuel CG limit is 25.5% to 26.5% MAC, gear down.

When using less than a 1035-gallon fuel load for a flight, the fuel should be carried in the main tanks. The primary reason for carrying the fuel in the main tanks is that in the event of the loss of air pressurization to the fuel tanks, **THE MAIN TANKS WILL GRAVITY FEED** but the auxiliary tanks **WILL NOT** gravity feed. Carrying fuel in the mains also makes servicing the aircraft easier, maintains a higher fuel level in the sump tanks and avoids an aft CG for landings. See figure 5-9.

FLIGHT STRENGTH DIAGRAM

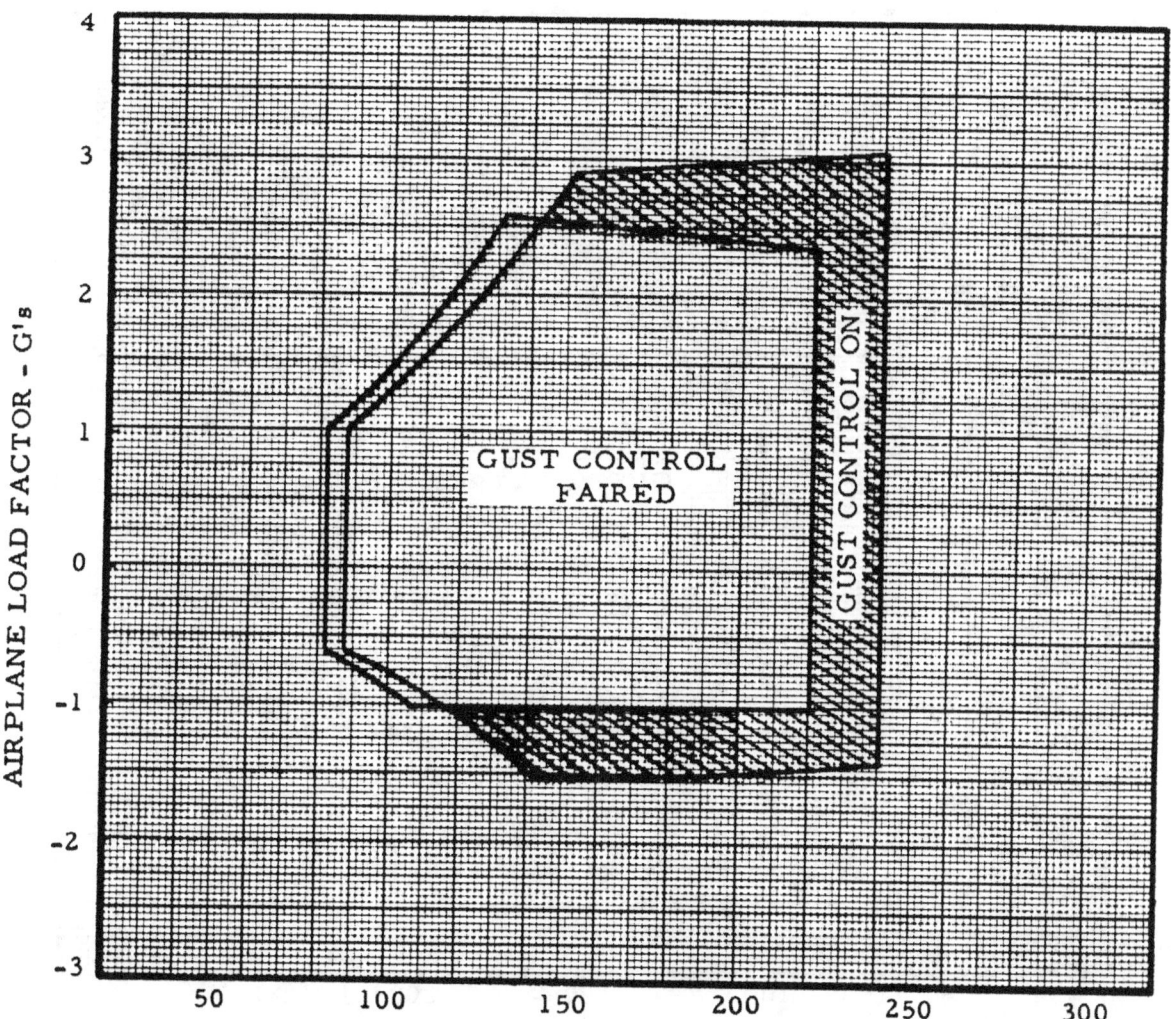

CAUTION
THE ABOVE V-n DIAGRAM DEFINES THE AIRPLANE CAPABILITY FOR SYMMETRICAL FLIGHT. HOWEVER, ALL AIRCRAFT ARE TRADITIONALLY LIMITED TO A REDUCTION IN THESE VALUES OF FROM 20 TO 33 PERCENT FOR ABRUPT ACCELERATED OR MAXIMUM VELOCITY ROLL.

Figure 5-3

FLIGHT STRENGTH DIAGRAM

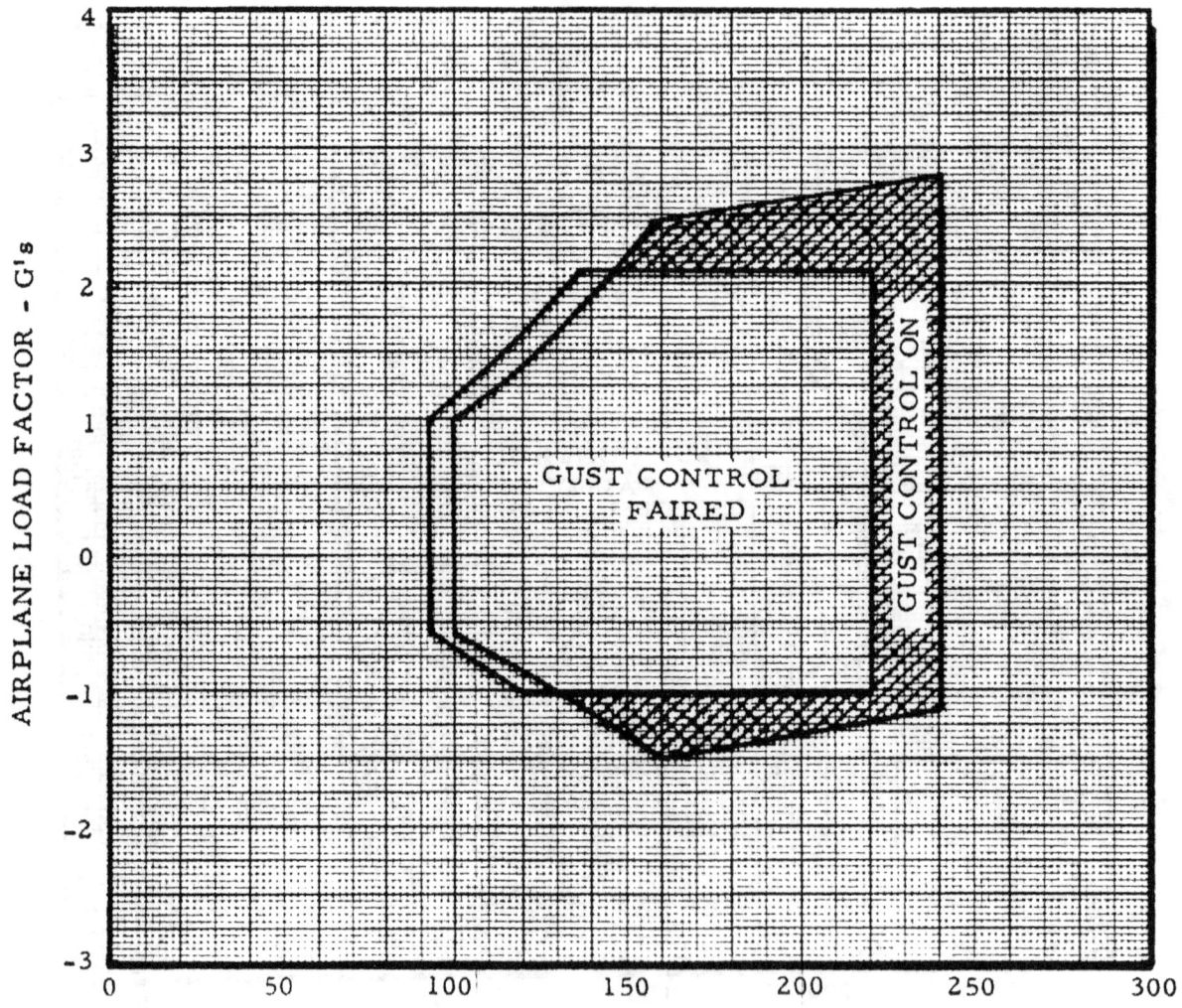

CAUTION

THE ABOVE V-n DIAGRAM DEFINES THE AIRPLANE CAPABILITY FOR SYMMETRICAL FLIGHT. HOWEVER, ALL AIRCRAFT ARE TRADITIONALLY LIMITED TO A REDUCTION IN THESE VALUES OF FROM 20 TO 33 PERCENT FOR ABRUPT ACCELERATED OR MAXIMUM VELOCITY ROLL.

Figure 5-4

FLIGHT STRENGTH DIAGRAM

WITHOUT SLIPPER TANKS

21,760 LB GROSS WEIGHT

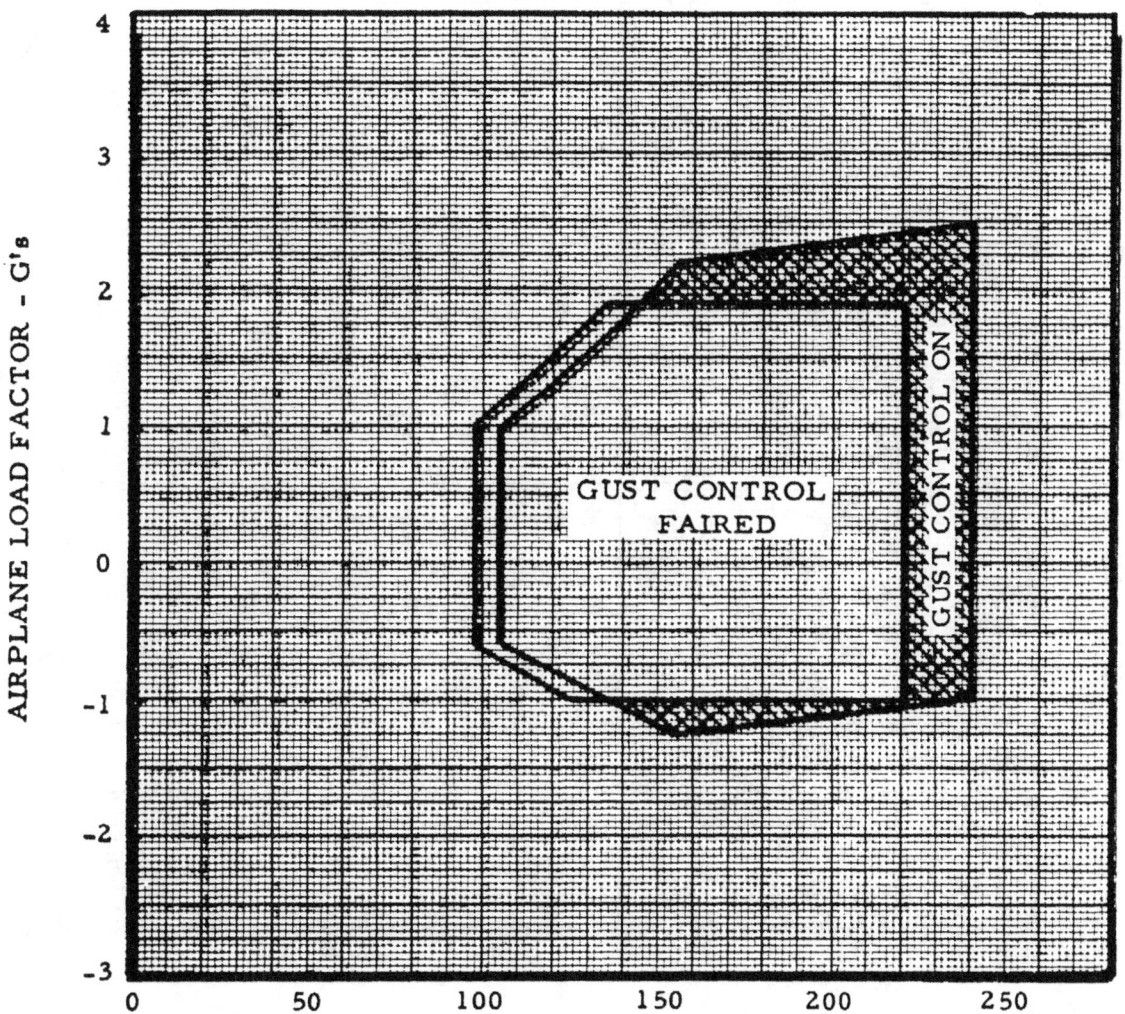

CAUTION

THE ABOVE V-n DIAGRAM DEFINES THE AIRPLANE CAPABILITY FOR SYMMETRICAL FLIGHT. HOWEVER, ALL AIRCRAFT ARE TRADITIONALLY LIMITED TO A REDUCTION IN THESE VALUES OF FROM 20 TO 33 PERCENT FOR ABRUPT ACCELERATED OR MAXIMUM VELOCITY ROLL.

Figure 5-5

FLIGHT STRENGTH DIAGRAM

WITH SLIPPER TANKS

23,240 LB GROSS WEIGHT

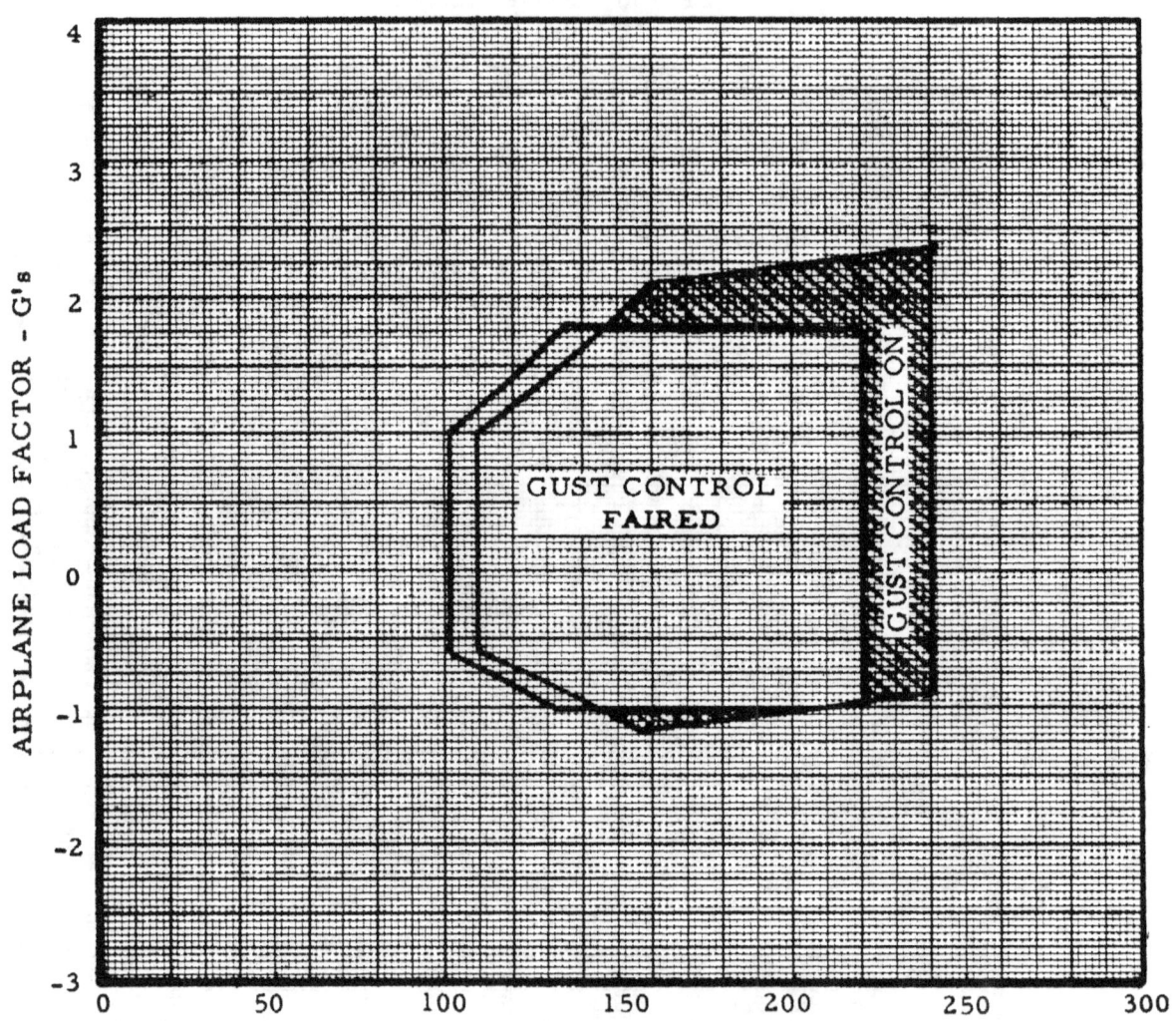

CAUTION

THE ABOVE V-n DIAGRAM DEFINES THE AIRPLANE CAPABILITY FOR SYMMETRICAL FLIGHT. HOWEVER, ALL AIRCRAFT ARE TRADITIONALLY LIMITED TO A REDUCTION IN THESE VALUES OF FROM 20 TO 33 PERCENT FOR ABRUPT ACCELERATED OR MAXIMUM VELOCITY ROLL.

Figure 5-6

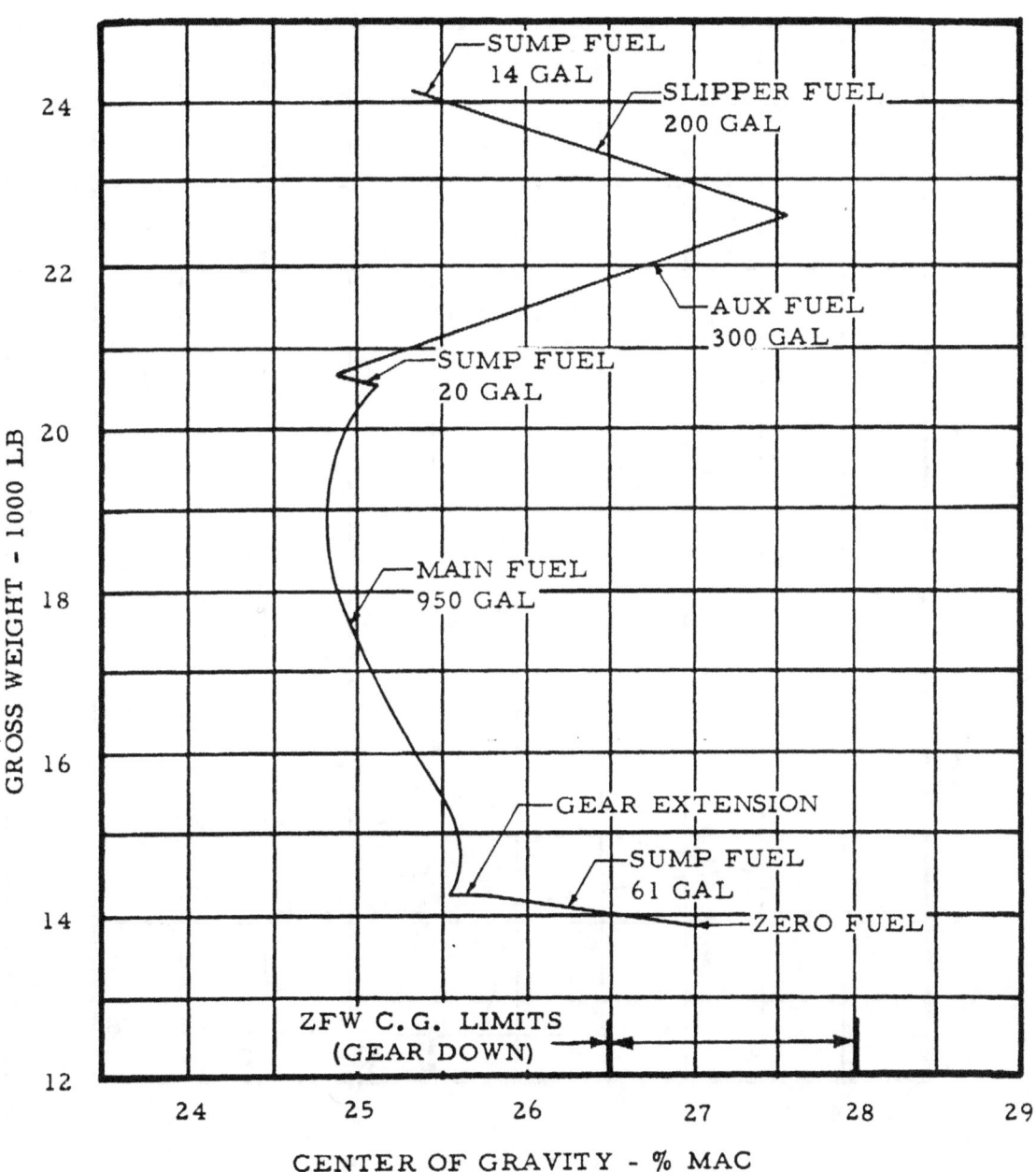

Figure 5-7

MODEL U-2F
CENTER OF GRAVITY VS GROSS WEIGHT

WITH SLIPPER FUEL

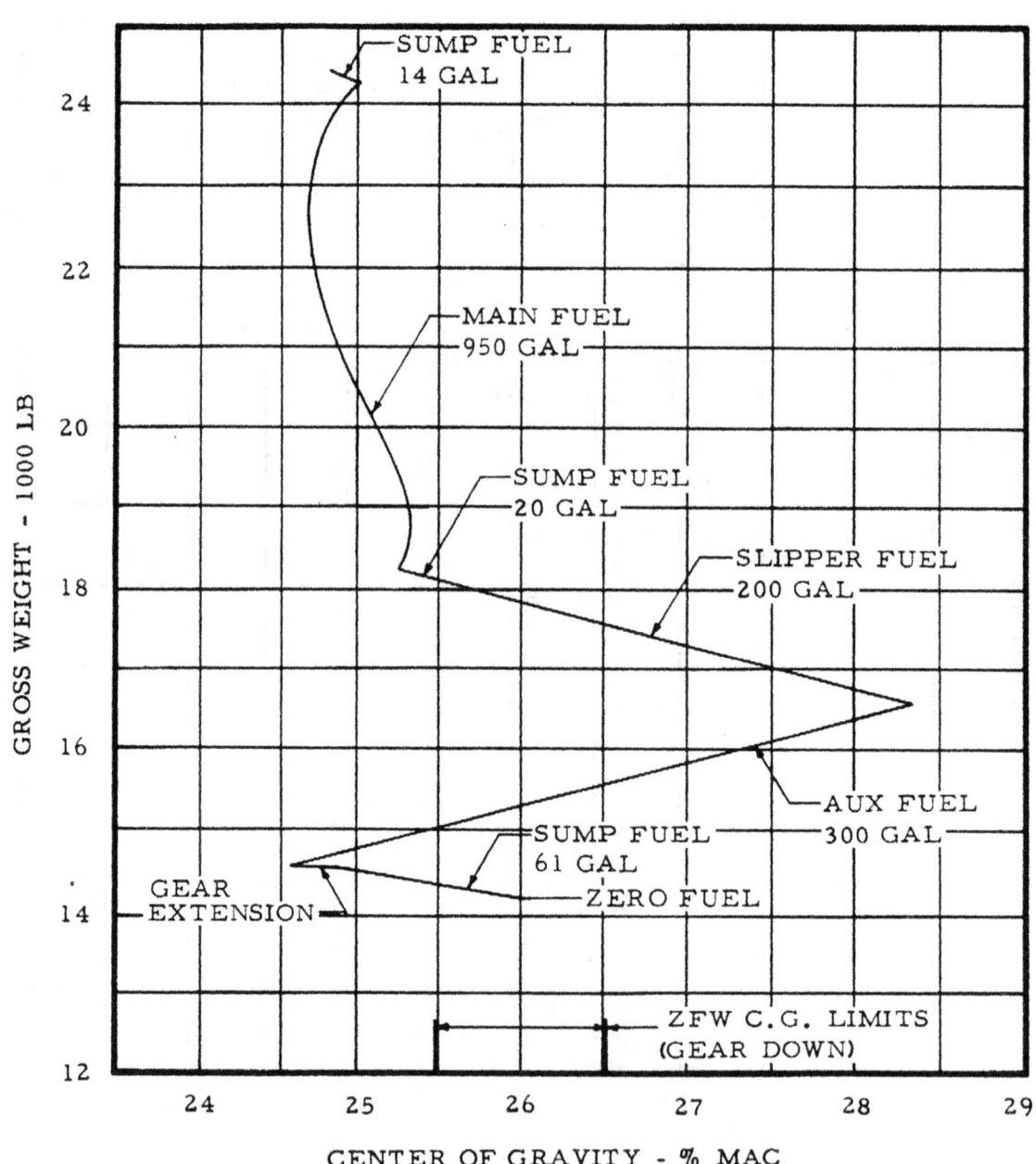

Figure 5-8

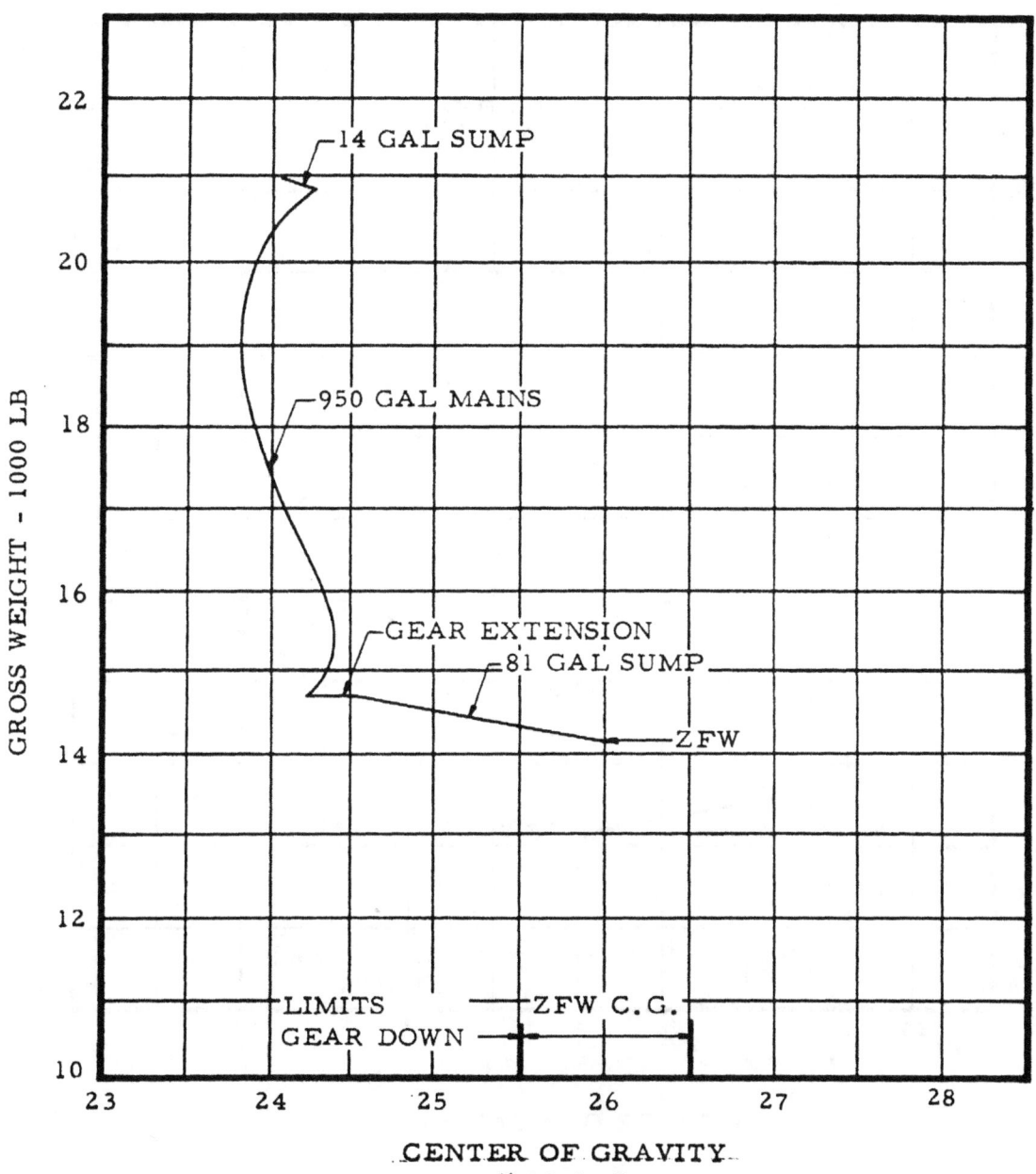

Figure 5-9

MODEL U-2F
CENTER OF GRAVITY VS GROSS WEIGHT

WITHOUT SLIPPER FUEL

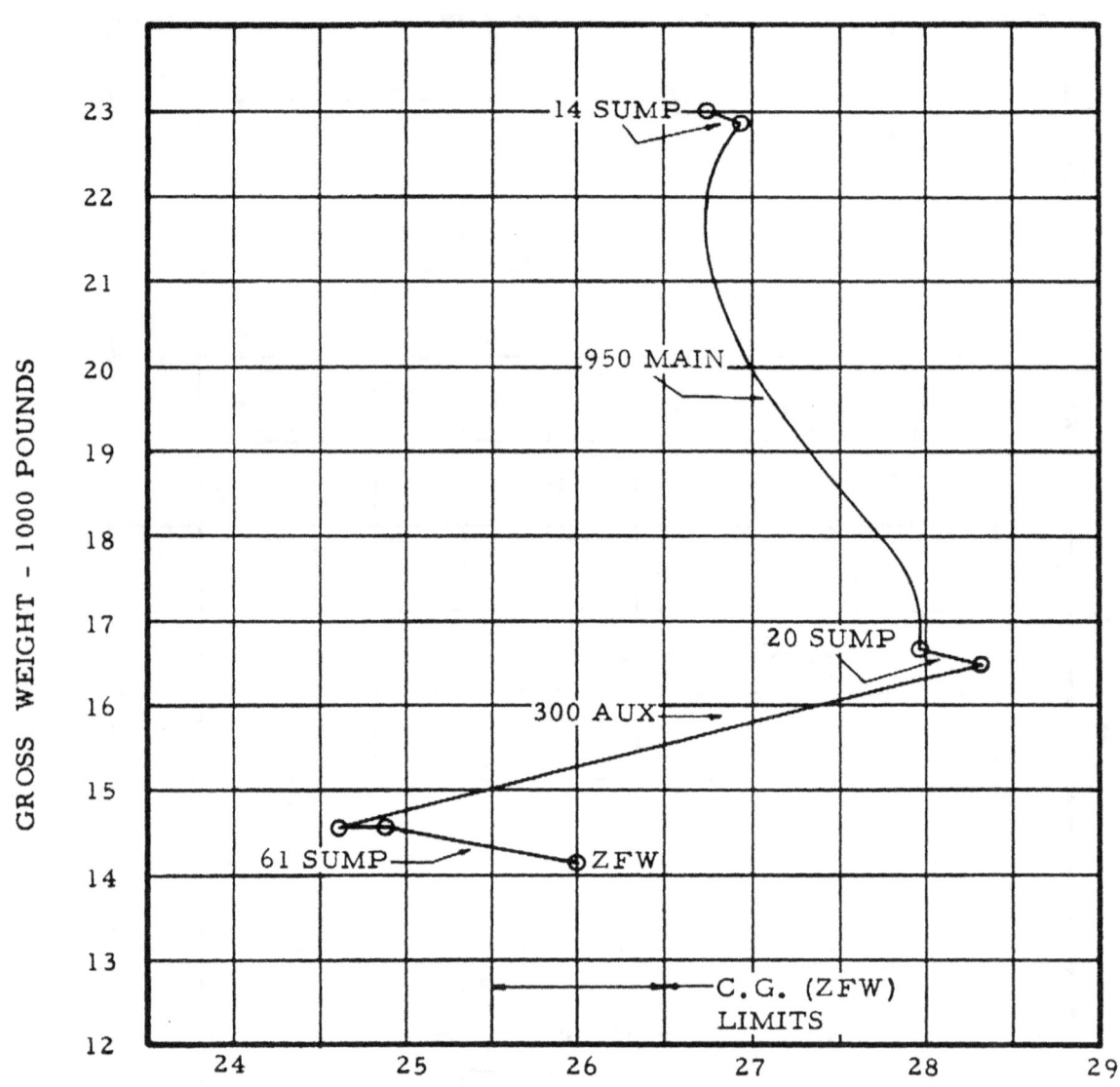

Figure 5-10

AF (C)-1-1   SECTION V

MODEL U-2C
CENTER OF GRAVITY VS GROSS WEIGHT

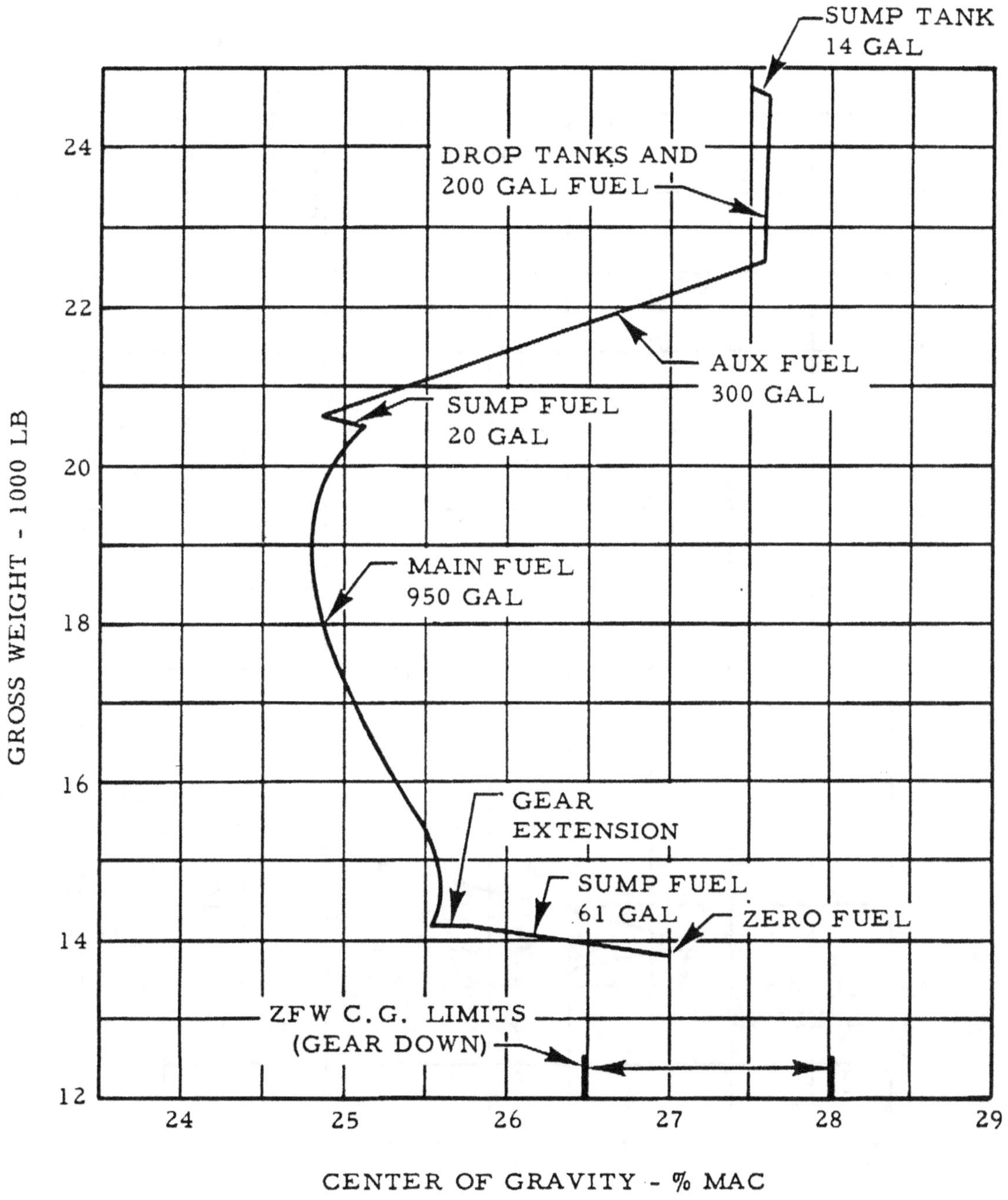

Figure 5-11

MODEL U-2F
CENTER OF GRAVITY VS GROSS WEIGHT

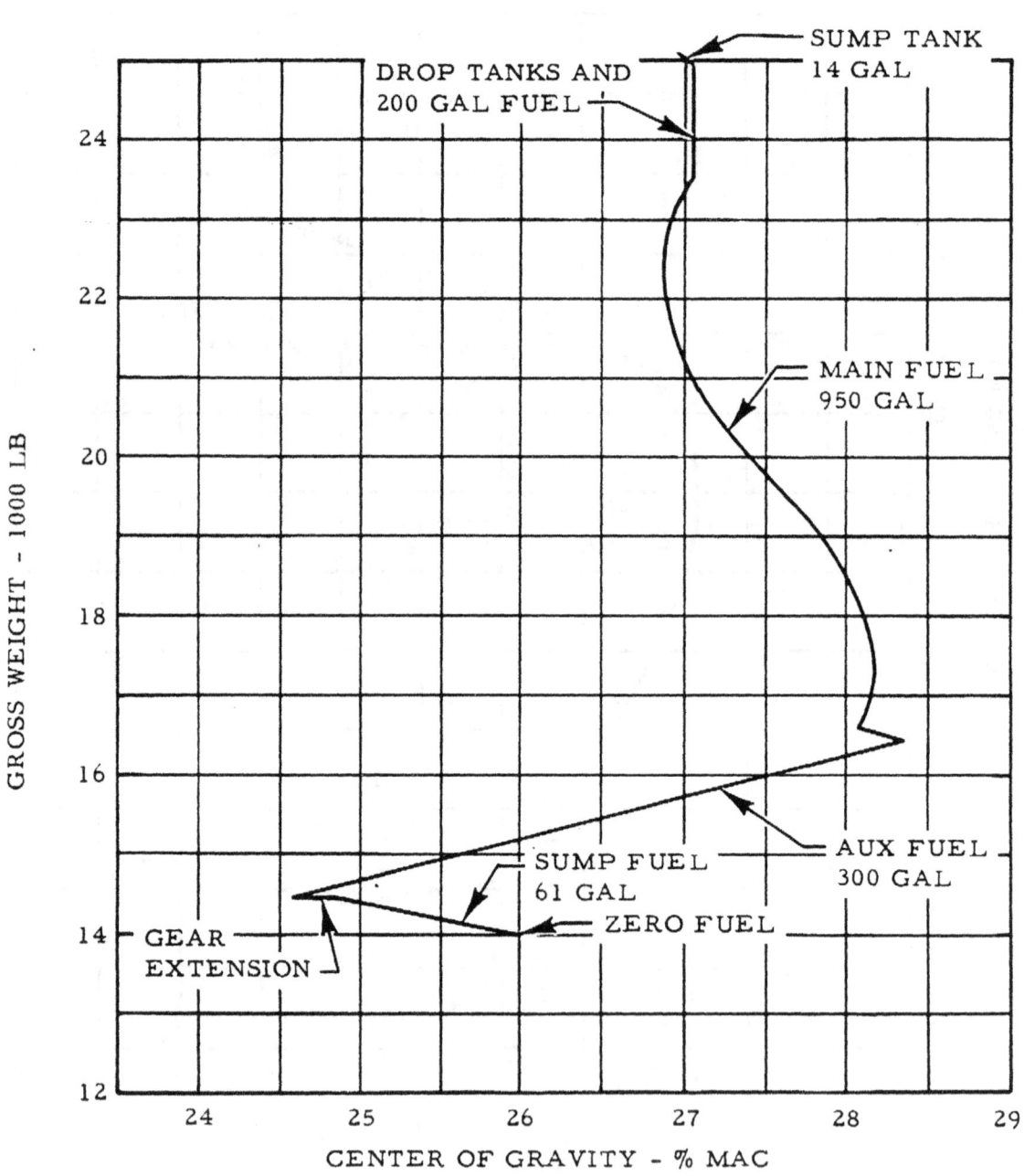

Figure 5-12

AF (C)-1-1

FLIGHT CHARACTERISTICS — SECTION 6

TABLE OF CONTENTS

	PAGE
INTRODUCTION	6-1
FLIGHT CONTROL CHARACTERISTICS	6-1
LONGITUDINAL STABILITY	6-2
LONGITUDINAL TRIM CHANGES	6-4
DIRECTIONAL STABILITY	6-4
ROLLING MANEUVERS	6-5
STALLS	6-5
LATERAL TRIM CHARACTERISTICS	6-6
AIRSPEED ACCELERATION	6-7
WING BUFFET	6-7
MACH NUMBER CHARACTERISTICS	6-8
SPINS	6-12
SPEED BRAKES	6-12
GUST CONTROL	6-12
CRUISE CLIMB	6-13
DESCENT	6-13
APPROACH	6-14
LANDING	6-14
GO-AROUND	6-15
FLYING WITH EXTERNAL LOADS	6-16
FORMATION FLYING	6-16

INTRODUCTION

The flight characteristics of this airplane are in general conventional and similar to other airplanes. Its design is predicated on maximum performance consistent with reliability. The bicycle landing gear with pogos is devised for minimum weight in order to gain maximum altitude performance. A complete familiarization with the aircraft is essential to full utilization of its capabilities. The information in this section is based on extensive flight testing and operational experience.

FLIGHT CONTROL CHARACTERISTICS

CONTROLS

The flight controls are conventional consisting of rudder pedals and a wheel mounted on a control column. All control surfaces are directly connected to the cockpit controls by means of cables. No power boost is provided.

ELEVATOR

The elevator control forces vary from light to heavy depending on CG position, airspeed, and other variables. At low speed the forces are light but the elevator is large and relatively effective. Therefore, during the approach and landing stages there should be no abrupt movements of the elevator control, but rather steady light pressures.

AILERON

In general, the aileron control forces will feel heavy, particularly at the higher speeds. This is due in part to the long wing span. At landing pattern speeds the control forces are much lower and the rate of roll is adequate. With the gust control in the GUST position, the aileron forces are lighter but the maximum rate of roll is lower because of the reduced travel of the ailerons. At low altitudes and GUST position, downward rolling manuevers should be limited due to the reduced aileron travel. However, holding full aileron for recovery from steep attitudes causes no structural problems.

RUDDER

The rudder control forces are moderately high. Rudder control is not used extensively during flight except during takeoff and landing when large amounts of rudder may be necessary to maintain runway lineup and to counteract crosswind drift. Since there is very little roll with yaw, only a small amount of aileron is required to compensate for a side slip.

TRIM

Rudder trim is provided by a ground adjustable bend tab. Once this tab has been properly set for a particular airplane, it rarely requires any further attention. However, if directional trim is not correct the left and right wings will not feed fuel evenly and chronic wing heaviness will result.

Aileron and elevator trim tabs are actuated electrically and controllable from the cockpit.

There should be no substantial lateral trim change with change in speed, as noted by wheel position. When actuating the gust control, the control wheel may rotate but should not hold a considerably different position. Ground adjustments can be made to minimize these effects.

LONGITUDINAL STABILITY

STATIC STABILITY

The airplane is statically stable, but at high altitude is subject to normal Mach number effects which result in "tuck" or increase of nose down pitch moment with speed. Generally, when the airplane is trimmed at a given speed, a push force is required to increase speed, and a pull force is required to decrease speed. The force is usually less than 10 to 15 pounds.

At or near placard speeds, attention should not be diverted from flying the airplane since control force change with speed is small. The speed may increase beyond limits due to inattention or outside upset before the pilot becomes aware of it.

At lower altitudes the static stability is satisfactory.

At high altitude and heavy weight, the airplane is operating close to the limit of its capabilities. Under these conditions the airplane is flying at very high wing lift coefficients, and during manual flight requires the pilot's full attention. It is not practical to fly exclusively by reference to the airspeed indicator since a comparatively large attitude change may occur before the change is reflected in the airspeed. Therefore, corrections should be by reference both to attitude and airspeed.

DYNAMIC STABILITY

If the airplane is flown hands off and the autopilot is not engaged, an attitude upset can cause a long period motion which may be either stable or unstable. Below 60,000 feet the aircraft will generally oscillate slowly from a speed above trim. Above 60,000 feet, if upset and left hands off, the aircraft will go into a climb or dive and the airspeed will continue to decrease or increase until the pilot corrects the situation.

During approach and landing, the dynamic stability is affected by flap setting, airspeed and power. The airplane will become less stable if full flaps are combined with higher airspeed and higher power. When these affects are combined, such as using full flaps at 130 knots with considerable power, an unstable motion can be developed. This combination is not normally encountered. If it were, an upset would have to be introduced with hands off, before any unusual motion would develop. This motion would take the form of increasing or decreasing speed, together with changing airplane attitude. This is fairly slow in developing after the upset is introduced and is easily corrected by the pilot. With the use of full flaps at ordinary approach speeds, these effects are not present.

In addition to the effects described, buffet from incipient stabilizer stall may be encountered when using 25° to 35° wing flaps and high approach speeds. The degree of buffet will vary with different airplanes and will increase with increasing speed, G's, or air turbulence. Hard pushover maneuvers will also induce or increase stabilizer buffet. If flap placard speed is observed, the buffet, if it occurs, will be mild and cause no control difficulties.

MANEUVERING FLIGHT

Maneuverability

The acceleration limits are shown in Section V. It is possible to reach these limits during maneuvering flight at low and intermediate altitudes. At high altitude the G-capability is limited because the airplane is operating at high lift coefficients. Under some conditions, buffet will be encountered at 1-1/10 G.

Maneuvering Stick Forces

The maneuvering stick force normally ranges from 20 to 40 pounds per G at aft center of gravity positions. (Aft service loading.) However, at slow speed with high power, the force is lighter. At forward center gravity positions, the force is normally 50 to 70 pounds per G. At high altitude, the force can be 100 pounds per G at speeds near limiting Mach number.

HIGH ALTITUDE TRIM

There is sufficient elevator trim tab travel to trim out all elevator forces during climb conditions.

During cruise conditions, the GUST position of the gust control should not be used to relieve a nose heavy condition since this will cause wing buffet.

EFFECT OF CG POSITION

Forward movement of the center of gravity is stabilizing and results in higher stick forces. As may be seen in figures 5-7 through 5-10, the CG will move aft due to fuel usage from the slipper and sump tanks. The CG will move forward about 3-1/2% when the auxiliary tanks are feeding and will remain essentially constant while expending fuel from the main tanks.

LONGITUDINAL TRIM CHANGES

LANDING GEAR EXTENSION

The trim change is very slight and may be nose up or down. The force required is 5 pounds or less.

WING FLAP EXTENSION

A nose down trim change occurs. However, the elevator trim tab rate is sufficient to keep up with or exceed the amount of trim change required by the operation of the wing flaps. An opposite effect is produced by flap retraction. If the flaps have been extended and full nose up elevator trim has been applied, do not add full power at the same time the flaps are being retracted until the elevator tab setting has been considerably reduced. High push forces will be encountered under this condition if speed is allowed to increase too rapidly.

SPEED BRAKE EXTENSION

The speed brakes cause a nose up trim change. At approach speeds, a 10-pound push force is required in about 5 seconds. Above 175 knots a 25- to 35-pound push force is required in about 6 to 7 seconds.

GUST CONTROL OPERATION

Actuation of the gust control to the GUST position causes a nose up trim change. If desired, the forces may be reduced by applying some countering trim tab before actuation of the gust control.

EFFECT OF POWER

Power application causes a nose up trim change. If the power is increased from idle to part power stop rpm, the trim change can be resisted by a push force of 25 to 30 pounds in the first 5 seconds. As the airplane accelerates at constant altitude, the force will build up to 50 or 60 pounds. If trim is used to reduce the force it probably will not have a significant effect until about 10-15 seconds. Power reduction causes a nose down trim change. A 10 to 20 pound pull force is required in 3 to 5 seconds.

DIRECTIONAL STABILITY

The maximum steady sideslip angle is about 12° with full rudder deflection. Two thirds rudder will develop about 10° sideslip. The rudder forces are relatively high; at approach and landing speeds about 150 pounds pedal force is required for maximum sideslip. At 250 knots, 150 pounds pedal force is required to develop 2° to 3° steady sideslip.

The aircraft has very little roll due to yaw. Right rudder is required to lift the left wing and vice versa. However, this action is very slow. Slipper tanks will slightly aggravate the yawing tendency. Slight yawing oscillations which might occur can be readily damped by using the rudder.

ROLLING MANEUVERS

The heavier weight of the U-2C and U-2F results in a decreased rate of roll with a resultant increase in force required to stop the roll. This comparison was made with the U-2A which displays characteristics as follows:

At approach and landing speeds the rate of roll is about 25° per second with full aileron deflection. The total wheel force is about 40 to 50 pounds. With the gust control shifted to the GUST position the rate of roll is about 10° per second.

The aileron forces become heavier as speed is increased. At 200 knots in the FAIRED position an 85-pound force is required for 15° per second rate of roll; however, in the GUST position the force is reduced to 70 pounds.

The rate of roll is considerably higher at high altitude; 45° per second at 67,000 feet for a 50-pound wheel force.

ADVERSE YAW

Rudder should be used to coordinate the roll, particularly at lower speeds and during landing approach. A full aileron deflection roll at 85 knots will develop about 15° adverse yaw if it is not coordinated. About 2/3 rudder and 150 pounds pedal force are required to reduce the yaw to zero.

STALLS

ONE-G STALLS

The airplane has a moderate stall with normal stall warning buffet. Small sharp edged spoilers located on the wing leading edge near the fuselage cause turbulence at high angles of attack which impinges on the horizontal surface and provides the stall warning. At heavy weights the buffet is less noticeable.

During the approach to the stall, some roll and yaw may be encountered. These characteristics should be controlled by use of aileron and rudder. If the wings are not held level and if there is uncorrected yaw just prior to the stall, the airplane may roll moderately to either the left or right. The pitch down varies from slight at aft CG position to abrupt at a forward CG position. Recovery is effected by easing the control column forward and controlling roll and yaw with aileron and rudder as necessary. Application of power is effective in stall recoveries at lower altitudes.

The retractable stall strips should be extended during stall evaluations and landings. These strips do not affect the stall speed but provide symmetrical stall characteristics on each wing and counteract yaw and roll, both during approach and at the stall.

Stalls should be avoided at high altitude. Recovery is extremely difficult due to the narrow range between high speed buffet and low speed stall. In addition it is possible to induce engine flameout with low airspeeds and abrupt aircraft stall recovery maneuvers.

EFFECT OF BANK ANGLE

Flying in a bank maintaining altitude produces an effect on stall speed which is similar to the effect of weight. Any steady turn requires that the vertical component of lift be equal to the weight of the airplane and the horizontal component of lift be equal to the centrifugal force. Therefore, the airplane in a steady turn develops a lift greater than the weight and stall speeds are increased, see figure 6-1.

ACCELERATED STALLS

Accelerated stalls are preceded by buffeting. The stall is normal and may show slight rolloff. Recovery is made by relaxing the back pressure.

LATERAL TRIM CHARACTERISTICS

If the basic aerodynamic trim of a particular U-2 aircraft is not properly adjusted, it will result in uneven fuel feeding. If an aircraft has chronic fuel feeding problems, they may be traced to this cause.

ELECTRICAL AILERON TAB

The controllable aileron tab should only be used to adjust for any basic aerodynamic trim characteristics.

If a takeoff is made with full main fuel tanks or with all fuel tanks full, it can be assumed that any lateral imbalance is aerodynamic and should be corrected with the aileron tab. However, later during the flight, any further wing heaviness is probably due to uneven fuel feeding. The fuel cross transfer system should be used to correct for this condition.

STALL SPEEDS - KIAS

WING FLAP - DEG	BANK ANGLE - DEG	GROSS WEIGHT - LBS					
		14,000	16,000	18,000	20,000	22,000	24,000
0	0	76	81	86	91	95	99
	30	81	87	92	98	102	106
	45	90	96	102	108	113	118
	60	107	114	121	128	134	140
15	0	73	78	82	87	91	96
	30	78	84	88	93	98	103
	45	87	93	98	103	109	114
	60	103	110	116	123	129	136
35	0	69	74	79	83	87	91
	30	74	80	85	89	94	98
	45	82	88	94	99	104	109
	60	97	105	111	117	124	129

Figure 6-1

AIRSPEED ACCELERATION

POWER

The combination of aerodynamic cleanness and high thrust gives this airplane an acceleration potential which makes possible the rapid attainment of speeds in excess of structural limitations. Power can be applied as fast as necessary at lower altitudes.

CAUTION

Extreme caution should be exercised to prevent exceeding airspeed limitations when large power increases are made. This condition is more critical on power application with landing flaps extended because of lower structural limitations.

DESCENT MANEUVERS

Care should be exercised during any descent maneuver due to the rapid acceleration of the aircraft. Do not allow any descent attitude to become so steep that you have difficulty observing the structural limitations during the recovery.

WING BUFFET

When the air no longer follows the wing contour, separation occurs. A separation is usually unstable and the resulting turbulent flow which shakes the airframe is called wing buffet.

Separation will occur at low Mach numbers as a result of angle of attack alone, and a fully developed separation is called stall. The localized separation which precedes wing stall results in wing stall warning buffet, and is a warning of impending stall.

Separation will occur at high Mach numbers at specific combinations of Mach number and angle of attack which are a characteristic of the particular wing. This separation results in wing Mach buffet.

The usable combination of angle of attack and Mach number are bounded by the wing stall buffet boundary and the wing Mach buffet boundary. The intersection of the stall buffet and Mach buffet boundary forms a "corner" in which the normal wing low speed stall is identical to its high speed Mach buffet limit. This is illustrated in figure 6-4.

An airplane can maintain level flight at the stall buffet or the Mach buffet boundary if sufficient power is available. A large amount of power is available in the U-2C and U-2F aircraft and level flight is possible at high angle of attack and moderately high Mach number simultaneously. The U-2C and U-2F thus operate near the "corner".

Figures 6-2, 6-3, and 6-4 are provided to describe graphically the buffet boundaries of the U-2C and U-2F aircraft. Figure 6-2 shows the Mach buffet boundary for maximum cruise power in terms of Mach number versus altitude. Figure 6-3 shows Mach buffet boundary for maximum cruise power in terms of IAS versus altitude. It also shows the stall warning buffet line for the maximum weight for a given altitude which is achieved in maximum power cruise climb.

In figure 6-4, the buffet boundaries at constant weight, which are independent of power, are superimposed upon the envelope created by the variation of weight with altitude data from figure 6-3 during a maximum power cruise profile. The particular constant weights shown are for maximum power cruising at 66,000 feet, 71,000 feet, and 74,000 feet. Referring to figure 6-4, it will be seen that for a given weight a reduced power setting with the accompanying reduced altitude will result in an increase in the speed range existing between the stall and Mach buffet boundaries. The recommended climb speed schedule, therefore, provides greater speed margins between the buffet boundaries when cruising at reduced power.

Figure 6-3 shows a slight widening of the corridor as altitude is increased. This is a result of the weight/altitude relationship for maximum cruise power.

Cold temperatures increase the thrust of a jet engine at a given altitude and EGT. On a cold day, an airplane can fly at higher altitudes and higher angle of attack at a given weight than at normal temperatures at the same EGT. This would result in a narrowing of the "corridor". Narrowing of the corridor does not occur when the aircraft is flown to the maximum EPR schedule listed in Section V, since EPR is a direct measure of thrust and is unaffected by variations in outside air temperature.

During maximum power cruise flight at the proper speed, the airplane should not encounter buffet. If the autopilot should allow the speed to increase by several knots and then make a correction by pulling the nose up too sharply, one or the other buffet boundary may be encountered.

With the gust control in the GUST position, the buffet becomes worse since a higher angle is necessary to maintain level flight.

MACH NUMBER CHARACTERISTICS

This is a subsonic aircraft and is subject to the same tuck and buzz tendencies associated with most subsonic aircraft. It is placarded to a nominal Mach number of 0.80. However, under conditions of heavy weight, high altitude, and faired gust controls, it is impractical to fly faster than 0.70 Mach number because of buffet, rolloff and/or sharp tucking tendencies. The rolloff and tucking can be minimized by placing the gust control in the Gust position. The speed can be increased to 0.80 Mach number and only buffet will still be present.

As Mach number is increased in the faired position, elevator control forces will increase rapidly. These high pull forces can be relieved by placing the gust control in the GUST position. At altitudes below 35,000 feet no difficulty will be encountered due to Mach effect if the pilot does not exceed the placard airspeed limitations for the aircraft. At all altitudes Mach characteristics can be aggravated by high gross weight or by extending the speed brakes, or by wing heaviness due to uneven fuel load.

MACH BUFFET BOUNDARY MAXIMUM CRUISE POWER

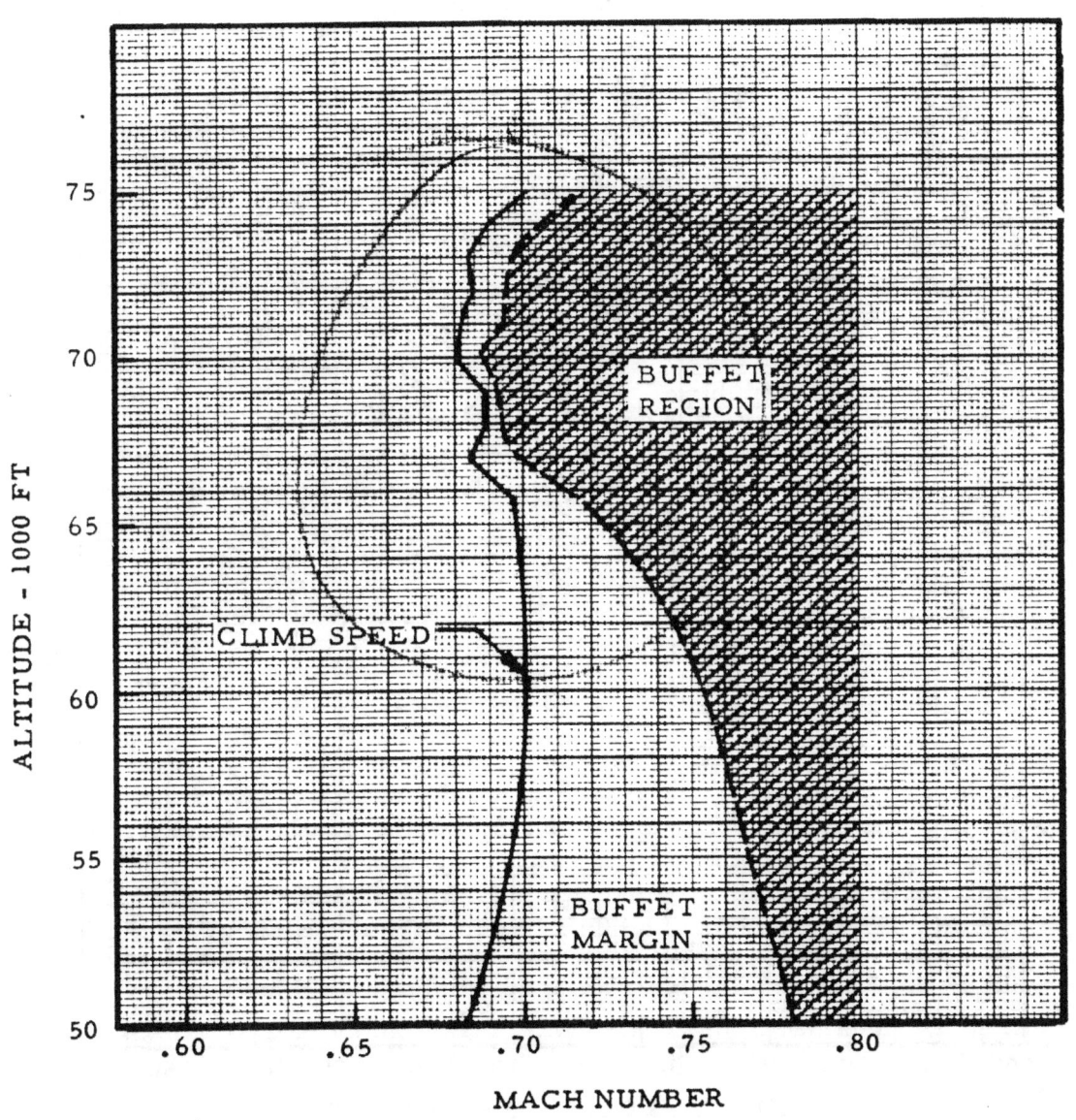

Figure 6-2

SECTION VI   AF (C)-1-1

1-G SPEED ENVELOPE vs ALTITUDE

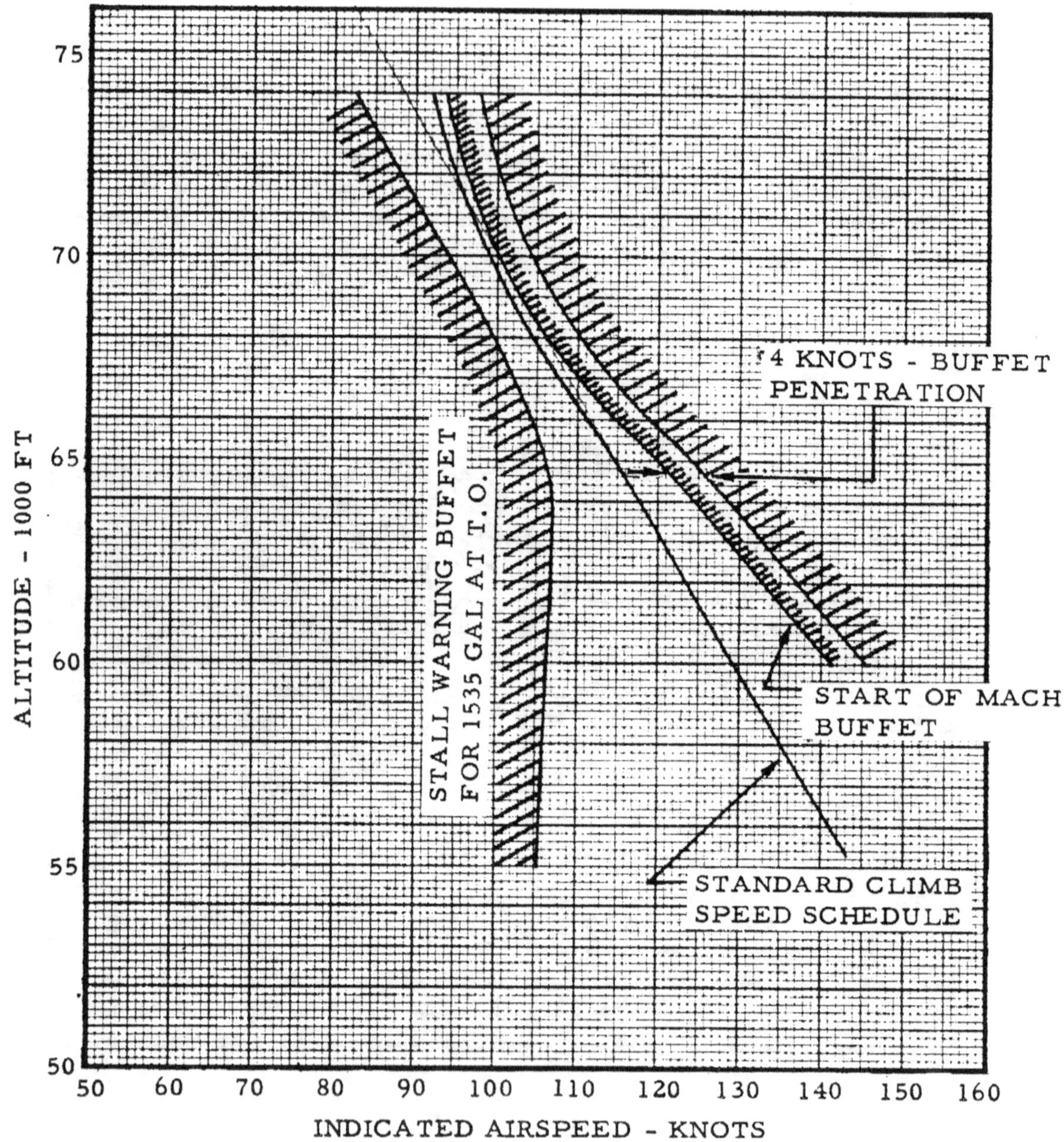

Figure 6-3

6-10

STALL AND MACH BUFFET BOUNDARIES

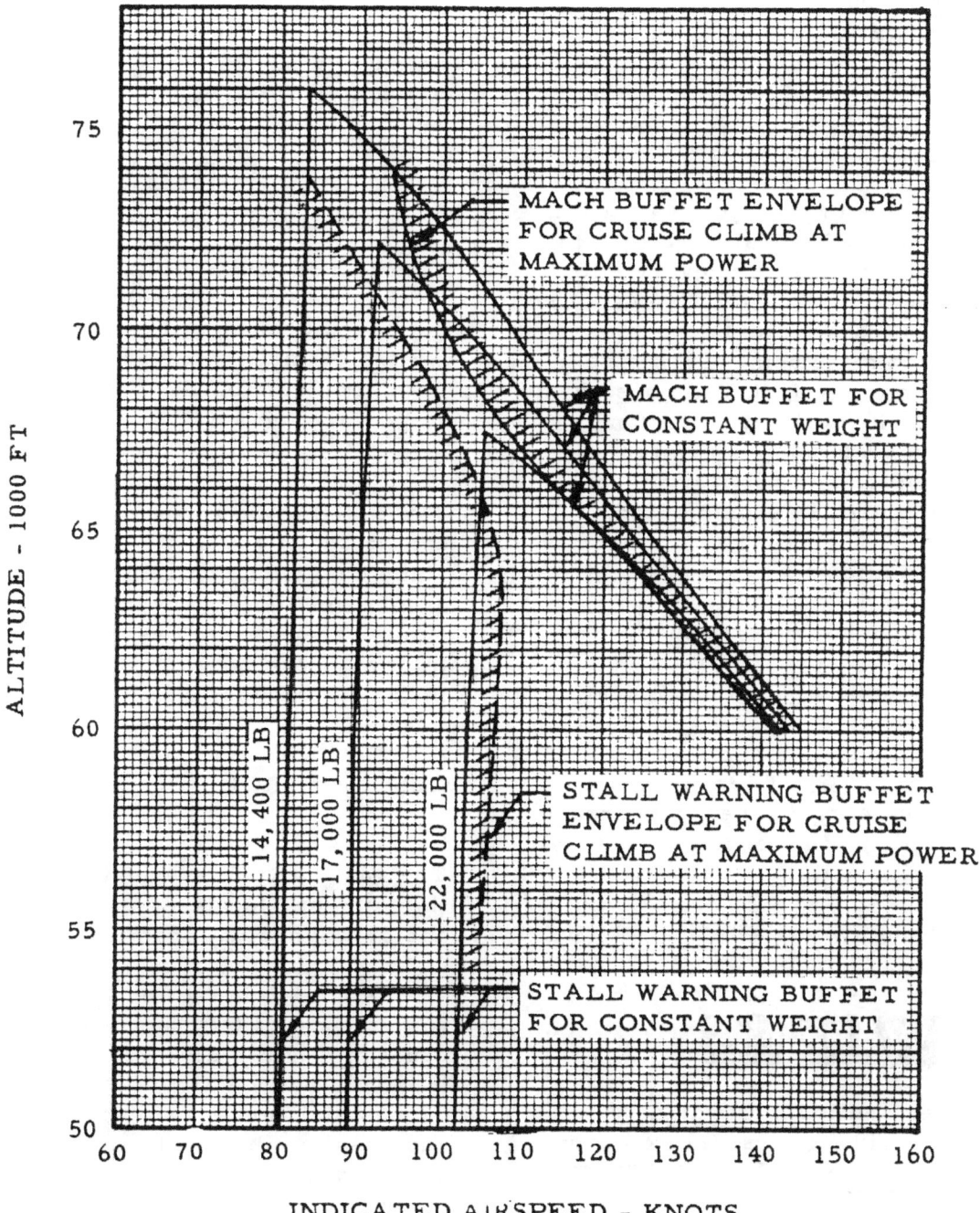

Figure 6-4

In order to stop the aileron or rudder buzzing, it will be necessary to slow down as much as 10 knots below the speed at which the buzz started. Therefore, it is imperative to start recovery as soon as buzzing is felt in the controls. If corrective action is taken as soon as the first Mach effect is noticed, there should be no difficulty making a cautious recovery.

SPINS

Although spins are prohibited, it is possible that the airplane could be allowed to enter a spin. If this should occur, every effort should be made to effect an early recovery since structural damage or failure may easily result from the spin. In the event that airplane enters a spin, the flaps should be retracted. The position of the landing gear and speed brakes is considered unimportant in effecting a recovery. The following procedure is recommended:

1. Throttle to idle.

2. Rudder against spin rotation.

3. Control column forward until rotation stops.

4. Gust control to GUST, speed brakes and landing gear extended to decrease the possibility of exceeding structural limits during pullout.

5. If drop tanks are installed, arm and release.

WARNING

The recovery pullout must be as gentle as possible to prevent exceeding maximum acceleration limits.

SPEED BRAKES

The airplane has conventional speed brakes which are moderately effective. Their primary use is as a drag producing device for descent, approach and landing. The speed brakes can be extended or retracted at any speed. They cause a moderate airframe and tail buffeting which is of no particular concern.

NOTE

At very high altitudes and at limit Mach number (0.80), fully extended speed brakes will cause sharp pitching oscillations. There is no loss of control, but for a smoother ride, reduce the angle of descent.

The speed brakes are fully variable and can be set at any desired position. In some cases they will creep closed from an intermediate position after a period of time. They cannot creep closed from the fully extended position if the switch is left in the extend position.

GUST CONTROL

The gust control is a device installed on this airplane to make possible its structural and aerodynamic capabilities.

The wing airfoil section was selected to have the best lift/drag characteristics for cruising flight. This resulted in a very highly cambered airfoil. This in turn causes high nose down pitching moments and resultant high balancing tail loads. As airspeed increases, these tail loads increase very rapidly and a structural weight penalty is imposed on the fuselage and horizontal stabilizer. The portion of the gust control which causes the wing flaps to be shifted up 4 degrees eliminates the high camber of the airfoil and greatly reduces the balancing tail load. The most important fact to remember is that tail loads increase rapidly with airspeed and the placard speed of 220 knots with gust control FAIRED must be carefully observed. With the gust control in the GUST position, the tail loads are materially reduced. However, at the placard speed of 240 knots, limiting wing bending loads are produced if a design gust is encountered. To summarize: tail loads are limiting at 220 knots FAIRED, wing bending turbulence loads are limiting at 240 knots in GUST.

The high aspect ratio of the wing was chosen for good efficiency at high altitude. The resulting long span imposes high structural loads on the wing. In order to accomodate higher gust loads at higher speeds, the center of pressure must be moved inboard. This is accomplished by the portion of the gust control which shifts the aileron up 10 degrees. This action relieves the wing tip area loads and thereby reduces structural loads on the inboard section of the wing.

The flight instruments should be closely monitored when actuating the gust control at night or in weather, due to the change in trim.

CRUISE CLIMB

The cruise climb comprises the major portion of a typical flight. The airplane will slowly climb as fuel is burned and weight decreases. Maximum allowable engine power must be used to obtain proper performance. During the cruise climb it is necessary that the normal airspeed climb schedule be closely maintained. Usually, the autopilot will be used to fly the airplane during this phase. If the autopilot is not available, the airplane can be flown manually in a normal manner except that it will require more attention than it does at low altitudes. Because of the proximity of the buffet region to the climb speed schedule, the airplane will have buffet, rolloff, and/or tucking tendencies, if the climb speed is exceeded. This condition is corrected by exerting a pull force on the elevator control and slowing the airplane back to trim speed. A slow airspeed increases the possibility of an engine flame-out, and a return to proper speed should be done slowly. No abrupt control movements should be made.

DESCENT

The descent in the high altitude area from 75,000 feet down to 70,000 feet is slow. The engine power, even on minimum flow, is still considerable at this altitude and the drag items do not produce much drag at indicated airspeeds at 90 to 100 knots.

A descent with the gust control FAIRED will probably encounter buffet, rolloff, and sharp tucking tendencies at speeds faster than the standard climb speed schedule. Elevator stick force will be high.

After a descent is established with gear down, speed brakes out, on climb speed schedule, the gust control may be placed in GUST.

The descent should be limited to the speed for light to moderate buffet. The buffet will be heavier and will start at a lower Mach number at higher altitudes and/or heavier gross weights. Recommended descent speeds with 100-200 gallons of fuel remaining are given below. These speeds are an approximate guide. Light to moderate buffet will be encountered at the higher altitudes for the GUST configuration.

FAIRED

Alt Range	Airspeed
75-55	Descent on standard airplane schedule.

GUST

Alt Range	Approx. IAS Below Needle
75-70	5-10 kn
70-65	0-10 kn
65-60	0-10 kn
60-55	10 kn

At medium and lower altitudes a descent speed of 150 knots is used if turbulence is present or anticipated. If the air is smooth, the rate of descent can be considerably increased by descending at 200 knots IAS.

Wing flaps should never be extended above 45,000 feet, since dangerous pitching moments are produced.

During descent, the pilot should be aware of the possibility of encountering high winds at medium altitudes. The descent should be planned so that such winds will not cause you to fall short of your destination. In many cases the descent can be started from a position which will ensure tail winds.

APPROACH

The approach is conventional. Make a lateral trim check before reaching the field as outlined in Section II. Initial approach with wing flaps up will require only idle power or slightly more. If full flaps are used after turning downwind, a conventional power approach can be made.

The threshold speeds given for 0° to 15° wing flaps are computed on a basis of 110% of stall speed for zero flaps and 115% of stall speed for 15° flaps. The threshold speeds given for 25° to 35° wing flaps are based on 112% of stall speed for 25° flaps and 114% of stall speed for 35° flaps.

On final approach make liberal use of the rudder to maintain proper alignment and to correct for drift. If it is necessary to pick up a wing before touchdown, use coordinated rudder to correct for adverse yaw tendency.

LANDING

This aircraft has a conventional landing gear arrangement with the center of gravity located a short distance behind the main gear. The fuselage ground attitude is approximately 4-1/2°. The most desirable method of landing is to contact the ground on both the main and tail wheels simultaneously. Skips and bounces are caused by allowing the main gear to touch down first.

LANDING WEIGHT

During the early landing stages of a pilots checkout, the landing weight should be kept reasonably low. This provides greater structural margin and, even more important, it simplifies lateral and directional control on the runway. When the amount of fuel in the wings is low, there is less sloshing and less lateral inertia.

After the early landing stages of training have been successfully completed, the nominal landing fuel load of 550 gallons is considered satisfactory.

After the pilot gains considerable experience, landings with higher fuel loads can be safely accomplished if necessary when landing conditions are optimum. Normally, the fuel dump system may be used to obtain landing weight.

WING FLAPS

The use of wing flaps lowers the stall speed and so lowers the touchdown speed. However, of equal or greater importance on this airplane is the added drag with wing flaps extended. With extensions down to 15°, very little drag is added. With 25° flaps, the drag is noticeable. With 35° flaps (full), the added drag is a major factor.

DIRECTIONAL CONTROL

It is important to touch down with the aircraft aligned with the runway and without drift. During the high speed portion of the landing roll, a steering correction should not be held in too long since it is possible to oversteer.

The position of the control column is very important in assuring adequate directional control. It should be held in the aft position in order to hold the tail wheel firmly in contact with the runway.

GO-AROUND

CAPABILITY

In this airplane, a large amount of power is available so that there is never a problem in easily reaching a safe altitude and airspeed as far as power is concerned. Fairly substantial changes in control force may be encountered, depending on the amount of power used. The method of application of the power and other pertinent factors are covered in succeeding paragraphs.

BEFORE TOUCHDOWN

If the decision to go-around is made early and power is applied before touchdown, there is no particular difficulty. If it is necessary to use full throttle, the wing flaps should not be retracted until elevator trim has been applied.

AFTER TOUCHDOWN

Wing Flaps

The primary reason for retraction of wing flaps is for better aileron control and to remove the necessity for observing the flaps extended placard speed, in the event of go-around. Therefore, if time permits, the flaps can be started up. If not, they can be retracted later.

Elevator Trim

Application of nose-down elevator trim should be started as the power is applied.

Elevator Forces

As power is applied, the control column should be moved forward slowly and firmly since a push force will be necessary to resist the nose-up trim change. This force will vary and depends on the amount of power application, CG position, airspeed and elevator trim tab position.

SECTION VI   AF (C)-1-1

Power Application

The power can be rapidly increased to approximately 85% RPM. This results in only a light nose-up trim change. If the power is rapidly increased to the maximum, the trim change is considerably larger and the push forces correspondingly much higher. Usually, 85% RPM is sufficient power for the go-around. If more power is required, it can be added as airspeed increases. If excessive push forces are encountered, they can be relieved by a slight reduction in power.

FLYING WITH EXTERNAL LOADS

SLIPPER TANKS OR DROP TANKS

Drop tanks or empty slipper tanks do not affect the flight characteristics of this airplane. The additional weight of fuel will decrease climb and altitude performance slightly. Slipper tanks cannot be dropped in flight.

CAUTION

Drop tanks and slipper tanks should not be carried at the same time.

POGOS

The airplane can be flown satisfactorily at low and medium altitudes with pogos attached. The drag created by the pogos requires slightly more power for level flight and the rate of climb is reduced. There is no change in aircraft maneuverability. They are usually retained on the initial checkout flight to aid the pilot with ground control during landings.

FORMATION FLYING

This airplane was not designed to participate in formation flying, although it presents no problem if the wing man maintains a position where the lead airplane is easily viewed and clearance is adequate. Close formation should be avoided due to the relatively high aileron control forces and low rates of roll, particularly at the higher speeds. Formation flying should be performed with wing flaps retracted and gust control faired to provide maximum aileron movement.

If close formation is encountered for any reason, be on guard for heavy control forces due to wing downwash.

AF (C)-1-1

SYSTEMS OPERATIONS — SECTION 7

TABLE OF CONTENTS

	PAGE
ENGINE OPERATION	7-1
FUEL SYSTEM OPERATION	7-3

ENGINE OPERATION

COMPRESSOR STALL

The high output, high pressure ratio, J-75 engine was designed to operate at high altitudes with superior fuel consumption. To do this it must operate as close to the stall region as possible. There is no evidence that compressor stall has ever caused damage to the J-75 engine. Nevertheless, compressor stall is disconcerting and can result in flameout. Compressor stall is very much like aircraft wing stall, and compressor blades may be thought of as miniature wings. Every airfoil is limited to a maximum angle of attack, which if exceeded, will cause the airfoil to stall. As with the wing stall experienced in the aircraft, the air separates from the airfoil section. With this separation, the lift is greatly reduced on the wing. In the compressor, this loss is evidenced as a loss in pressure ratio and therefore a reduction in pressure level at the compressor discharge.

During acceleration or deceleration compressor stall may be encountered due to improper fuel scheduling to the burner. For instance, if the fuel flow is too high, temperature and pressure in the burner become higher than design limits, thereby causing abnormal back pressure on the compressor. This decreases the airflow for a given RPM. The effective angle of attack increases beyond the airfoil's critical angle of attack and causes the airfoil section to stall. This results in reduced airflow for an instant in the compressor and pressures are greatly reduced.

Another condition of possible compressor stall occurs at high altitude, particularly with high power operation. This condition is brought about by low temperature and Reynolds Number effect. When the air gets thinner, it has difficulty in following the contours of the airfoil section of the compressor blade, thereby reducing the stall margin. As the air gets colder, the engine tends to operate closer to stall. Therefore, more care must be exercised when operating at altitude where the air is both thin and cold.

Stalls vary in severity, depending on whether the stall involves only a portion of a stage, a stage, several stages, or an entire compressor. Partial stall may produce roughness with or without audible accompaniment of rumble, drone, etc. More complete stalls may produce noises varying in intensity from pistol shots to cannon fire, and can be very disconcerting if the pilot doesn't know what to expect.

7-1

To avoid compressor stalls at high altitude with the J-75, the pilot must remember the following:

1. Make no abrupt throttle movements.

2. Climb at recommended airspeed and control the engine power to stay within EPR limits.

3. Avoid abrupt or uncoordinated maneuvers.

To get the engine out of a compressor stall, the pilot should slowly retard the throttle until the compressor stall stops or the throttle reaches idle. If the stall persists in the idle position, reduce altitude and increase airspeed to maintain an even pressure distribution at the compressor face. The chances of stall recovery improve as altitude decreases because of higher compressor inlet temperatures and the Reynolds number effect. Since there is a possibility of overtemping the engine during a compressor stall, the pilot should be prepared to shut off the engine if corrective action does not break the stall. Compressor stall at high altitude may result in flameout. In case of flameout a descent must be made to air start altitude.

ENGINE ROUGHNESS

Engine roughness is generally obtained at high EGT's in the altitude range between 40,000 and 60,000 feet. This is commonly known as the "BADLANDS". The roughness is easily distinguished from the customary high frequency, low amplitude "BUZZ" which is present during climb.

In order to avoid engine roughness during climb, the following EGT schedule is recommended: 630°C from takeoff to 40,000 feet, 485°C from 40,000 to 60,000 feet. Above 60,000 feet, the power can be increased to 665°C EGT or limiting EPR.

Engine power should be reduced starting at 35,000 feet in such a way that the EGT schedule described above is obtained at 40,000 feet.

ENGINE BANGING

Engine banging may occur in the 40,000- to 60,000-foot altitude range with EGT above 485°C. The banging is not necessarily associated with engine roughness. The banging can cause extreme shaking of the airplane and possible flameout.

If banging occurs, power should be reduced immediately. There have been no cases of engine overtemping during the banging. The banging can be avoided by climbing at the 485° EGT power schedule.

ENGINE OIL SYSTEM

Engine oil temperature should be monitored closely during maximum altitude cruise. With the dual-air-oil coolers the temperature should remain within limit. However, if the 121°C temperature limit is reached, the power should be reduced. Level cruise flight can be maintained at reduced power until the oil cools. If desired, power can then be increased and altitude regained.

Generally, the oil temperature will increase faster if the airplane is flown with a light load and a fast climb is immediately made to maximum altitude.

Engine oil pressure should be monitored during flight and the limits shown in Section V must be observed.

POWER REGULATION AT ALTITUDE

The primary engine operating variables are engine speed, engine pressure ratio, and exhaust gas temperature. Limits for these variables must be observed carefully at altitudes above 40,000 feet in order to avoid engine compressor stall. (See Section V for specific limitations.) An altitude may be reached where the minimum fuel flow setting equals that required for maximum exhaust gas temperature. This is called "minflow".

In this case, retarding the throttle to idle will not change the power condition. In the event the maximum EGT is exceeded, it can be reduced by increasing speed or lowering the gear and/or extending speed brakes to lose altitude.

INTERCOMPRESSOR BLEED

Intercompressor bleed is required at altitude and during descent conditions when operating at engine pressure ratios below the minimums listed in figure 5-1. When bleed valves are closed and engine pressure ratios are below minimum, a low compressor stall and engine flameout will occur. The resultant flameout is immediate and can not be avoided by the pilot.

The bleed valves are left open during landing to avoid compressor instability during acceleration if a missed approach is executed. The small loss in thrust with the bleed air on during landing is not significant.

FUEL SYSTEM OPERATION

The fuel system ordinarily requires little of the pilot's attention because of its simplicity. One cockpit switch controls the auxiliary boost pump and continuous ignition; a second operates the main tank fuel cross transfer pump. Four dump switches on the lower left console control fuel dumping.

SLIPPER TANKS

A small AC electrical fuel pump is provided in each slipper tank to ensure complete utilization of slipper fuel. These slipper strippers are controlled by a switch on the lower left console. The pumps should be turned on only when it is known that sufficient fuel has been used from the auxiliary tanks to provide space to accommodate the slipper tank residual fuel. This can be ensured by turning the slipper pumps on only after fuel remaining is less than 800 gallons on the U-2C aircraft or less than 300 gallons on U-2F aircraft.

WARNING

Excessive pressures in the auxiliary tanks can be caused by operating the pumps when the auxiliary tanks are full.

The pumps should be operated in level or cruise climb flight (after auxiliary tank level is low enough) for approximately thirty minutes to transfer the fuel from the slippers. Following this procedure, there can be as much as 10 gallons left in each slipper due to nose up attitudes. If it is necessary to strip the slipper tanks completely, the stripper pumps should be operated again during the first part of descent for five minutes, holding a minimum of 80% engine RPM so that the AC generator frequency will be within the correct operating range.

AUXILIARY BOOST PUMP

The auxiliary boost pump (electrically-driven) is installed as a backup for the normal boost pump (hydraulically-driven) in the event of hydraulic system or pump failure. The auxiliary pump should be ON above 5,000 feet to eliminate possibility of engine flameout due to pump or hydraulic failure. Satisfactory operation of the auxiliary pump is indicated by the absence of an amber warning light while the auxiliary boost switch is ON. Failure of the auxiliary pump is indicated by the amber light which is on the Master Caution Annunciator Panel. Failure of the normal pump will be evidenced by a decrease in fuel pressure.

The auxiliary pump should be on before takeoff to eliminate possibility of engine flameout due to pump or hydraulic failure.

CROSS TRANSFER PUMP

The cross transfer pump switches control small reversible pumps. Main tank fuel cross transfer on both models is used to level the load in the left and right main tanks. The auxiliary tanks in the U-2F start feeding after the main tanks are emptied, and a fuel cross transfer pump is provided on that model to level the fuel load in the left and right auxiliary tanks.

NOTE

To preclude pumping fuel overboard, do not use auxiliary tank cross transfer pump unless fuel load is below 300 gallons.

Cross transfer of fuel from one wing to the other may be necessary due to uneven fuel feed in flight. If an aircraft has chronic wing heaviness requiring fuel transfer for lateral balance, it is usually due to a basic trim problem which should be corrected.

The fuel cross transfer system is very simple and reliable. There may be a slow-down in transfer rate during descent. This is due to a drop in the wing tank air pressure differential caused by the change in altitude.

MAIN TANK PRESSURE SWITCH

A float switch in the sump tank shuts off main tank pressurization air when the fuel level in the sump tank drops below the main tank float valves. The MAIN TANK pressure switch overrides the float switch allowing the main tanks to repressurize and raise the fuel level in the sump tank above the float switch. The MAIN TANK pressure switch may then be returned to NORMAL for normal main tank feeding. The air pressure switch is actuated for takeoff and after refueling.

NOTE

The MAIN TANK pressure switch must be returned to the normal position after 15 minutes or in any event, before the main tanks are emptied.

AUXILIARY TANK PRESSURIZATION VALVE

A pressurization air shutoff valve controls the flow of air to the auxiliary tanks. The shutoff valve should be closed by placing the AUX TANK PRESS switch in the OFF position when the auxiliary tanks are empty. The fuel tank pressurization air will then bypass the empty auxiliary tanks and the added flow of air into the main tanks will improve the fuel flow from these tanks.

FUEL DUMP OPERATION

The fuel dump system may be used to dump each of the four wing tanks individually. The system should not normally be used for partial dumps since no indication of fuel remaining exists unless fuel is dumped to the standpipes and the counter is properly reset.

NOTE

Dump switches must be closed before the fuel tanks can be pressurized.

DROP TANKS

When drop tanks are used, the drop tank panel should be checked before power is applied to the aircraft to be assured that the ARM and RELEASE switches are both OFF and that the indicator light switch is ON. Also, the left and right drop tank pressure circuit breakers should be pushed in.

Drop tank fuel will then feed automatically, when the engine is started and builds up sufficient pressure in the tanks.

NOTE

The drop tank low pressure lights will normally illuminate and may stay on until shortly after takeoff.

When both drop tanks empty (green) and both low pressure (amber) lights illuminate, turn the indicator light switch OFF.

Normally the drop tanks are dropped only after they are empty. The drop tanks can be dropped at any time however (if airspeed and altitude limitations are observed, see Sections II and III) by placing the ARM switch to the ARM position, and the RELEASE switch to RELEASE. Actual release occurs about three seconds later because it takes about that length of time for the motor-driven release hook to move the necessary distance.

SECTION VIII

CREW DUTIES

Not applicable to this airplane.

AF (C)-1-1

ALL WEATHER OPERATION

TABLE OF CONTENTS

	PAGE
INTRODUCTION	9-1
INSTRUMENT FLIGHT PROCEDURES	9-1
ICE AND RAIN	9-4
TURBULENCE AND THUNDERSTORMS	9-6
NIGHT FLYING	9-6
COLD WEATHER OPERATION	9-7
LANDING ON ICY RUNWAYS	9-7
HOT WEATHER AND DESERT PROCEDURES	9-7

INTRODUCTION

Except where some repetition is necessary for emphasis, clarity, or continuity of thought, this section contains only those procedures that differ from or are in addition to the normal operating instructions.

INSTRUMENT FLIGHT PROCEDURES

This airplane has the same stability and flight handling characteristics during instrument flight conditions as when flown under VFR conditions. However, like most jet airplanes, it is sensitive to changes of control pressures and requires constant attention to flight instrument indications. For best results, make all changes in pitch, bank, and power smoothly and keep the airplane properly trimmed.

It is especially important that recommended airspeeds be adhered to on the final part of an instrument landing approach. An approach at speeds exceeding those recommended may result in overshooting, because the low drag characteristics make it difficult to quickly reduce airspeed unless full wing flaps are used. Turns should not exceed 30 degrees of bank. Turns during standard instrument approaches are 2-needle width and made at the rate of 3 degrees of turn per second.

NOTE

At high altitude, a 30° bank will be a 1-needle width turn and made at a rate of approximately 1-1/2 degrees of turn per second.

PREFLIGHT

To ensure a successful flight under instrument conditions, complete the normal preflight inspection as given in Section II. Particular attention must be placed on the proper functioning and settings of flight instruments, radios, and any electronic navigational equipment aboard.

TAXI

No taxi checks other than those listed in Section II are required.

BEFORE TAKEOFF

Accomplish the Before Takeoff Check as listed in Section II. After aligning the aircraft visually on the runway, adjust the miniature aircraft of the attitude indicator level with the 90-degrees index marks on the indicator case. Check compass heading against known runway heading. Turn on the pitot heat if a weather penetration is expected.

INSTRUMENT TAKEOFF AND INITIAL CLIMB

1. Maintain aircraft alignment by visual reference to runway during takeoff roll. The takeoff roll is the same as a normal VFR takeoff.

2. Retract landing gear as soon as aircraft is definitely airborne and a positive climb is established by reference to the vertical speed indicator and altitude - a positive rate of climb should be maintained during gear retraction.

INSTRUMENT CLIMB

1. Below 40,000 feet, maintain no more than 90 percent RPM. This will result in a reduced climb attitude and provide a better attitude indicator presentation. The attitude indicator presentation with full power and 160 knots IAS is difficult to accurately interpret.

2. Maintain 160 knots for climb until climb schedule requires a reduction of indicated airspeed.

3. Shift gust control to GUST position prior to entering cloud formations, or as soon after as possible when a low ceiling exists, but not lower than 1000 feet above the terrain. Return to FAIRED position at 40,000 feet or when clear of clouds.

INSTRUMENT CRUISE FLIGHT

Rarely will a maximum altitude flight in the U-2 be conducted under other than visual flight conditions. Experience has shown that only an occasional cumulus cloud in tropical regions will reach the U-2 cruising altitudes. Cruising flight under instrument conditions below maximum altitude is the same as cruise flight under visual conditions, except aircraft should be flown in GUST configuration.

STEEP TURNS

Bank angles of more than 30 degrees in the U-2 are not considered a normal maneuver in instrument conditions and should only be used in an emergency.

HOLDING

Holding speed is 120 knots.

DESCENT

Descent under instrument conditions is performed as outlined in Section II.

TYPICAL JET PENETRATION PROCEDURES

For a typical Jet Penetration procedure, see figure 9-1. Prior to beginning penetration, note fuel quantity for computation of pattern and threshold speeds.

TYPICAL JET PENETRATION

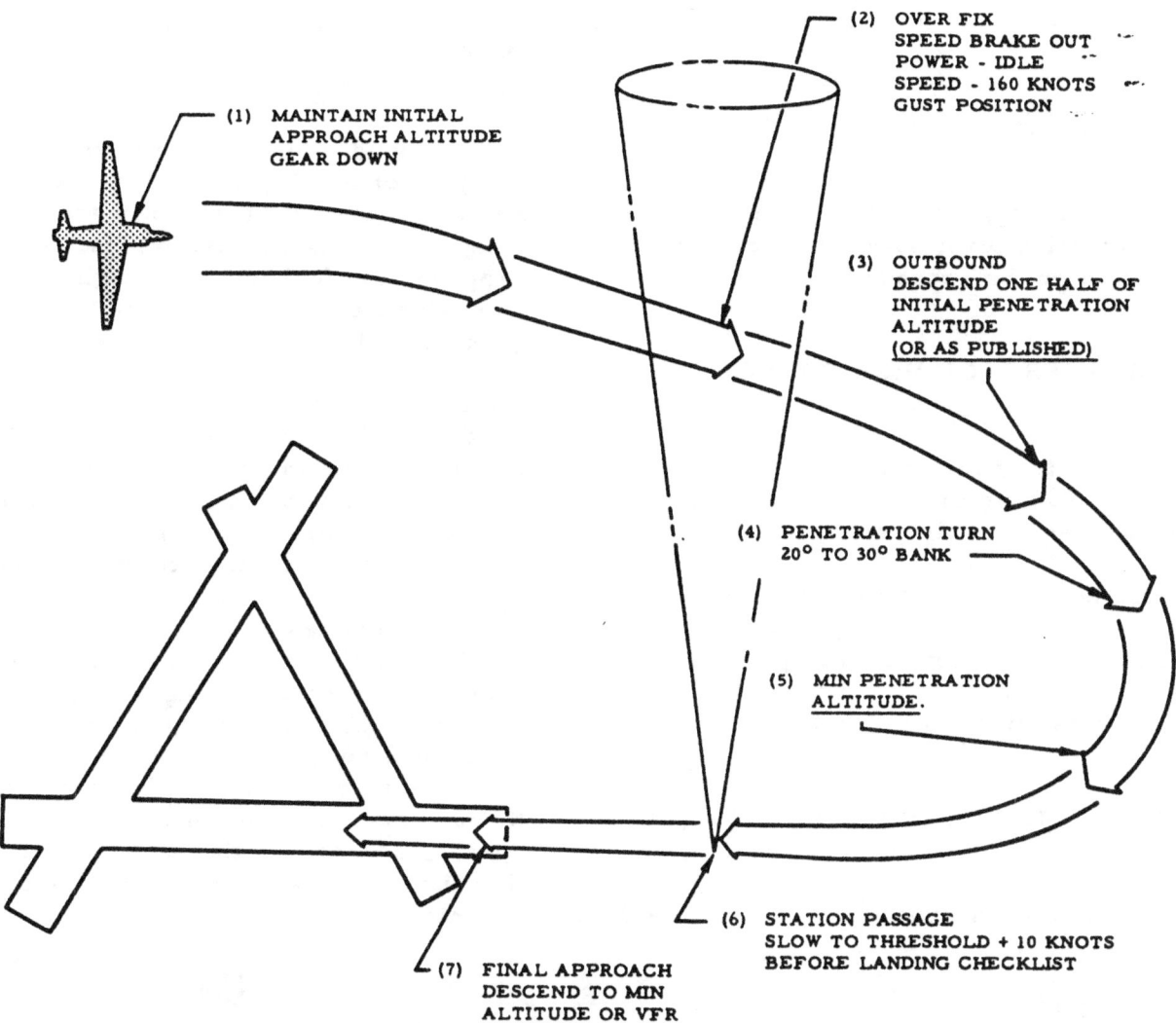

Figure 9-1

SECTION IX　　　　　　　　　AF (C)-1-1

Initial Penetration Altitude

Over the fix for an instrument penetration, the gear is down and the gust control is in GUST position. The throttle is retarded to idle, speed brakes extended and airspeed held to 160 knots as in a normal penetration.

Penetration Turn

A bank of 20 degrees is recommended in the penetration turn; however, a bank of 30° should not be exceeded.

MISSED APPROACH PROCEDURE

When necessary to execute a missed approach from an instrument letdown, or GCA, proceed as follows:

1. Add power as required (80-90% RPM).

2. Raise flaps before reaching 130 knots.

3. Execute missed approach.

GROUND CONTROLLED APPROACH (GCA)

For a typical GCA procedure, see figure 9-2.

ICE AND RAIN

Ordinarily rain gives no appreciable trouble in flight other than to restrict visibility. It is possible for heavy rain to cause incorrect airspeed readings.

ICING

The greatest concern when operating in icing conditions in the U-2 is structural icing in climb or descent phases of a mission. Before any visible moisture is encountered in either climb or descent, the defroster should be set and the pitot heat turned on.

If icing conditions are encountered during any part of the flight, ice may have formed and remained on the ram air scoops of the wing tank suction relief valves. If this happens, the fuel low level light may come on during descent and the pilot should proceed according to instructions in Section III.

Engine icing can be experienced but apparently requires such precise conditions that it is quite uncommon. If fog is present or the dew point is within 4°C of the ambient temperature, conditions exist under which jet engine icing can occur without wing icing. Ice can build up on the inlet guide vanes of the engine when the airplane is flown through areas where icing conditions prevail. Icing of the guide vanes affects the flow of air, causing loss of thrust and roughness. Extreme conditions could cause a flameout. If, after flying through engine icing conditions, and engine operation indicates that icing has occurred, find the best power setting and land as soon as practicable.

NOTE

Do not switch to EMER fuel system if engine roughness is thought to be due to icing. The EMER system altitude compensation probe (not used by the NORM system) may be iced and the EMER fuel flow schedule will not be compensated for altitude.

If the EPR inlet probe is iced, the gage will indicate erroneous values and the EGT gage should be used for continuing the flight.

GCA PATTERN

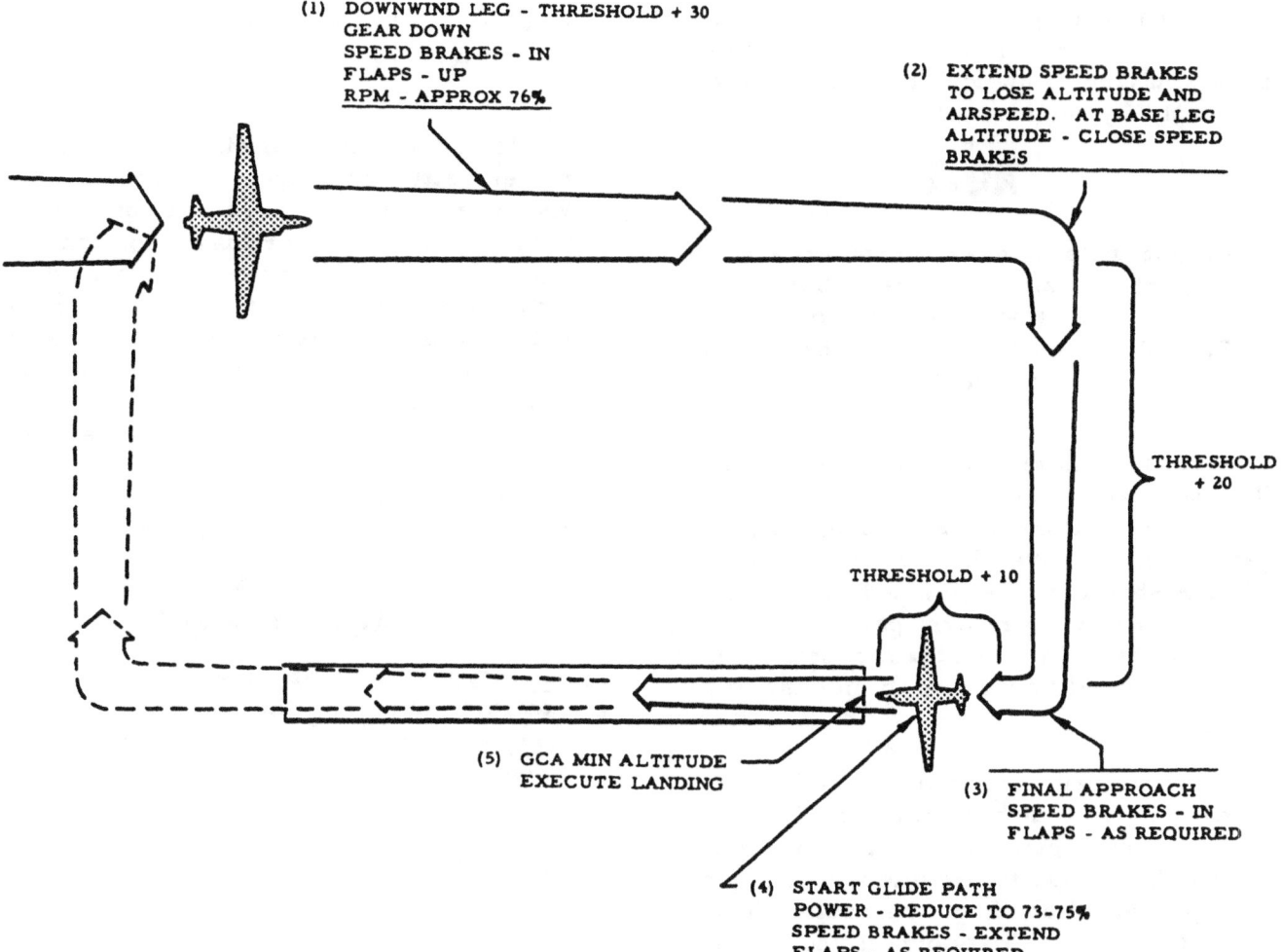

Figure 9-2

TURBULENCE AND THUNDERSTORMS

It is possible, in this aircraft to encounter light to heavy clear air turbulence at normal cruising altitudes. Light turbulence will not prevent use of the autopilot. Moderate to heavy turbulence may cause the autopilot to overcontrol and will necessitate the pilot's having to switch off the autopilot and manually fly the aircraft.

NOTE

In moderate to heavy turbulence at normal cruising altitudes above 60,000 feet, leave the gust control FAIRED since the stall will be reached before limit load factor.

Thunderstorms should be avoided in the U-2 because of its low structural load limitations. Normal cruise flight will generally be above all thunderstorms. Flights should not be planned through an area where thunderstorm penetration might be possible. Thunderstorm penetration in the U-2 should be made only in case of an emergency.

If it is impossible to avoid entry into a thunderstorm, the aircraft should be prepared prior to penetration. Power setting and pitch attitude to provide a good attitude gyro presentation should be established prior to entry. Gust control should be placed in the GUST position.

NOTE

It is possible to place the gust control in the GUST position even when the aircraft is flamed-out if hydraulic pressure and electrical power are available.

The recommended airspeed for thunderstorm penetration is 130 knots IAS, with gear and speed brakes extended. The pitot heater switch should be turned ON. Tighten safety belt and shoulder harness. Turn off any radio equipment rendered useless by static and turn cockpit lighting full bright to minimize blinding effect of lightning.

While in the storm, maintain a constant power setting and fly pitch attitude. Expect turbulence, precipitation and lightning. Concentrate primarily on maintaining a level attitude by reference to the attitude indicator. Do not chase the airspeed indications or extreme aircraft attitudes may result. A heavy rain may interfere with the indicated airspeed by partial blocking of the pitot tube pressure head. Use as little elevator and aileron control as possible in order to minimize the stresses imposed on the aircraft.

Although a constant power and pitch attitude is flown, the airspeed should be monitored so as not to exceed placard speed.

NIGHT FLYING

Check the lighting equipment thoroughly and be familiar with the location of all lighting switches in the cockpit. For night taxiing, the landing lights will be used to enable the pilot to observe and avoid obstacles. On night weather flights the NAV lights can induce spacial disorientation. The pilot should turn them off if symptoms of spacial disorientation arise. The use of oxygen on all night flights is recommended.

COLD WEATHER OPERATION

Cold weather operation of the U-2 does not present abnormal difficulties; however, certain precautionary procedures should be employed to provide safe operational conditions and afford pilot comfort.

The majority of cold weather operating difficulties are on the ground. Fresh or melted snow must be removed to insure necessary traction. To prevent possible structural damage to the aircraft during ground operation, all snow or ice covered areas must be level and free of ruts. Non-skid material, such as sand, should be available for application on the runway when adverse crosswinds exist.

The U-2 should be hangared during cold weather operations to minimize fuel, oil, hydraulic and air leaks.

All aircraft ground checks are as listed in Section II. The aircraft is normally prepared for flight prior to towing from hanger to the starting area. Station time must be adequate for performance of all required checks.

The performance of the U-2 in cold weather is generally similar to the normal performance of the aircraft. Takeoff roll is shorter and initial rate of climb greater.

LANDING ON ICY RUNWAYS

The procedures and techniques used for approach and touchdown on an ice or snow covered runway are the same as for a dry runway. It is very important to touch down in perfect alignment with the runway and to maintain directional control until coming to a complete stop.

WARNING

Any overcontrolling or allowing aircraft to veer may result in an out of control slide.

Immediately after touchdown, if runway length permits, raise the wing flaps. This makes it easier to hold the wings level and also prevents damage to the flaps from slush and ice. When brakes are used, care must be exercised to avoid sliding the wheels.

Landings on slippery runways under relatively high crosswind conditions should be avoided. The crosswind component acceptable to the U-2 aircraft is reduced due to the decrease in effectiveness of the tailwheel in maintaining directional control.

The drag chute should not be used with crosswind conditions on an icy runway unless absolutely necessary. If it is used, it should be jettisoned at about 50 knots.

HOT WEATHER AND DESERT PROCEDURE

Hot weather and desert operation is virtually identical with normal operation with these exceptions:

SECTION IX   AF (C)-1-1

PILOT COMFORT

Pilot comfort prior to takeoff and at low altitudes is a necessary consideration. It is recommended that an air conditioned vehicle be employed to transport a pilot from the point of prebreathing to the aircraft. The pilot should exert himself as little as possible, when exposed to the heat, to prevent overheating. After the pilot is in the aircraft, it is recommended that airconditioning be piped into the cockpit, and/or a protective sun shade be installed to shield the pilot from the sun.

EXTERIOR INSPECTION

Intake ducts should be inspected for evidence of sand. If excessive sand is found, do not start the engine until a more thorough check has been made and sand is removed.

ON ENTERING THE AIRPLANE

Excessive dust accumulation on instrument dials and blown sand on movable flight controls. dials, and switches must be cleaned away. Care must be exercised in any contact with the canopy as plexiglas is damaged easily in hot weather.

JUST PRIOR TO STARTING ENGINE

The external sun shield (howdah) over the cockpit should be removed.

The ground airconditioning should be the last thing removed before closing the canopy. Turn on the fan to circulate air in the cockpit.

STARTING THE ENGINE

Normal starting procedures are used in hot weather.

TAXI

Taxi carefully with low power to prevent blowing dust on ground personnel and equipment. Braking should be kept at a minimum, although brake overheating is not normally a problem in the U-2.

BEFORE TAKEOFF

Normal procedure is used.

TAKEOFF

The U-2 will take off quickly under any temperature condition, although there is a marked increase in the takeoff roll in hot weather as compared to cold weather. Be prepared for thermal turbulence after takeoff.

AFTER TAKEOFF - CLIMB

After takeoff, the normal climb schedule will be followed. The gust control should be placed in the GUST position to prevent excessive loads on the aircraft if thermal turbulence is encountered.

CRUISE

Hot weather has no critical effect on normal cruise.

DESCENT

Descent is normal. Expect turbulence at lower altitudes.

LANDING

Use a normal landing technique, but expect the thermals to cause the aircraft to wallow on approach. Avoid steep angles of bank as thermal turbulence could suddenly increase the angle of bank to a dangerous degree.

AFTER LANDING

Hot weather operation requires the pilot to be cautious of gusts and wind shifts on the ground. Ground roll is slightly longer than normal.

ENGINE SHUTDOWN

Engine shutdown is normal.

POST FLIGHT

All ducts and openings should be covered soon after landing to prevent possible dust accumulation.

APPENDIX 1
PERFORMANCE DATA

TABLE OF CONTENTS

PART	PAGE
1 - INTRODUCTION	A1-1
2 - TAKEOFF	A2-1
3 - CLIMB	A3-1
4 - NORMAL CRUISE	A4-1
5 - ALTERNATE CRUISE	A5-1
6 - MAXIMUM ENDURANCE	A6-1
7 - DESCENT	A7-1
8 - LANDING	A8-1
9 - STALL SPEEDS	A9-1

PART 1 - INTRODUCTION

Subject	Page
Airspeed and Altitude Corrections	A1-1
Mechanical Instrument Error	A1-1
Installation Errors	A1-2
True Airspeed	A1-2
Outside Air Temperature Correction	A1-2
Standard Temperature vs Altitude	A1-3
Fuel and Fuel Density	A1-3
Performance Data Basis	A1-3

Charts	Page
Airspeed Position Error Correction	A1-4
Altimeter Position Error Correction	A1-5
TAS - Altitude - Outside Air Temperature Relationships	A1-6
Outside Air Temperature Corrections	A1-7

INTRODUCTION

General information for the pilot is provided in this section. Information is given on airspeed and altitude corrections, and true airspeed variation with altitude and temperature.

AIRSPEED AND ALTITUDE CORRECTIONS

Mechanical Instrument Error

The mechanical errors in the airspeed indicator and altimeter are extremely important because of the small buffet margin during cruise at altitude. If an airspeed indicator reads slow by 2-3 knots, the airplane will actually be flying faster than the recommended speed schedule and probably will encounter Mach buffet. The same situation will probably occur if the altimeter reads low by an appreciable amount. Therefore, indicators with large errors or an unusual amount of 'stickiness' should be replaced.

The mechanical error in the airspeed indicator can be checked by referring to the correction card for the instrument. It is recommended that the correction be applied so that the proper cruise airspeeds are used.

Altimeter error should also be checked by referring to the correction card for the instrument and the corrections applied so that the proper corrected airspeeds are flown at altitudes which have been corrected for instrument mechanical error. Above 60,000 feet the altimeter is operating much closer to the design limit of the mechanism than at lower altitudes, and for this reason specification tolerances permit instrument errors as great as 1500 feet. Any time the airplane cruise altitude does not appear to

Appendix I
Part 1 - Introduction

be correct for the fuel on board, the instrument error should be rechecked. However, it must be remembered that the cruise altitude will also be affected by outside air temperature and airplane gross weight.

Installation Errors

There is an airplane angle of attack and wing flap effect on the static pressure source which introduces small errors in the airspeed and altimeter readings. These are commonly known as installation or position errors.

The correction to airspeed and altitude readings for this effect are shown in figures A1-1 and A1-2. For this airplane, the error is small; but has been accounted for in determining the optimum climb speed schedule shown in Section II. In the normal cruise configuration, gust control faired, the airspeed correction is only about one knot or less and the altitude correction is correspondingly small, about 100 to 150 feet at cruise altitude. For the landing configuration, the airspeed correction is less than one knot below 100 knots IAS. With the gust control in the gust position there is no correction.

TRUE AIRSPEED

True airspeed (TAS) is obtained by dividing equivalent airspeed ($V_T \sigma^{1/2}$) by the square root of the atmospheric density ratio ($\sigma^{1/2}$). At sea level, with standard day pressure and temperature conditions of 29.92 inches of mercury and 15° C, true airspeed is equal to calibrated airspeed.

Airplanes are equipped with a system for measuring outside air temperature. Figure A1-3 can be used to obtain true airspeed directly from the observed temperature readings.

Figure A1-3 applies only when flying at the standard climb speed schedule given in Section II and should not be used at other airspeeds.

OUTSIDE AIR TEMPERATURE CORRECTION

Knowledge of the outside air temperature is valuable for evaluating airplane and engine performance and for obtaining true airspeed for navigational purposes. Figure A1-4 may be used to correct the observed temperature readings to true outside air temperature. Figure A1-4 applies only when flying at the standard climb speed schedule given in Section II. However, if the airspeed is merely one or two knots off the standard schedule, figure A1-4 can still be used with only a small loss in accuracy.

The corrections shown in the chart apply only to a system with 100% total temperature recovery. The system installed is an improved modification of a special Franz-Howland probe which has 100% total temperature recovery. This system is unaffected by radiated heat or direct sunlight as long as air is moving through the probe.

NOTE

Do not use the temperature indication when the airplane is on the ground. With no airflow through the probe, an erroneous temperature reading will be obtained.

STANDARD TEMPERATURE VS ALTITUDE

The performance shown in this book is not based on an NASA standard day unless specifically stated. The standard day variation of outside air temperature with altitude is not always representative of actual conditions. Usually the region of coldest temperatures is from 50,000 to 60,000 feet and at higher altitudes the temperature increases about 5°C to 15°C. However, this pattern will vary in different parts of the world and should not be interpreted as being a universal temperature variation. The effect of nonstandard outside air temperature on cruise performance is discussed in Parts 4 and 5 of Appendix I.

FUEL AND FUEL DENSITY

All of the data are based on tests with JPTS fuel corresponding to MIL-F-25524A.

The standard fuel weight is 6.58 pounds per gallon at a temperature of 15°C. However, as in the case of the U-2 airplane, the fuel-oil heat exchanger and engine-driven fuel pump heat the fuel. This heating of the fuel results in an increase in fuel volume of about 3% as measured by the totalizer.

PERFORMANCE DATA BASIS

The performance charts are based on flight data rather than calculations.

Data should be used for mission planning purposes only for aircraft at or near the zero fuel weights for which the curves were made. The flight time will be increased approximately 2.4 minutes for each 100 pounds decrease in zero fuel weight and decreased 2.4 minutes for each increase in zero fuel weight.

Range changes approximately 16 nautical miles for each 100 pound incremental change in zero fuel weight.

Altitude, for a given totalizer reading, changes approximately 100 feet for each 100 pound incremental change in zero fuel weight.

Appendix I
Part 1 - Introduction

AF (C)-1-1

AIRSPEED POSITION ERROR CORRECTION

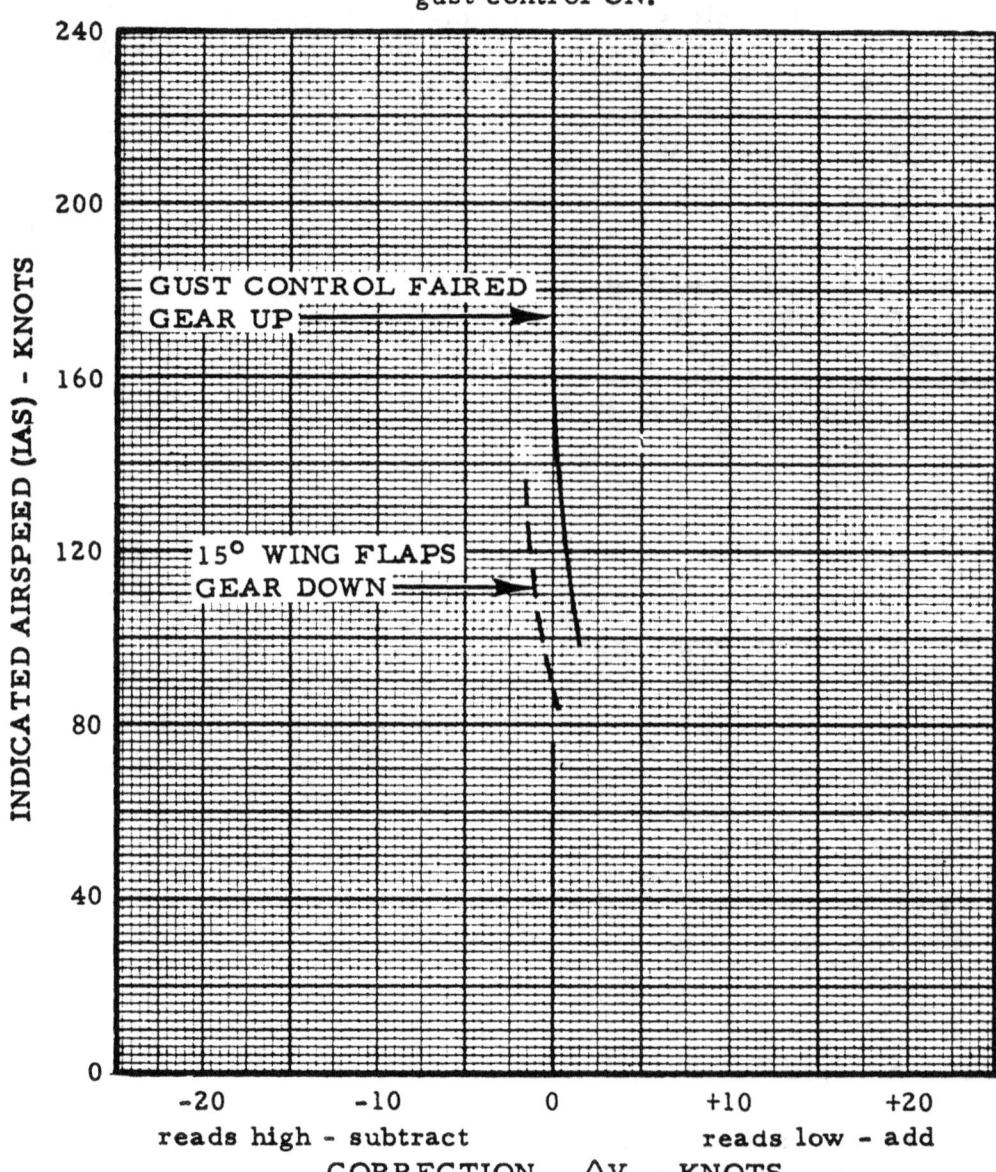

NOTE: Zero correction with gust control ON.

CAS = IAS + ΔV_i

CAS = Calibrated airspeed
 = Airspeed corrected for instrument and position error.

IAS = Airspeed corrected for instrument error.

Figure A1-1

ALTIMETER POSITION ERROR CORRECTION

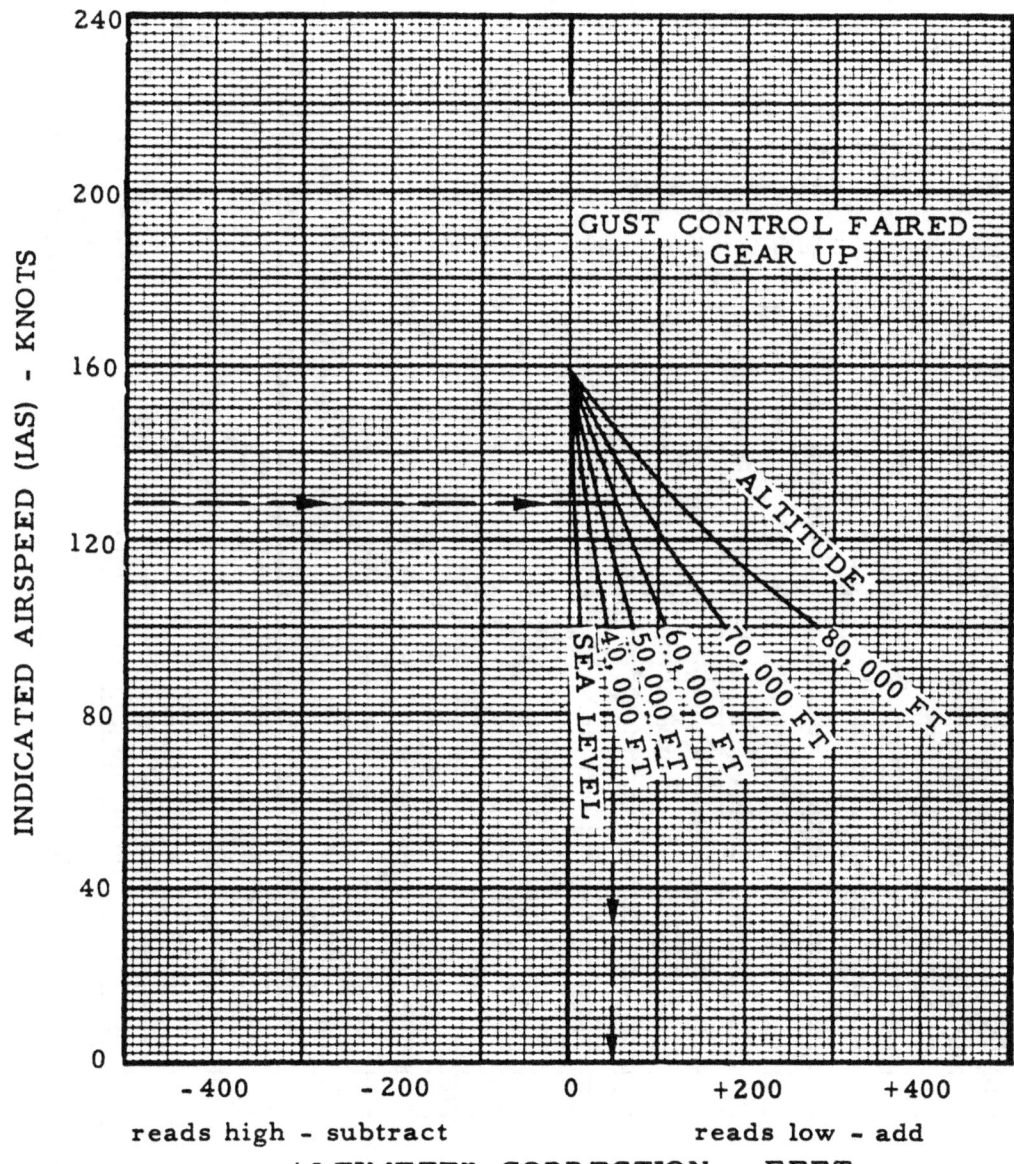

Example:

At an indicated airspeed of 130 knots and observed altimeter reading of 60,000 feet, the correction is +50 feet. Therefore, the corrected reading is (60,000 + 50) = 60,050 feet.

Figure A1-2

Appendix I
Part 1 - Introduction

TAS - ALTITUDE - OUTSIDE AIR TEMPERATURE RELATIONSHIPS

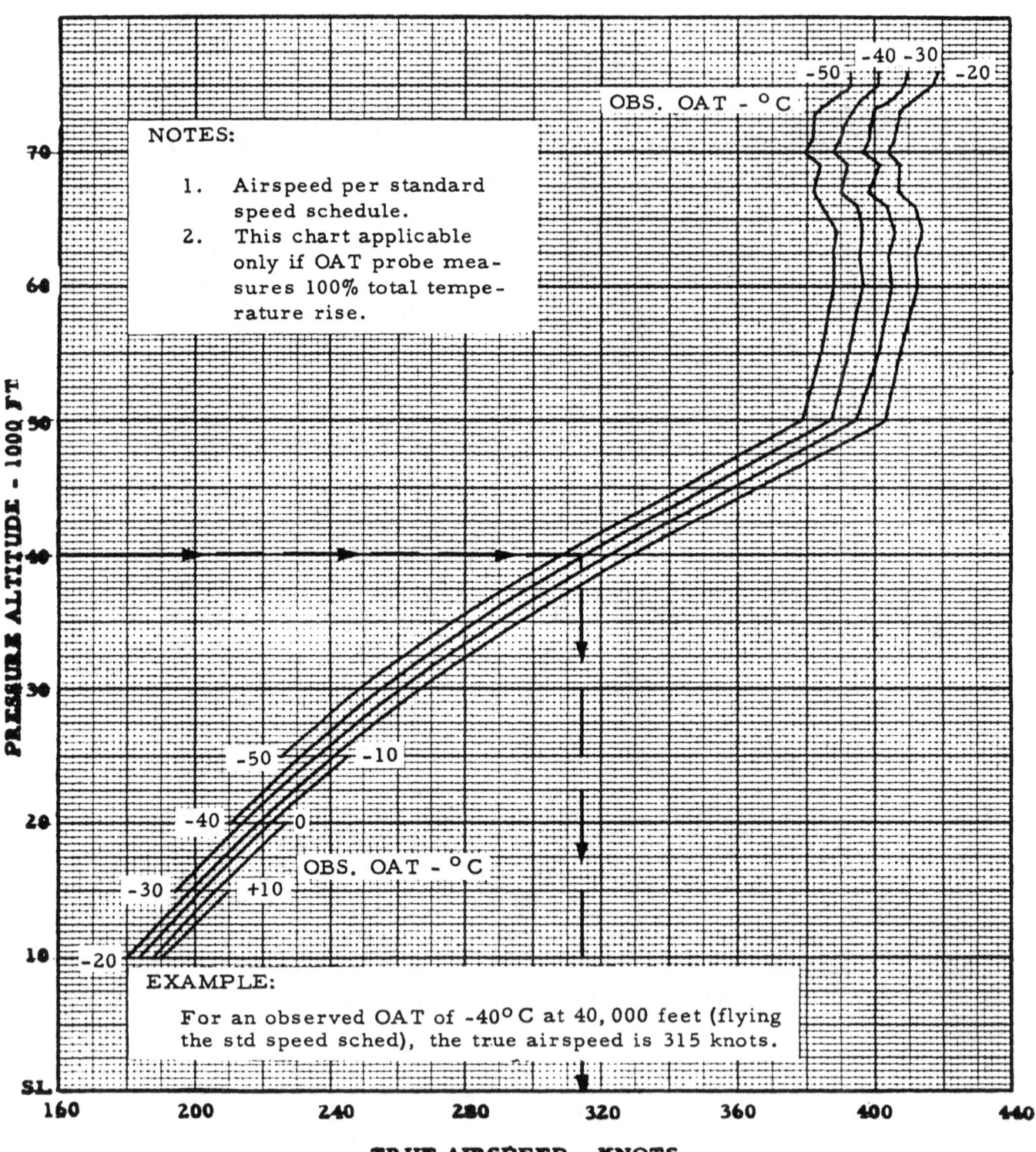

Figure A1-3

AF (C)-1-1

OUTSIDE AIR TEMPERATURE CORRECTIONS

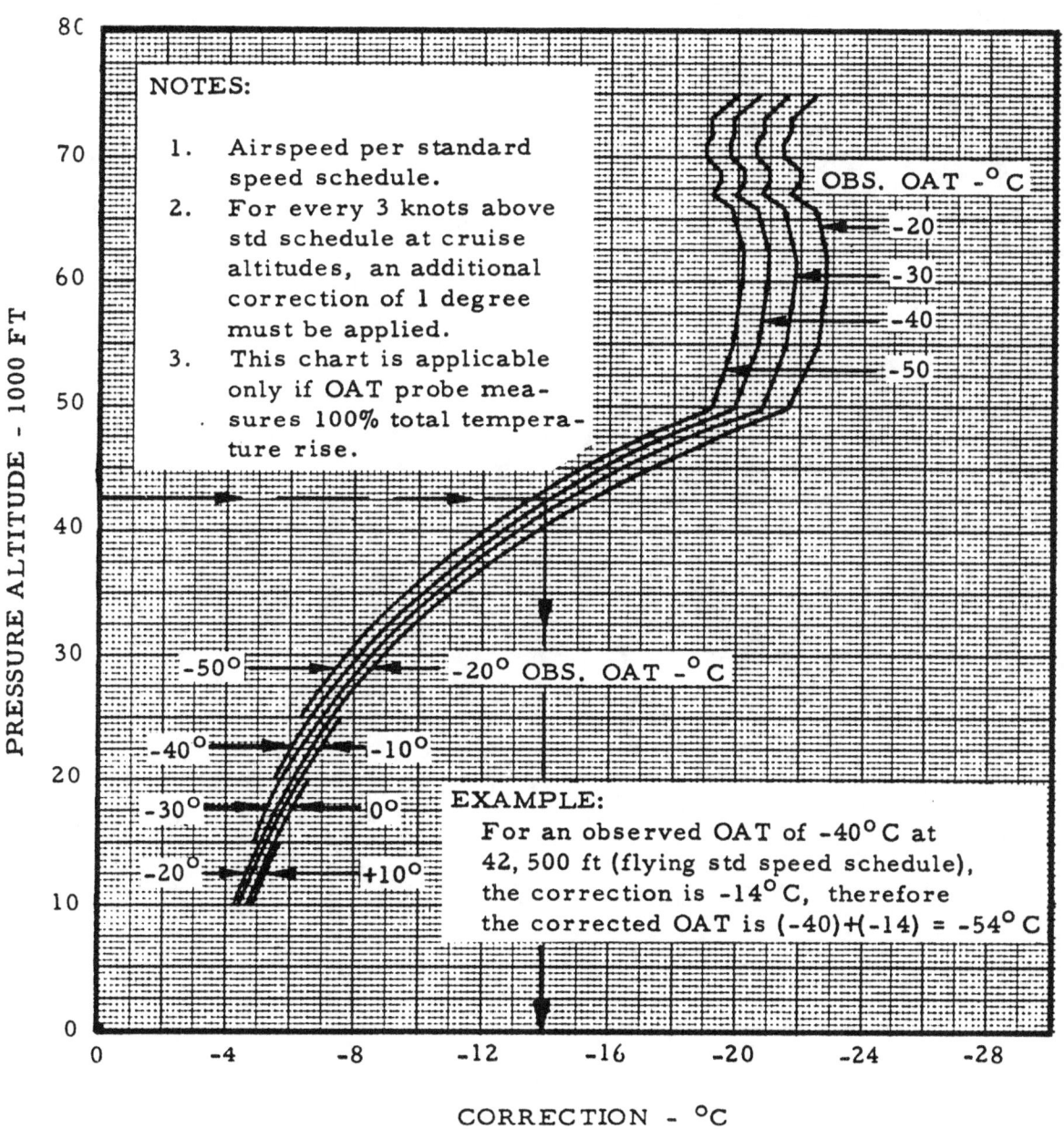

Figure A1-4

AF (C)-1-1

Appendix I
Part 2 - Takeoff

PART 2 - TAKEOFF

TABLE OF CONTENTS

Subject	Page
Takeoff	A2-1
Takeoff Speeds	A2-1
Effect of Wing Flaps	A2-1
Effect of Wind and Runway Slope	A2-1

Chart	Page
Takeoff Speeds	A2-2

TAKEOFF

Acceleration is rapid and takeoff distances are very short. Therefore, line speed and refusal speed are not significant for this airplane. The major variables which have an effect on the takeoff are as follows:

(a) Engine thrust.

(b) Gross weight.

(c) Outside air temperature.

(d) Field altitude.

Both engine thrust and air density are reduced as air temperature or altitude is increased. This results in longer takeoff distances on warm days or at higher field elevations.

The sea level standard day takeoff ground roll distance with a 1520-gallon fuel load and at throttle stop RPM is 1300 feet.

TAKEOFF SPEEDS

Figure A2-1 shows the stall speeds and recommended takeoff speeds with wing flaps faired. The takeoff speeds are about 115% of the stall speeds.

EFFECT OF WING FLAPS

It is recommended that normal takeoffs be made with flaps faired. The normal takeoff airspeeds shown in Section II are about 115% of stall speeds with flaps faired. Tests have shown that distances are not reduced by using wing flaps if this stall speed margin is maintained at all flap settings. Distances are about the same with flaps faired or with flaps down 15 degrees. With 25 degrees flaps, the takeoff distance is increased.

EFFECT OF WIND AND RUNWAY SLOPE

Because of the short takeoff distances, the effects of wind and runway slope are of less significance than for airplanes requiring longer runways.

Changed 6 November 1967

Appendix I
Part 2 - Takeoff

TAKEOFF SPEED

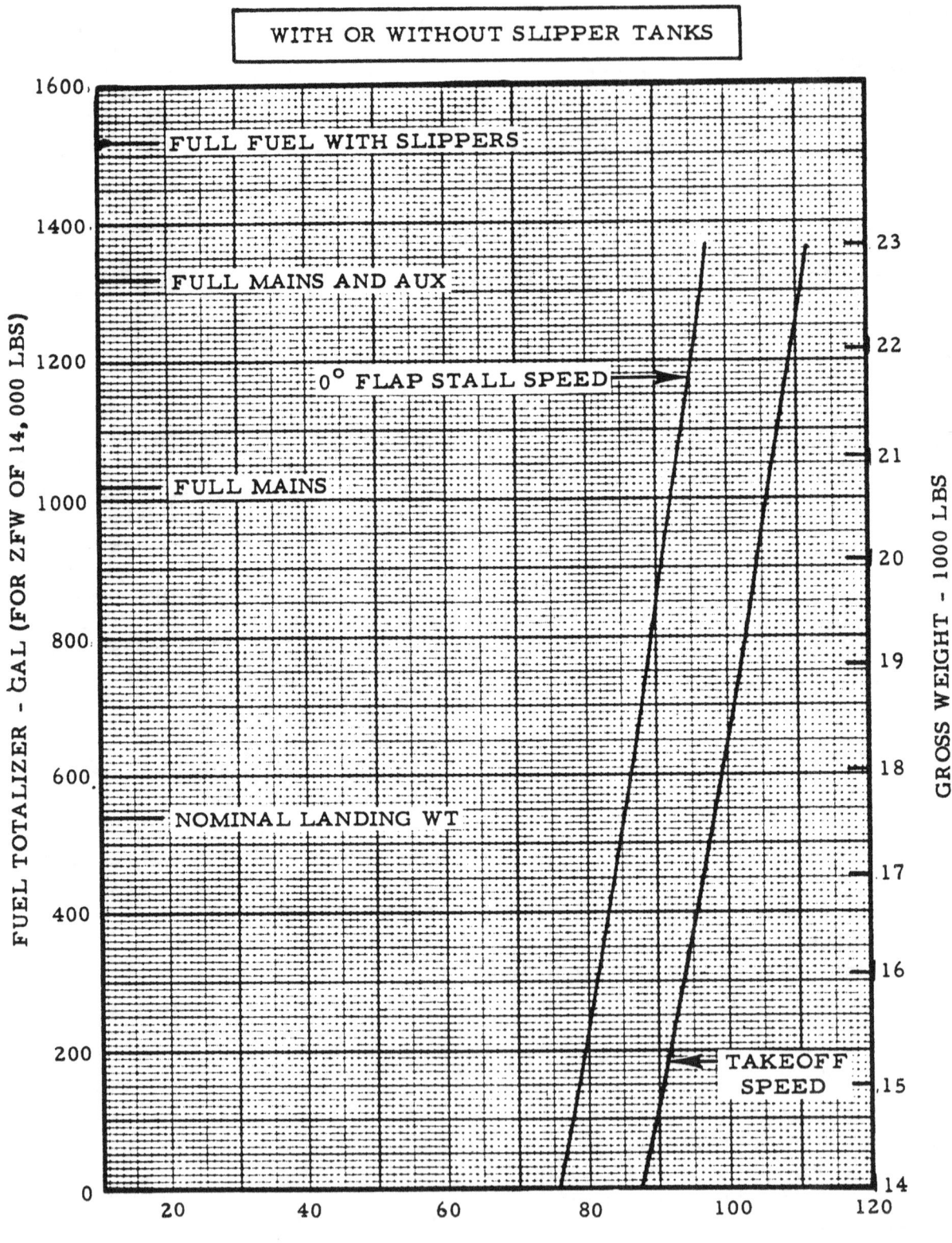

Figure A2-1

PART 3 - CLIMB

TABLE OF CONTENTS

Subject	Page
Climb	A3-1
Effect of Outside Air Temperature	A3-2

Charts	Page
Time, Distance, and Fuel to Climb With Slipper Tanks	A3-3
Time, Distance, and Fuel to Climb Without Slipper Tanks	A3-4
Climb from 35,000 Feet to 64,000 Feet Maximum Range Cruise Altitude - Empty Main Tanks - With Slipper Tanks	A3-5

CLIMB

Figures A3-1 and A3-2 show the climb performance in the form of time, fuel, and distance required to climb from one altitude to any other altitude up to start of cruise climb. Figures show the performance when using 665° EGT/Limiting EPR above 60,000 feet.

Figure A3-3 shows the climb performance to maximum range cruise altitude (for that gross weight) after an unsuccessful refueling attempt. This information is useful in case of a missed refueling attempt which may require a climb to maximum range cruise altitude to return to base.

A fifteen gallon fuel allowance is provided for start, takeoff and acceleration to climb speed. This allowance assumes that the engine is started on the runway and the takeoff is made immediately. If this procedure is not followed and taxiing is required, more than 15 gallons of fuel will be expended. Approximately three gallons per minute of fuel is expended during ground idle. This extra fuel burned on the ground will decrease the fuel available for cruise and will penalize the range. Therefore, it is recommended that every effort be made to minimize the amount of fuel consumed prior to takeoff.

Changed 6 November 1967

Appendix I
Part 3 - Climb

AF (C)-1-1

EFFECT OF OUTSIDE AIR TEMPERATURE

Climb performance will vary on different days due to changes in outside air temperature. However, the effects on fuel to climb and distance to climb tend to cancel.

On a hot day the cruise ceiling is lower but more time and fuel are required to climb to ceiling. However, the climb distance is also greater and this will compensate for the increased fuel consumption. On a cold day the opposite effects occurs; the cruise ceiling is higher, but less time and fuel are required to climb to ceiling. However, the climb distance is also shorter, which compensates for the decreased fuel consumption. These variations from standard performance can be approximated as follows:

Hot Day

1. Cruise ceiling - decrease 160 feet per degree centigrade warmer than standard temperature.

2. Fuel to ceiling - increase two gallons per degree centigrade warmer than standard temperature.

3. Distance - increase two miles per degree centigrade warmer than standard temperature.

Cold Day

1. Cruise ceiling - increase 135 feet per degree centigrade colder than standard temperature.

2. Fuel to ceiling - decrease one gallon per degree centigrade colder than standard temperature.

3. Distance - decrease one mile per degree centigrade colder than standard temperature.

TIME, FUEL, AND DISTANCE TO CLIMB

WITH SLIPPER TANKS

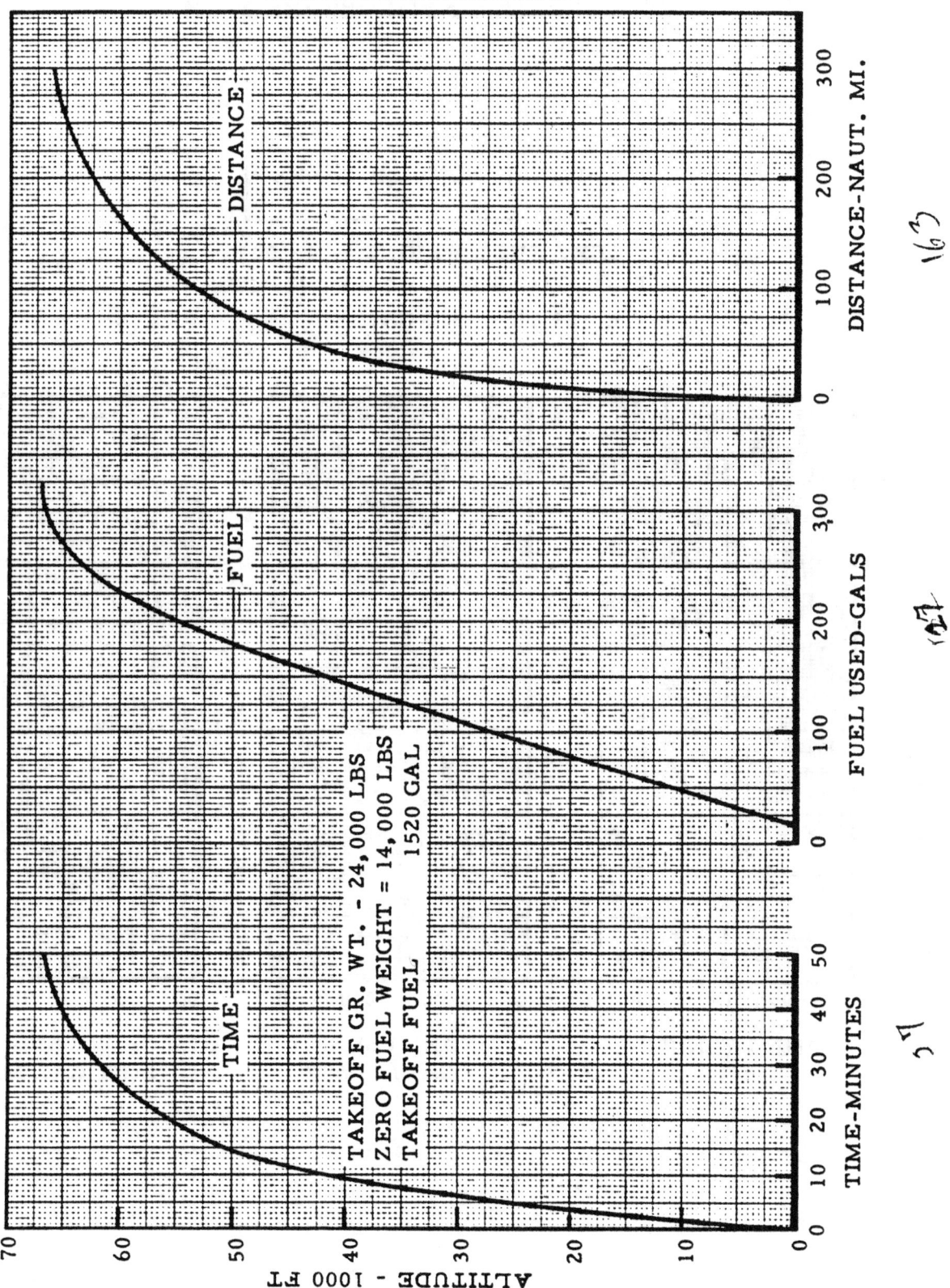

Figure A3-1

TIME, FUEL AND DISTANCE TO CLIMB

WITHOUT SLIPPER TANKS

ZERO FUEL WEIGHT = 14,100
T. O. GROSS WEIGHT = 22,777 LB
T. O. FUEL = 1320 GAL.

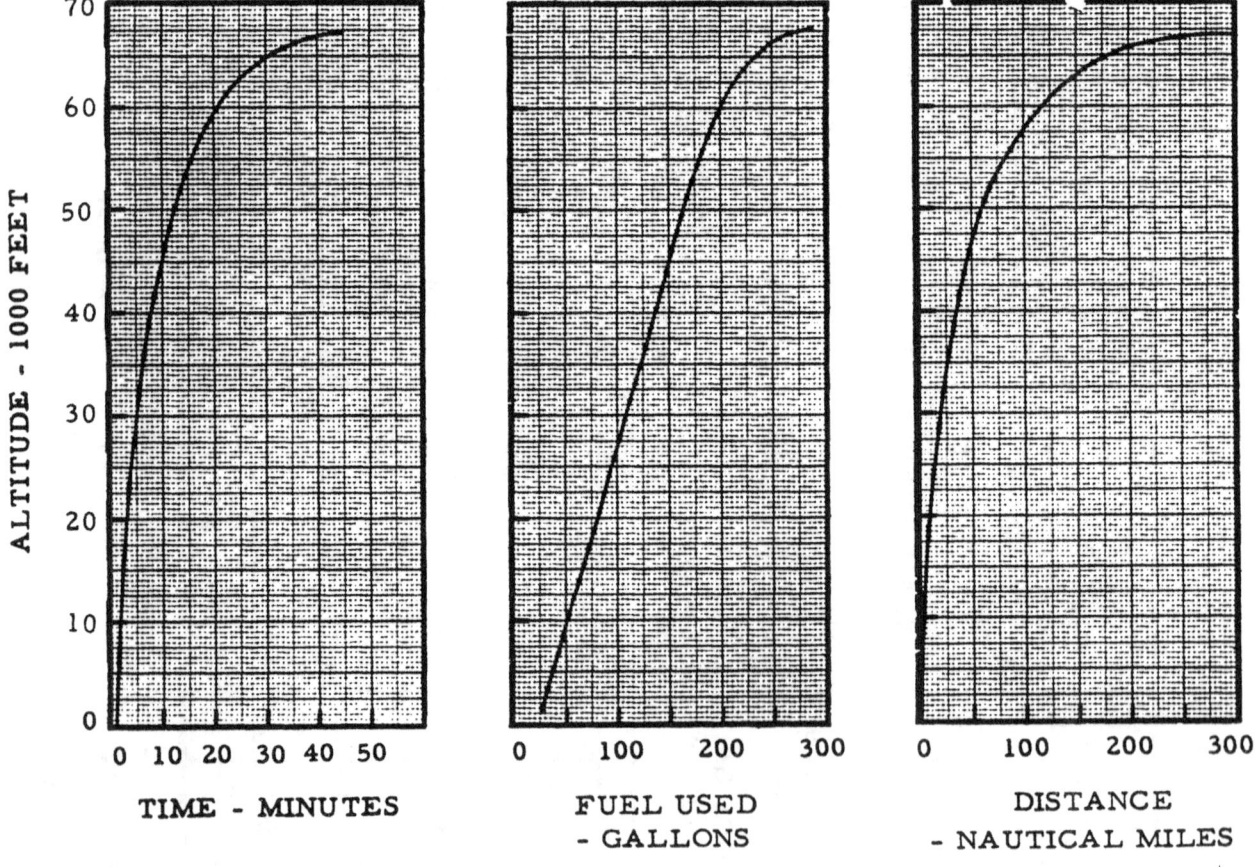

Figure A3-2

CLIMB FROM 35,000 FT TO 64,000 FT MAXIMUM RANGE CRUISE ALT

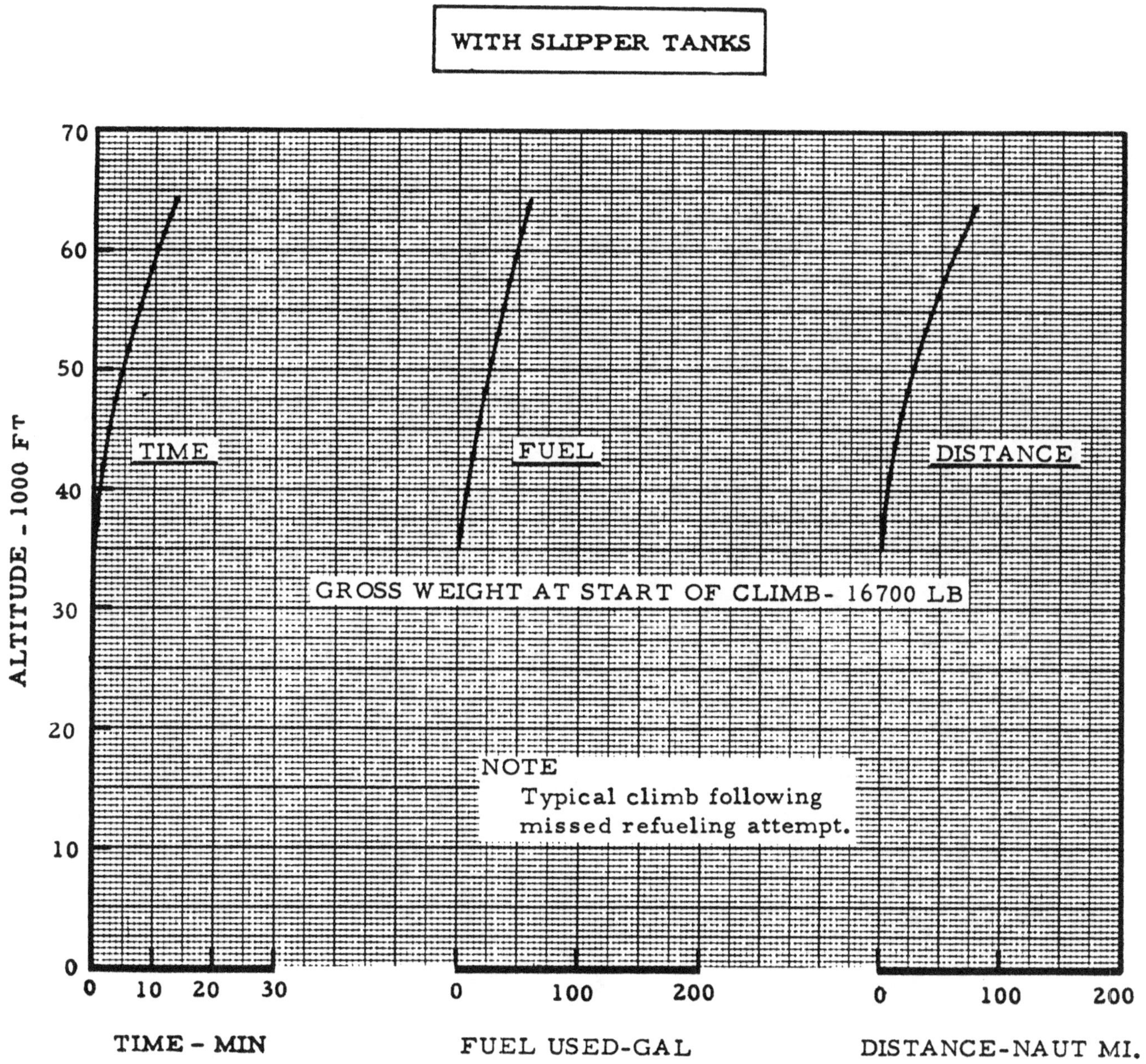

Figure A3-3

AF (C)-1-1

Appendix I
Part 4 - Normal Cruise

PART 4 - NORMAL CRUISE

TABLE OF CONTENTS

Subject	Page
Normal Cruise	A4-2
Maximum Altitude Cruise	A4-3
Level Cruise	A4-3
Maximum Range Cruise	A4-3
Alternate Maximum Range Cruise	A4-4
Split Profile Missions	A4-4
External Tanks	A4-4
Aerial Refueling Missions F	A4-5
Outside Air Temperature Effects	A4-6
Use of Fuel-Time Curve	A4-7

Charts	Page
Maximum Altitude Cruise Profile Without Slipper Tanks	A4-8
Maximum Altitude Cruise Profile With Slipper Tanks	A4-9
Maximum Altitude Cruise Profile With Empty Slipper Tanks	A4-10
Maximum Range Cruise Profile With Slipper and Drop Tanks	A4-11
Maximum Range Cruise Profile With Slipper Tanks	A4-12
Fuel Totalizer vs Time, 1320 Gallons at Takeoff, Without Slipper Tanks	A4-13
Range & Time vs Fuel Totalizer, 35,000 Feet Cruise With Slipper Tanks	A4-14
Return Profile after Unsuccessful Refuel Attempt, With Slipper Tanks	A4-15
Oxygen Consumption	A4-16
Oxygen Consumption, Model U-2F	A4-17

Changed 6 November 1967

NORMAL CRUISE

Cruise is along a climbing flight path at a very low rate of climb. The cruise climb is very sensitive to airspeed and the schedule shown in Section II should be used for best performance. As the weight decreases at a given altitude there is an increase in the excess thrust. By maintaining the speed schedule, this excess thrust is converted to the proper rate of climb. If the speed is too fast, Mach number effects and reduced rate of climb can decrease the miles per gallon.

Performance data are presented for maximum altitude cruise, level off cruise and maximum range cruise missions.

Altitude versus Time curves show the variation in altitude with time for various mission profiles. The nominal totalizer reading for each hour after takeoff is also shown on the curve. For a particular cruise climb type mission (either maximum altitude or maximum range) a given standard day altitude is always reached at the same gross weight (zero fuel weight plus fuel). However, differences in fuel loadings and/or zero fuel weights will result in arriving at a given altitude with different amounts of fuel or at a different elapsed time after takeoff.

Variations in outside air temperature will affect the altitude and fuel readings versus time as discussed under OUTSIDE AIR TEMPERATURE EFFECTS. The curves include the fuel used in climbing to altitude. No allowance is made for the apparent increase in fuel capacity due to "manufactured" fuel.

If additional time due to "manufactured" fuel were shown, the time would be increased by about 15 to 20 minutes maximum. These curves can be used during flight to keep a running check on fuel consumption. However, it must be emphasized that the fuel-time curve can vary for a number of reasons such as:

1. Fuel burned on ground prior to takeoff.

2. Differences in engines.

3. Difference in airplane gross weight.

4. Variation of outside air temperature from standard.

5. Variation from the standard cruise airspeed schedule.

The ground fuel consumption prior to takeoff is not included in fuel totalizer versus time curves because it can vary with different operational procedures. However, if possible, it should be minimized because the fuel reserve will be decreased for a mission of given duration. For example, if 25 gallons of fuel are burned prior to takeoff, the fuel reserve will be decreased by about 20 gallons for an 8-hour mission under standard conditions. The range would be unchanged. However, if the mission requirements specify a given landing reserve, then range must be sacrificed if the ground fuel consumption is high. For a mission with full internal fuel and a specified 50-gallon landing reserve, 25 gallons burned on the ground can reduce the range about 75 miles.

The standard fuel-time curves are based on cruising at the altitudes shown. If the cruise altitude is low the fuel consumption will be increased. This can be caused by temperatures that are hotter than standard, improper cruise speeds or heavy airplane gross weight. If the cause is entirely due to temperature there is a negligible effect on range because the increased true airspeed compensates for the higher fuel consumption. However, heavy airplane gross weight or improper cruise speeds can cause a loss in range. The cruise climb procedures outlined in Section II should be followed for best performance.

MAXIMUM ALTITUDE CRUISE

The maximum altitude mission is a conventional cruise climb using maximum power. This mission sacrifices range for altitude and results in the shortest range and flight time. The speed schedule shown in Section II must be used to attain the range. Even small increases in airspeed will decrease the fuel economy and result in loss of range.

LEVEL CRUISE

If desired, the range can be increased by leveling off and cruising at constant altitude and airspeed. The initial climb to the desired cruise altitude is made at maximum power. As fuel is burned and weight reduces, the power is decreased to hold the altitude and airspeed constant. The standard airspeed for the altitude is used.

Near the end of cruise the weight is reduced considerably and as the power is decreased to hold altitude, minimum EPR may be encountered. When this occurs, altitude should be held constant and the airspeed allowed to increase as fuel is consumed. Since the airplane is light, the speed can be increased without encountering a one-G buffet. Allowing the airspeed to increase at constant altitude will provide slightly better miles per gallon than climbing at minimum EPR at the standard speed schedule.

MAXIMUM RANGE CRUISE

Maximum range is obtained by flying a constant EPR, constant Mach 0.72 cruise climb starting at 59,000 feet without slipper or drop tanks, 57,000 feet with slipper or drop tanks, or 56,000 feet with both slipper and drop tanks. The initial climb is made to 56,000, 57,000 or 59,000 feet where power is reduced to 2.30 EPR and Mach 0.72 speed set up per speed schedule shown Section II. This EPR and Mach is then held constant and as fuel is burned and weight reduces, the airplane will cruise climb.

On a standard day, the cruise climb should follow the altitude time relationship shown in the chart if the proper EPR setting is established and the Mach 0.72 airspeed schedule is followed. This is the optimum gross weight-altitude relationship for cruise and will provide the best miles per gallon. However, large variations in outside air temperature are common in this altitude range and can affect the performance noticeably. This can be caused by changes in temperature with increasing altitude or can be due to flying into geographical areas of hotter or colder temperatures. The airplane will cruise climb more rapidly when flying into areas with colder temperatures. When flying into warmer temperatures the airplane will climb slower, may not climb at all, or may actually descend if the temperature increases enough. A temperature increase of 10°C may cause the airplane to fly at constant altitude or descend slightly. Then, when colder temperatures are encountered the airplane will climb rapidly until it is back on the fuel-altitude curve. The warmer temperatures will also increase the fuel consumption rate, but this is partially offset by the faster true airspeed as the temperature increases.

Appendix I
Part 4 - Normal Cruise

The standard fuel totalizer setting for a full internal fuel load is 1320 gallons. However, most of the airplanes will hold somewhat more than this in varying amounts. In addition, the fuel is heated by the fuel-oil heat exchanger and engine-driven fuel pump. The temperature of the fuel measured by the totalizer is therefore relatively hot compared to the fuel temperature when filling the tanks. This heating results in an increase in fuel volume. The increase in volume will vary, depending upon the temperature of the fuel when the tanks are filled. The maximum increase which can be expected with unrefrigerated fuel is about three percent. This is equivalent to about a 40 gallon increase with a full internal fuel load and a 30 gallon increase with full main tanks. To achieve maximum range, advantage should be taken of the additional fuel capacity and "manufactured" fuel. Since both of these factors are variable, they are ignored when the fuel counter is set. Therefore, the counter may read below zero when the fuel low warning light comes on, and full reliance must be placed in the light.

ALTERNATE MAXIMUM RANGE CRUISE

If range and altitude are both important for a mission, an alternate maximum range cruise procedure can be used which will result in approximately 3% less range, but at a cruise climb altitude an average of 1,500 feet higher than that of a normal maximum range cruise. Establish and maintain Mach 0.72 and 2.45 EPR starting at 59,000 feet if flying with slipper tank fuel. (No tests have been conducted with slipper tank fuel and 2.45 EPR.) If flying without slipper tanks (or with empty slipper tanks) establish and maintain Mach 0.72 and 2.45 EPR starting at 61,000 feet.

NOTE

The engine bleed valves must be closed or a loss of range will result. Observe the minimum EPR limits of figure 5-1.

SPLIT PROFILE MISSIONS

At the start of the cruise climb, the airplane is heavy and the maximum power profile is the least economical cruise condition. The most economical cruise is the maximum range profile. If this profile is used during the heavy weight portion of the mission, the miles per gallon are 35% to 40% better than the maximum power fuel economy. Therefore, the range can be increased by first cruising at the maximum range condition and then climbing to the required mission altitude. To take full advantage of this, it is desirable to fly the maximum range profile for as long as possible.

EXTERNAL TANKS

Slipper Tanks

All U-2 airplanes have provisions for slipper tanks. The total fuel capacity with slipper or drop tanks is 1520 gallons. The speed schedule is the same with or without slipper tanks.

With full internal fuel and full slipper tanks the flight time is increased about three quarters of an hour. With the same amount of fuel on board the maximum altitude cruise is about 500 feet lower than that of an airplane without slipper tanks.

The cruise altitude is lower, even with empty tanks, because of the tank installation weight and the added drag.

AERIAL REFUELING MISSIONS

On some aerial refueling missions it is desireable to "buddy" fly with the tanker airplane to the refueling site. For this reason, low level performance data were obtained for two low altitude conditions. The speed and altitude in each condition were chosen so as to be compatible with speed and altitude capabilities of a particular tanker airplane.

This information was obtained for a P-13A engine, however, it is considered valid for use with the P-13B engine.

LEVEL CRUISE AT 35,000 FEET

Level cruise performance at 35,000 feet is shown in figure A4-7. Cruise speed is 220 knots IAS with the gust control ON. Zero wind range with slipper tanks is 1540 nautical miles to empty main tanks; 540 gallons remaining on the fuel counter.

Longer overall range can be obtained by cruising out at maximum range cruise altitude and then descending to 35,000 feet for rendezvous and refueling. Allowing 100 gallons for descent and rendezvous, maximum range cruise provides 1940 miles to 640 gallons remaining on the fuel counter.

LEVEL CRUISE AT 20,000 FEET

Level cruise performance without slipper tanks at 20,000 feet is shown in figure A4-6. Cruise speed is 170 knots IAS (231 knots true speed). Zero wind range to empty main tanks (540 gallons remaining 3 1/2 hours after takeoff) is 810 nautical miles.

MISSED REFUELING

If the refueling attempt is unsuccessful, 540 gallons of fuel is sufficient to return to base from a point 1540 miles out. The profile is shown in figure A4-8. As shown, a climb to maximum range cruise altitude is required for the return flight.

If the refueling is to be made at a range greater than 1540 miles, an additional fuel allowance is required for return to base, or an alternate base closer than the takeoff point is required.

MISSION PROFILES

The performance shown in the preceding charts may be combined in various ways to determine overall mission profiles. For example, the profile shown in figure A4-8 was based on the range at 35,000 feet from figure A4-7, the climb performance in figure A3-3, the maximum range cruise performance in figure A4-5, and the descent performance in figure A7-1.

OUTSIDE AIR TEMPERATURE EFFECTS

Outside air temperature variations generally affect the cruise climb range more than the level cruise range. This is due to the fact that the range for a cruise climb mission is dependent on maintaining the proper relationship of gross weight to altitude, which is not the case for a level mission. For example, during a hot day level altitude mission a little more power is required to fly the scheduled indicated airspeed for the altitude. This increases the fuel consumption; however, the true airspeed is proportionately faster and the cruise miles per gallon are essentially the same as a standard day. On a cold day the effect is opposite. Less power is required for the same indicated airspeed and the fuel consumption is decreased. However, the true airspeed is proportionately slower and the cruise miles per gallon are still the same as a standard day.

This same effect is true to a large extent for the maximum power cruise climb mission. However, on a hot day the cruise altitude is lower for a given fuel load and the gross weight-altitude relationship is not the same as a standard day. Therefore, the change in true airspeed is not quite proportional to the difference in fuel consumption. Extreme temperature variations are required, however, to have a significant effect on the range.

The nature of the maximum power mission is such that it is flown at the most uneconomical gross weight-altitude relationship. Basically, the airplane is flying at too high an altitude and too heavy a gross weight for the best miles per gallon, which is the reason this mission has the shortest range. Consequently a lower cruise climb altitude improves the gross-weight-altitude relationship and the cruise miles per gallon. For this reason, the range may be slightly better on a hot day than a standard day. A cold day will tend to shorten the range slightly because the cruise altitude is higher, making the gross-weight-altitude relationship more unfavorable. However, it is emphasized that extreme temperature variations are required for this to occur and for practical purposes the maximum power range can be considered nearly independent of outside air temperature.

The maximum range mission is a cruise climb flown at the optimum gross-weight-altitude relationship. The best cruise miles per gallon are obtained at this condition. However, large variations in outside air temperature are common in this altitude range and can effect the performance noticeably. This can be caused by rapid changes in temperature with increasing altitude or can be due to flying into geographical areas of hotter or colder temperatures. The airplane will cruise climb more rapidly when flying into areas with colder temperatures. When flying into warmer temperatures the airplane will climb slower and actually may not climb at all, or may descend, if temperatures are very warm. If the temperature increases to the point where it is about $15°C$ warmer than standard, the airplane will not climb until colder temperatures are encountered. This will also increase the fuel consumption and if four to five hours are spent cruising at these conditions the fuel-time relationship can be about 25 to 30 gallons below the handbook curve. This, of course, is partially offset by the faster true airspeed as the temperature increases, which nearly compensates for the higher fuel consumption. However, when temperature variations of this order are encountered, the maximum range can be obtained by setting the power so that the airplane will climb at the optimum gross-weight-altitude relationship. This relationship is maintained by adjusting power so as to be at each altitude with the appropriate totalizer value shown for that altitude on the Altitude versus Time curve. On a cold day reduced power will be required and on a hot day increased power will be required to climb at the fuel-altitude schedule shown in the charts. When flying into areas of changing temperature it will be necessary to change the power setting in order to climb at the proper fuel-altitude relationship for maximum range.

USE OF THE FUEL-TIME CURVE

The fuel-time curve has two primary uses: preflight planning and checking the fuel consumption during a mission. The curve is very useful for these purposes; however, it is also commonly used to evaluate the aircraft and engine performance. Basically, the curve was not intended for this purpose and should not be used unless the variables which effect the fuel-time relationship are taken into consideration. Variations in outside air temperature cause the fuel-time curve to vary and often these are difficult to take into account. On a hot day the fuel consumption will be higher and the fuel-time plot will generally fall short of the handbook curve. On a cold day the effect is opposite. Therefore, the same aircraft and engine combination can show varying fuel consumption on different days. However, this does not necessarily mean that the range is different for these days. The temperature effects on true airspeed tend to compensate for the variations in fuel consumption, making the cruise miles per gallon and range essentially independent of temperature variations.

There are also other affecting factors which must be accounted for when evaluating the mission performance. These will affect both the time and range and are as follows:

1. Taxi fuel - discussed in the CLIMB section, Part 3.

2. Operational procedures involving low altitude level-off before climbing to cruise altitude.

3. Flight line turns during the mission.

4. Variations in gross weight with different equipment loads.

5. Drag due to external configurations.

6. Engine deterioration.

7. Fuel totalizer calibration.

If at all possible, taxi time and low altitude level-offs should be kept to a minimum because the resulting fuel consumption will decrease the fuel available for cruise.

Items 3 and 7 above are often difficult to account for; however, a method is suggested for quantitatively evaluating the combined effects of these variables. First, the amount of fuel burned before starting the climb to cruise altitude is checked, items 1 and 2 above. Then the outside air temperature during the mission is checked to determine whether there were any large variations in temperature which would cause the fuel consumption to vary appreciably from the handbook fuel-time curve. The mission fuel versus time is then plotted on the handbook curve for comparison. If the mission fuel does not fall more than 1-1/2% below the handbook curve, the handbook performance is considered to have been met. Note, that in calculating the allowable 1-1/2% deviation this must be based on the fuel consumed and not the fuel remaining. This method has been used to evaluate the performance of U-2 aircraft in service operation and generally is quite satisfactory. It is especially useful when range information is not available.

Appendix I
Part 4 - Normal Cruise

AF (C)-1-1

MAXIMUM ALTITUDE CRUISE PROFILE

WITHOUT SLIPPER TANKS

ZFW - 14100 LB
TAKEOFF FUEL 1320 GAL

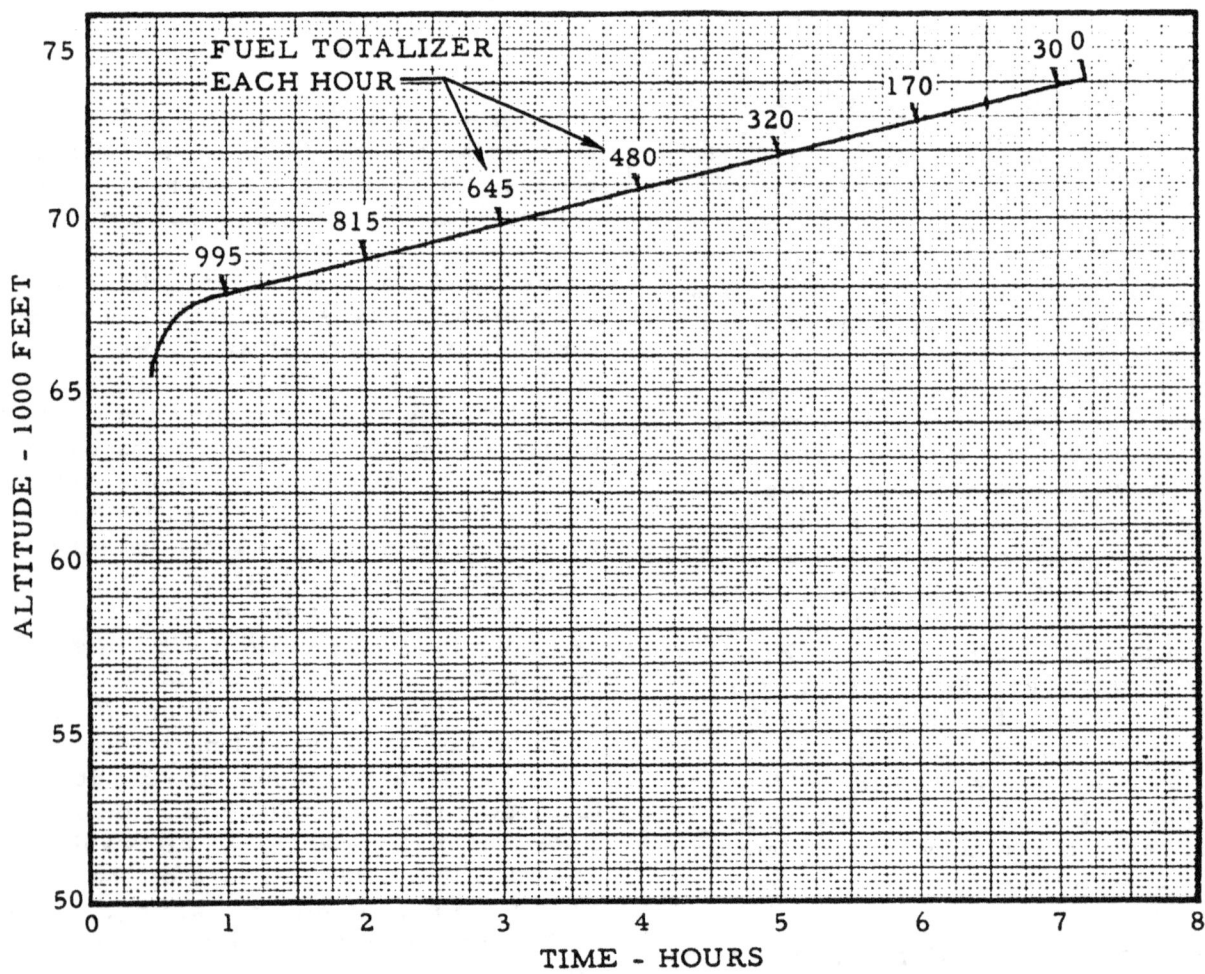

Figure A4-1

AF (C)-1-1
Appendix I
Part 4 - Normal Cruise

MAXIMUM ALTITUDE CRUISE PROFILE

WITH SLIPPER TANKS

ZFW - 14300 LB
TAKEOFF FUEL 1520 GAL

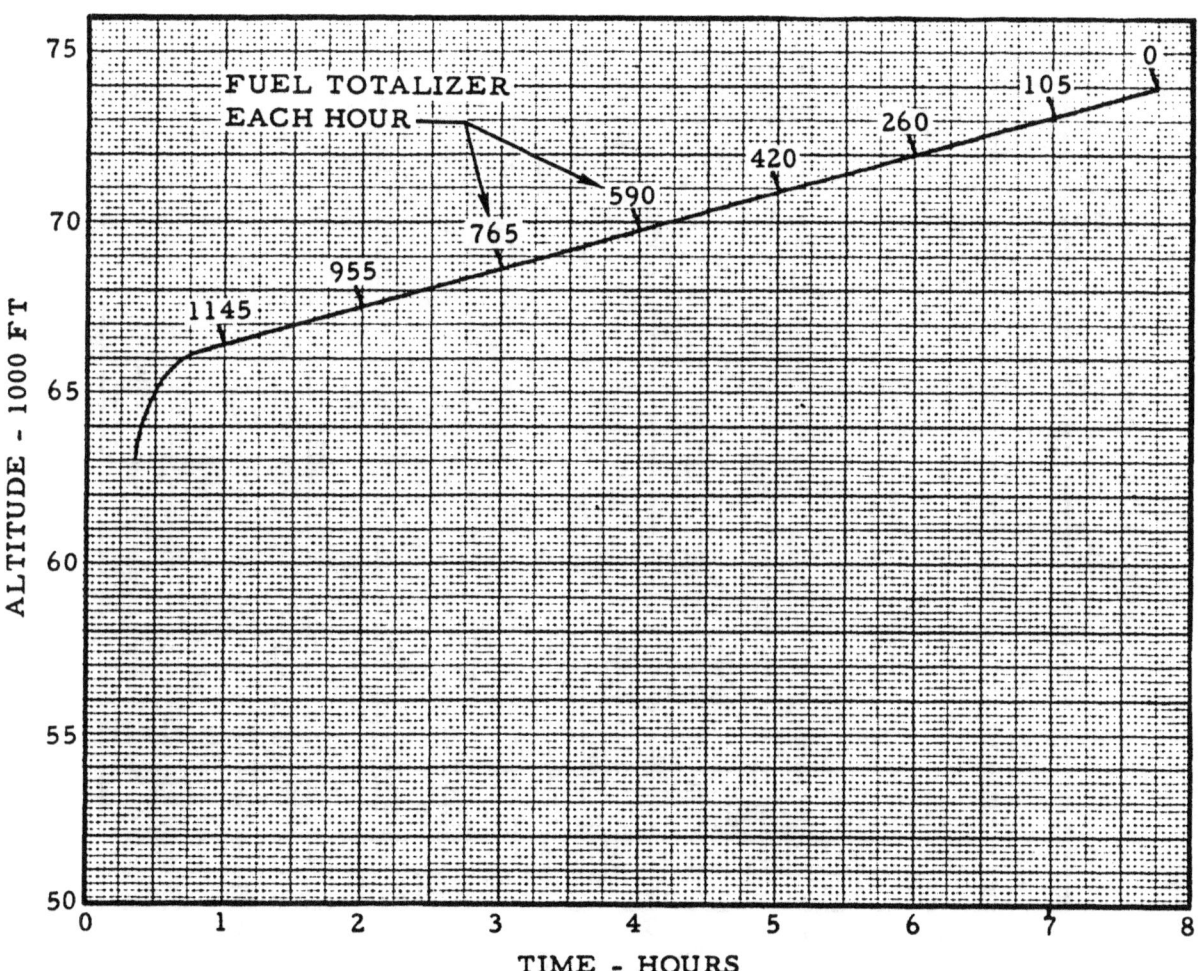

Figure A4-2

Changed 6 November 1967

A4-9

Appendix I AF (C)-1-1
Part 4 - Normal Cruise

MAXIMUM ALTITUDE CRUISE PROFILE

WITH EMPTY SLIPPER TANKS

ZFW - 14700 LB
TAKEOFF FUEL 1320 GAL

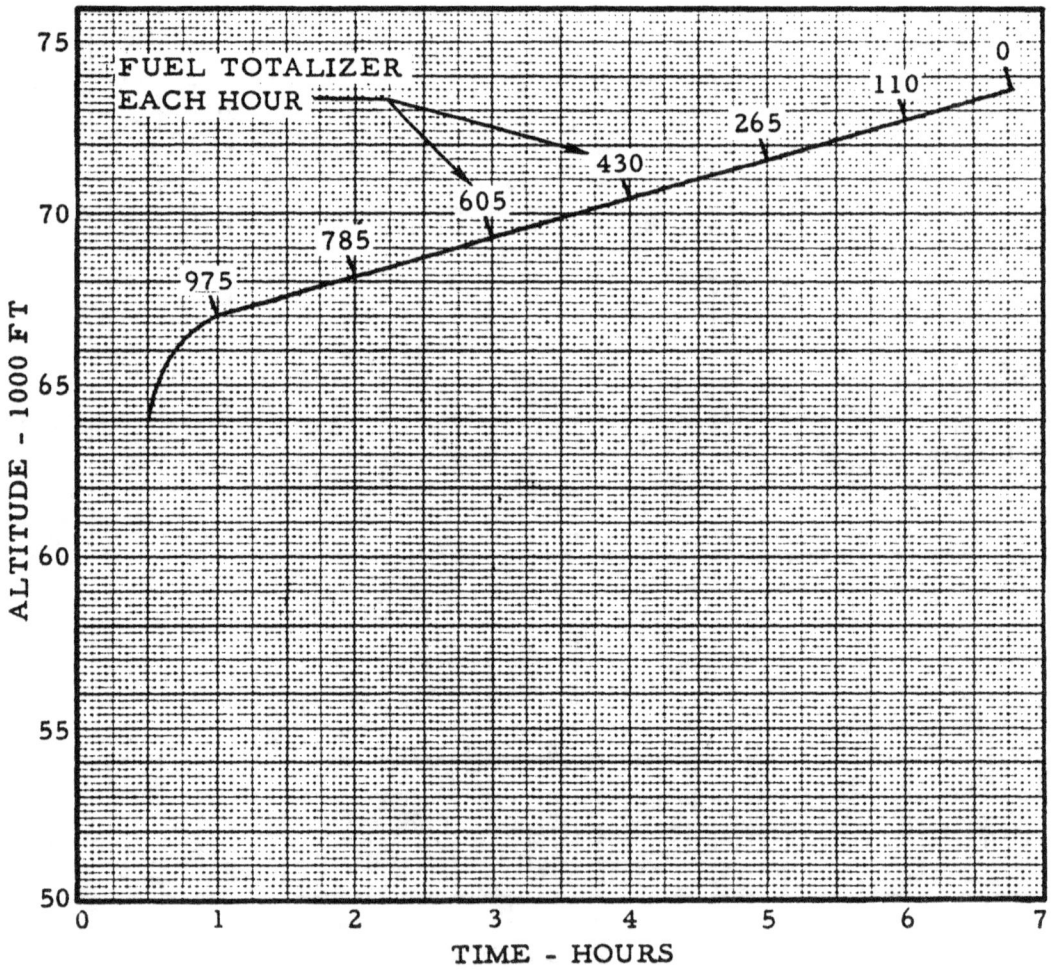

Figure A4-3

AF (C)-1-1
Appendix I
Part 4 - Normal Cruise

MAXIMUM RANGE CRUISE PROFILE

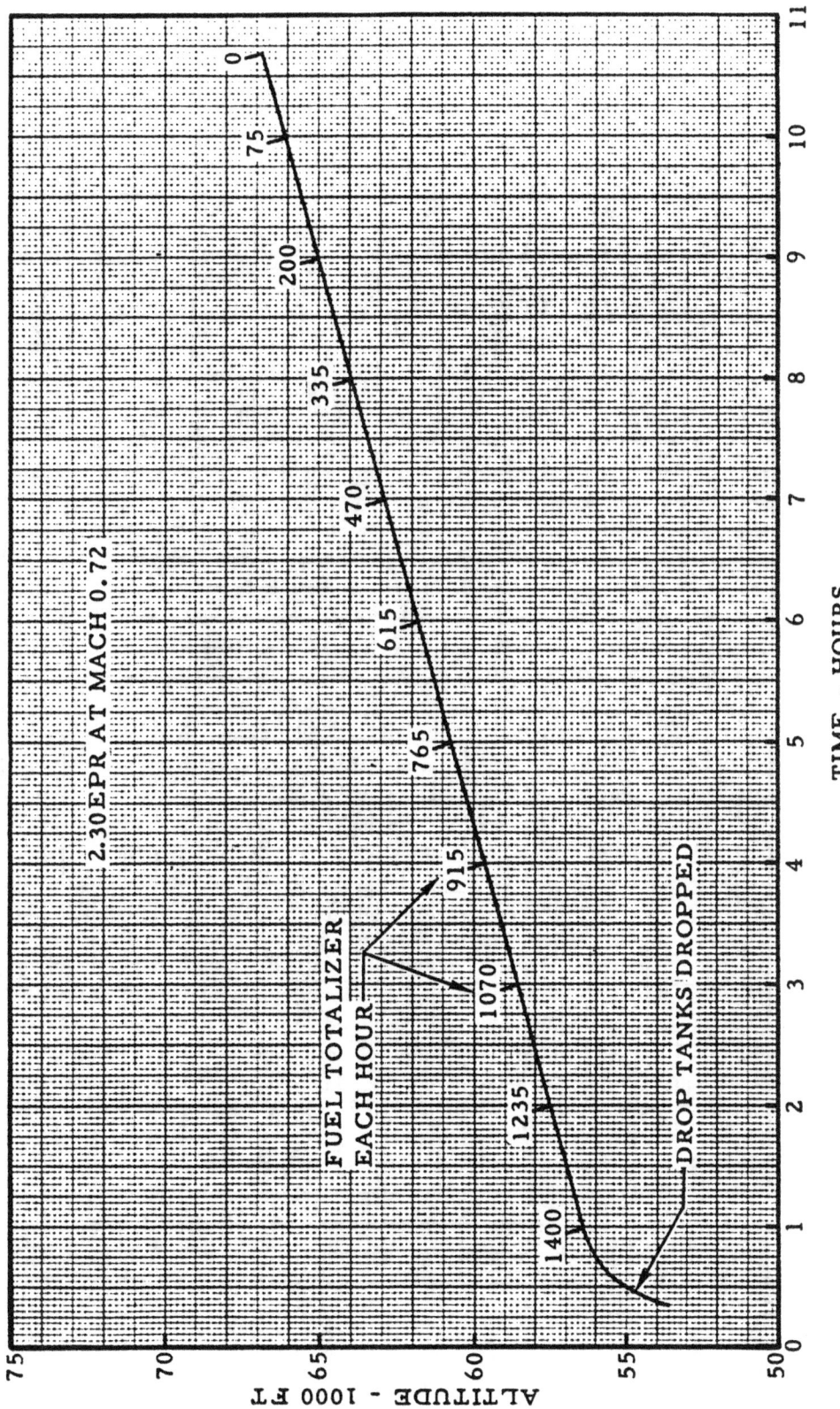

Figure A4-4

Appendix I
Part 4 - Normal Cruise

AF (C)-1-1

MAXIMUM RANGE CRUISE PROFILE

WITH SLIPPER TANKS

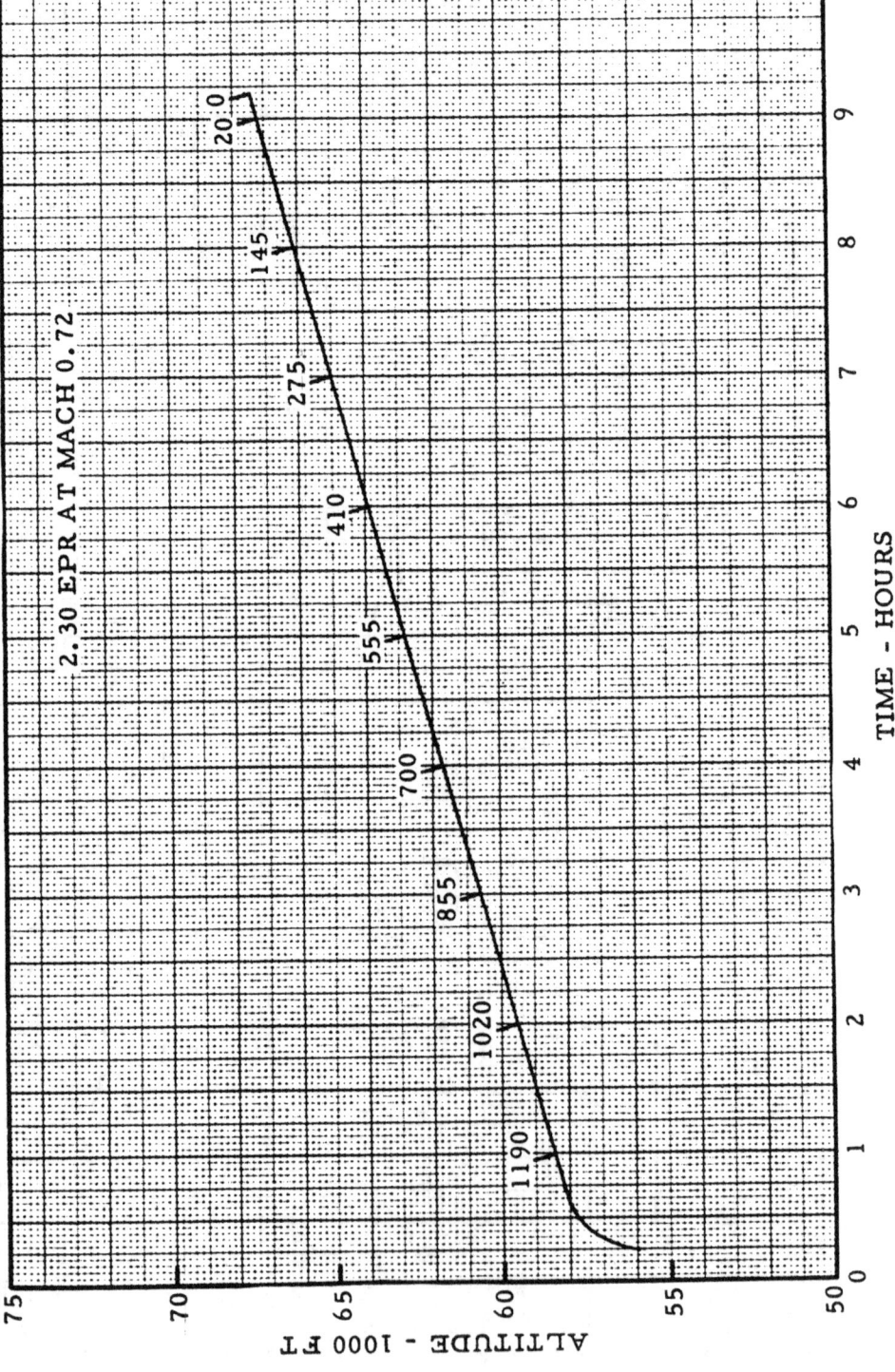

Figure A4-5

AF (C)-1-1
Appendix I
Part 4 - Normal Cruise

FUEL TOTALIZER VS TIME

WITHOUT SLIPPER TANKS

GROSS WEIGHT AT TAKEOFF = 22784 LB
LEVEL 20,000 FT   170 KIAS

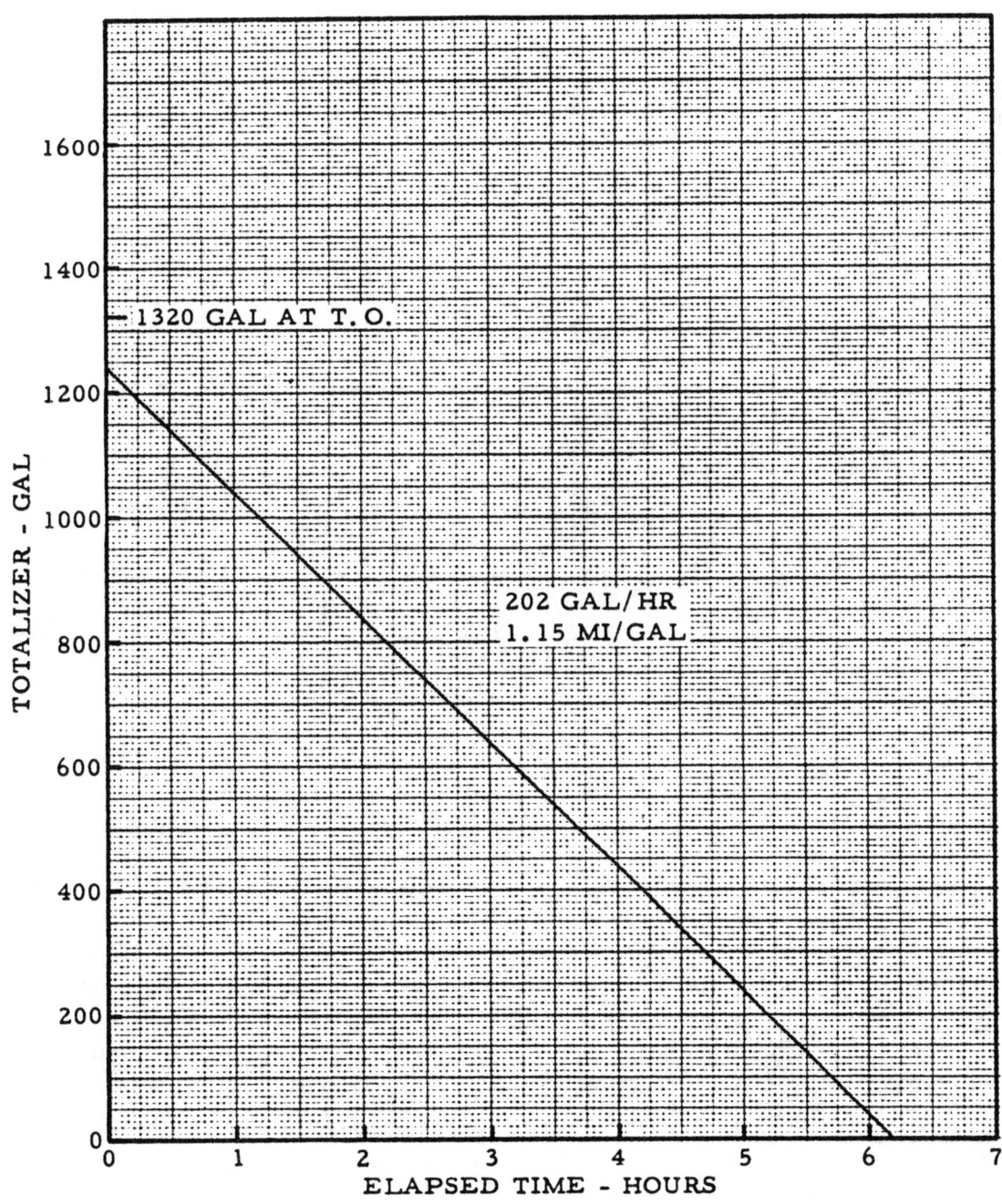

Figure A4-6

Changed 6 November 1967

A4-13

Appendix I
Part 4 - Normal Cruise

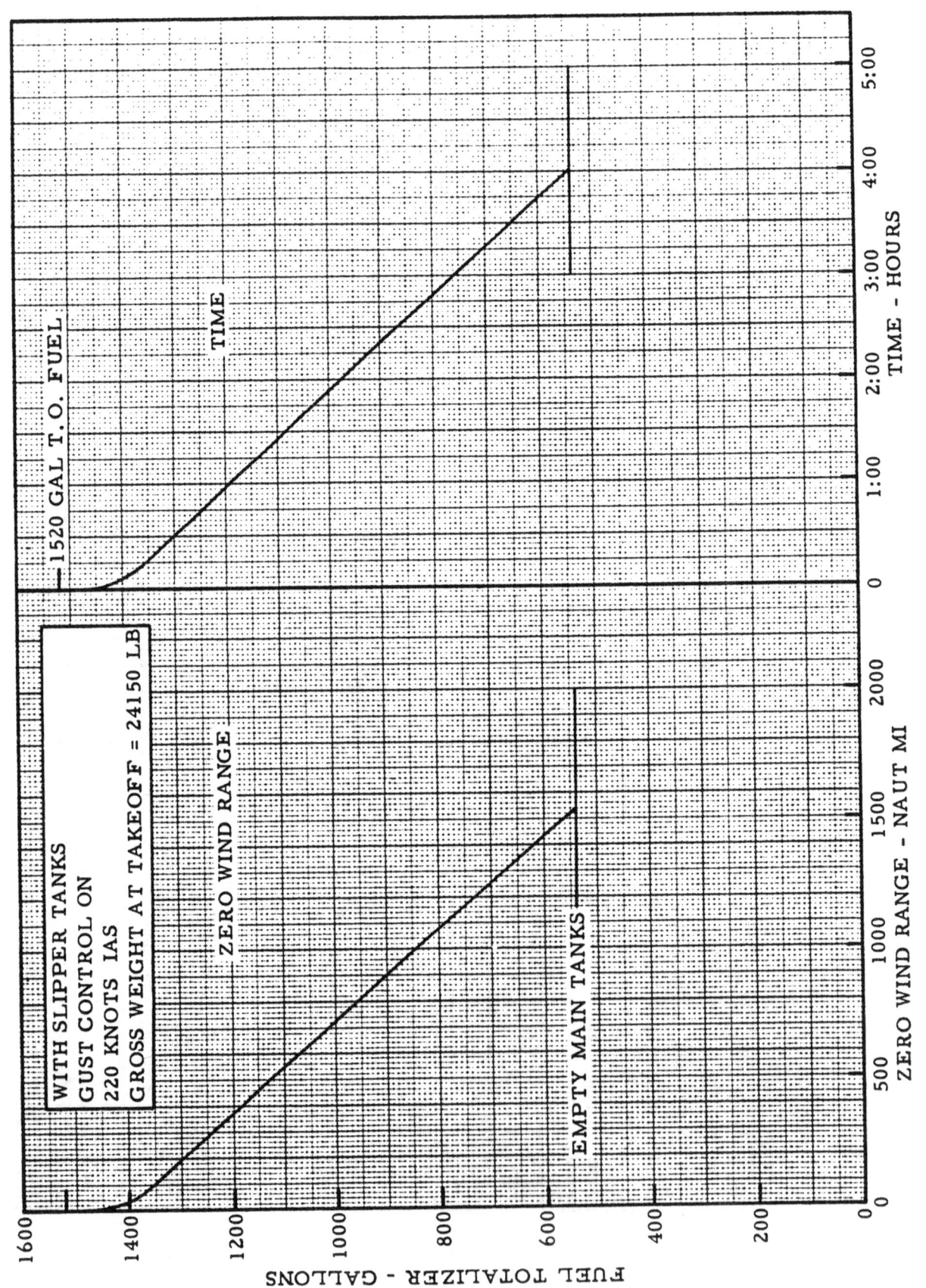

Figure A4-7

AF (C)-1-1
Appendix I
Part 4 - Normal Cruise

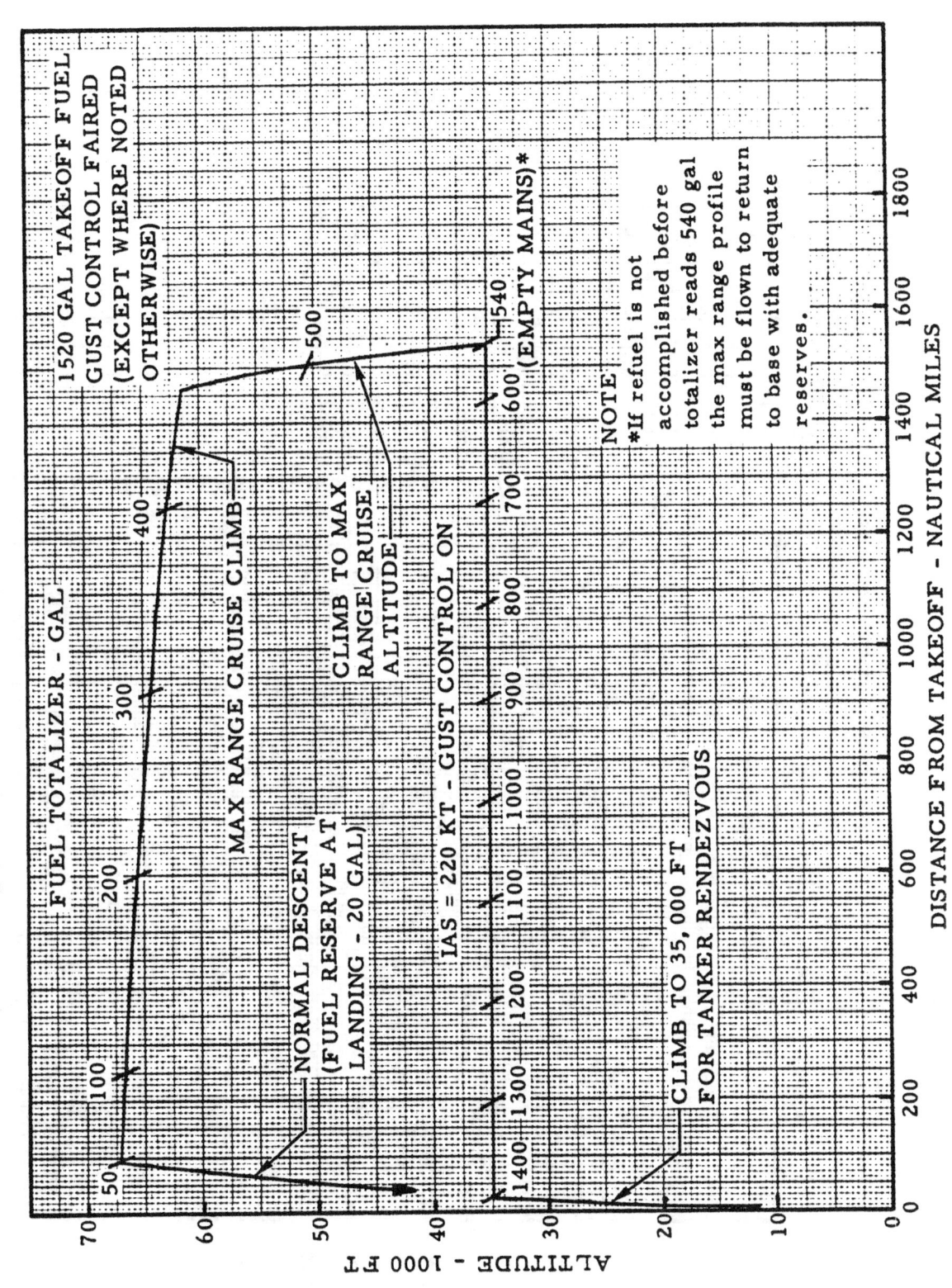

Figure A4-8

OXYGEN CONSUMPTION

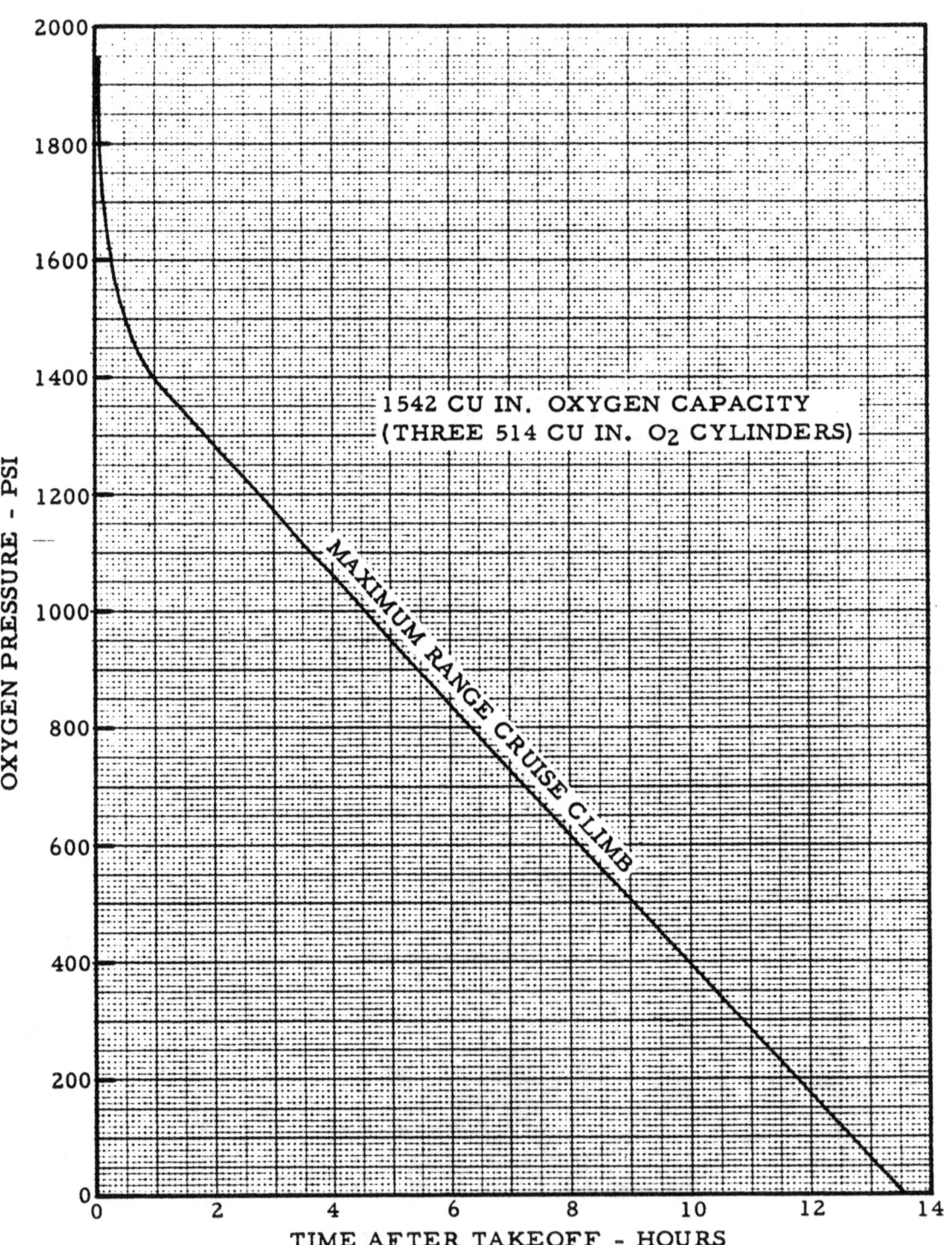

Figure A4-9

AF (C)-1-1

Appendix I
Part 4 - Normal Cruise

OXYGEN CONSUMPTION

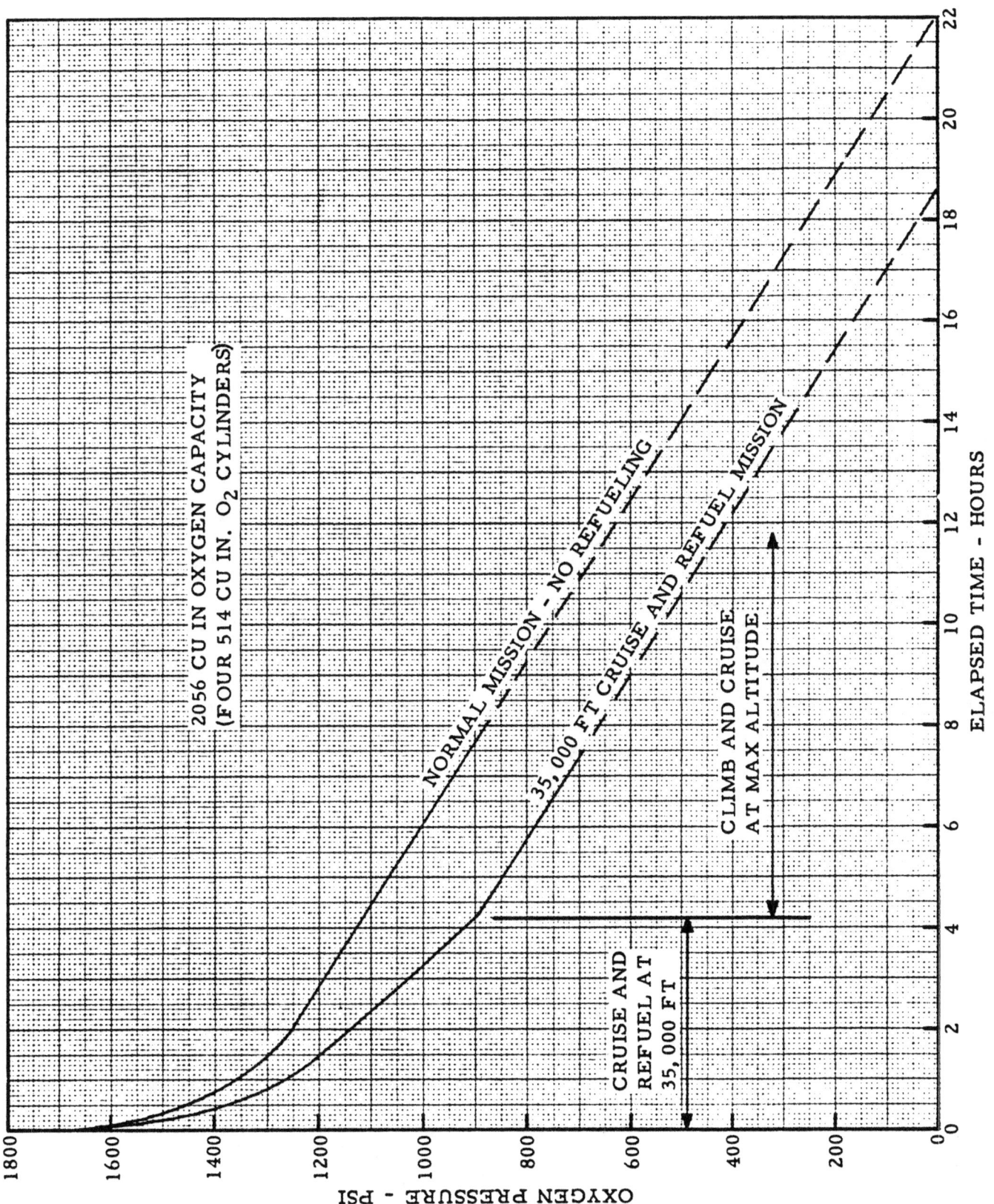

Figure A4-10

PART 5 - ALTERNATE CRUISE

TABLE OF CONTENTS

Subject	Page
Alternate Cruise	A5-1
Effect of Altitude	A5-1
Effect of Airspeed	A5-1
Effect of Gust Control	A5-2
Range Calculation	A5-2
Effect of Outside Air Temperature	A5-2
Effect of Gross Weight	A5-2
Effect of Wind	A5-2
Missed Approach	A5-3

Charts	Page
Nautical Miles per Gallon of Fuel, Altitude 10,000 Feet	A5-4
Nautical Miles per Gallon of Fuel, Altitude 20,000 Feet	A5-5
Nautical Miles per Gallon of Fuel, Altitude 30,000 Feet	A5-6
Nautical Miles per Gallon of Fuel, Altitude 45,000 Feet	A5-7
Nautical Miles per Gallon of Fuel, Altitude 60,000 Feet	A5-8
Optimum Range from Missed Approach - Starting at 10,000 Feet	A5-9

ALTERNATE CRUISE

Alternate cruise information with no slipper tanks is shown in figures A5-1 through A5-5 for altitudes from 10,000 feet to 60,000 feet. Cruise information at 45,000 feet is provided for flights without the pressure suits. Information at 20,000 and 30,000 feet is provided because this altitude range is about the highest that is practical for cruising without oxygen. This corresponds to a cabin altitude of 9,000 to 15,000 feet.

Information for this was obtained with a P-13A engine but should be reasonably accurate for the P-13B engine.

The charts are based on a gross weight of 15,380 pounds.

EFFECT OF ALTITUDE

Within limits, fuel economy improves with increased altitude. As can be seen from the charts, the maximum miles per gallon at 60,000 feet is about 4-1/2 times better than at 10,000 feet.

EFFECT OF AIRSPEED

The cruise information is presented in terms of fuel economy (miles per gallon) versus indicated airspeed with a subscale of standard day true airspeed. Maximum range is obtained by flying at the airspeed for maximum miles per gallon. However, at low and intermediate altitudes it is possible to fly at speeds appreciably faster than the maximum range speed with only a very small decrease in miles per gallon.

The entire speed range may be flown with a relatively small RPM variation, particularly above 45,000 feet. Therefore, careful RPM adjustments are required to establish the proper cruise speed. Due to differences in engine trim speeds and variations in airplane gross weight and outside air temperature, the RPM for maximum range speed may vary by several percent. Vary the RPM as required to establish the speed.

EFFECT OF GUST CONTROL

With the gust control in the GUST position, the maximum miles per gallon are obtained at a faster speed than with the gust control faired. Therefore, it is advantageous to cruise with the gust control in GUST position at altitudes of 55,000 feet and below. This permits a faster cruise speed in either smooth or rough air with no sacrifice in fuel economy. The airspeed limits specified in Section V should be observed.

RANGE CALCULATION

The range can be calculated as follows, using the miles per gallon from the appropriate chart:

Range = Mi/gal x gal remaining.
= Mi/gal x totalizer reading.

EFFECT OF OUTSIDE AIR TEMPERATURE

The miles per gallon will not change appreciably due to variations in outside air temperature. On a hot day, higher RPM and increased fuel flow will be required to maintain a given indicated airspeed. However, true airspeed will also be faster and will compensate for the increased fuel flow. On a cold day, lower RPM and decreased fuel flow will be required for the same indicated airspeed. However, true airspeed will also be slower and the miles per gallon will not change significantly.

EFFECT OF GROSS WEIGHT

The miles per gallon charts are based on a gross weight of 15,380 pounds. At heavier weights, higher power is required for a given indicated airspeed and miles per gallon are reduced. As weight decreases, the power must be reduced to hold constant airspeed and miles per gallon will increase. If desired, the effect of decreasing gross weight can be estimated by increasing the fuel economy 3% for each 1500 pounds of gross weight.

EFFECT OF WIND

The cruise information is presented for the zero wind condition. With zero wind, the air miles per gallon will equal the ground miles per gallon. At a constant indicated airspeed winds will not affect the air miles per gallon, but the ground miles per gallon will change. The ground miles per gallon will be increased by a tailwind and decreased by a headwind. The range will be changed by the same percentage as the ground miles per gallon. The following factors can be used to approximate these effects.

Tailwind

For each 10 knot increase in tailwind:

1. Increase miles per gallon 5% at 10,000 feet altitude.

2. Increase miles per gallon 3-1/2% at 30,000 feet altitude.

3. Increase miles per gallon 2-1/2% at 60,000 feet altitude.

Headwind

For each 10 knot increase in headwind:

1. Decrease miles per gallon 5% at 10,000 feet altitude.

2. Decrease miles per gallon 3-1/2% at 30,000 feet altitude.

3. Decrease miles per gallon 2-1/2% at 60,000 feet altitude.

The effect of wind should always be checked using the distance, time and fuel consumed between known check points.

MISSED APPROACH

When a landing at the planned destination is not possible and a missed approach is made starting at 10,000 feet, figure A5-6 can be used to determine range capability at various conditions of available fuel and at various level off altitudes. The figure also lists IAS for best fuel economy at each 5000-foot altitude increment, and descent distances. "Available fuel" is defined as totalizer reading at start of missed approach minus the desired amount of reserve fuel at the alternate. Figure A5-6 can be used in either of two ways: to determine optimum range at optimum altitude for the available fuel aboard, or to determine range at any altitude. The following examples illustrate use of the figure.

Example No. 1

After arrival over the planned destination at 10,000 feet with 200 gallons on the totalizer, it is found necessary to fly to an alternate. How far can you fly to an alternate and land with 50 gallons of reserve fuel? Available fuel is 150 gallons (200 - 50 gallon reserve). Figure A5-6 shows that the most range for 150 gallons of available fuel is 214 nautical miles and is obtained by climbing to 50,000 feet, flying level at an IAS of 136 knots, and starting a normal descent 55 miles from the alternate.

Example No. 2

Under the same conditions as Example No. 1, how far can you fly to an alternate if you climb to and level off at 25,000 feet (still landing with 50 gallons reserve)? Available fuel is still 150 gallons. Figure A5-6 shows that the range at 25,000 feet with 150 gallons of available fuel is 174 nautical miles and is obtained by maintaining an IAS of 176 knots in level flight at 25,000 feet and starting the descent 25 miles from the alternate.

Example No. 3

Under the same conditions as Example No. 1, how far can you fly if landing at the alternate is to be made with 25 gallons of reserve fuel? Available fuel is now 175 gallons (200 -25). Figure A5-6 shows that a range of 300 nautical miles can be obtained by climbing to the optimum altitude of 60,000 feet, flying level at an IAS of 122 knots, and starting descent 75 miles from the alternate.

Appendix I
Part 5 - Alternate Cruise
AF (C)-1-1

NAUTICAL MILES PER GALLON OF FUEL
ALTITUDE 10,000 FEET

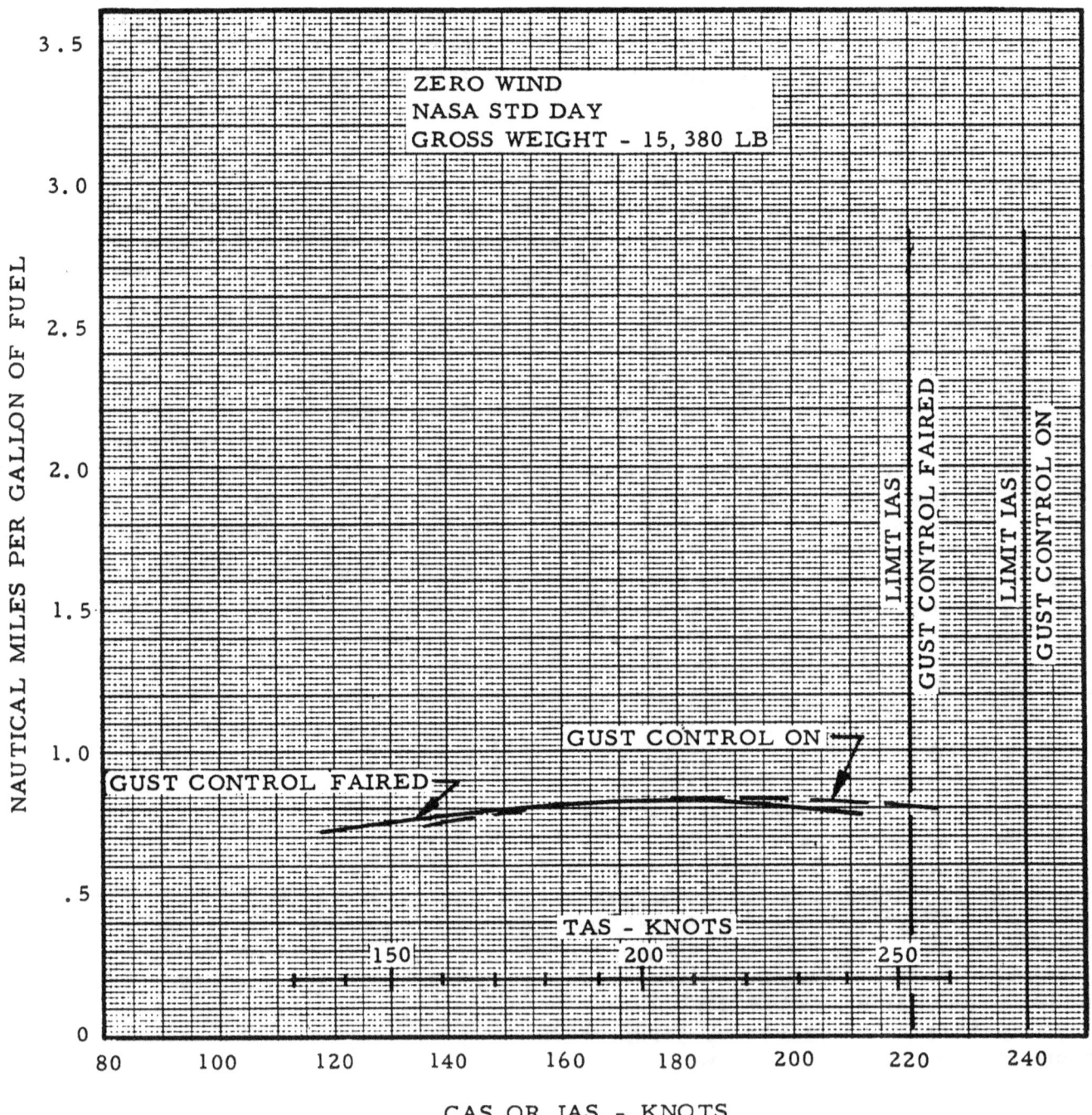

Figure A5-1

AF (C)-1-1 Appendix I
Part 5 - Alternate Cruise

NAUTICAL MILES PER GALLON OF FUEL
ALTITUDE 20,000 FEET

WITHOUT SLIPPER TANKS

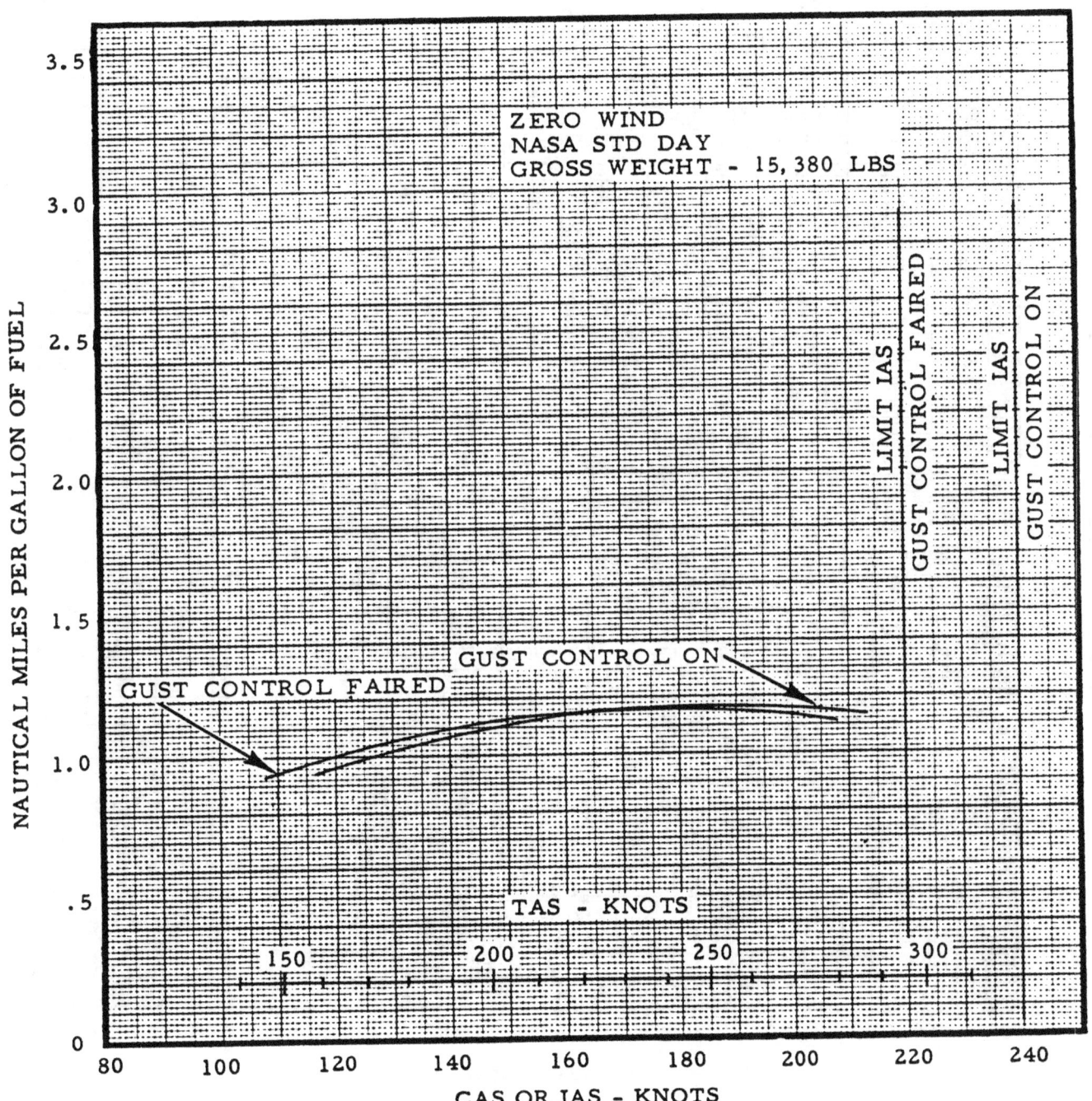

Figure A5-2

Appendix I AF (C)-1-1
Part 5 - Alternate Cruise

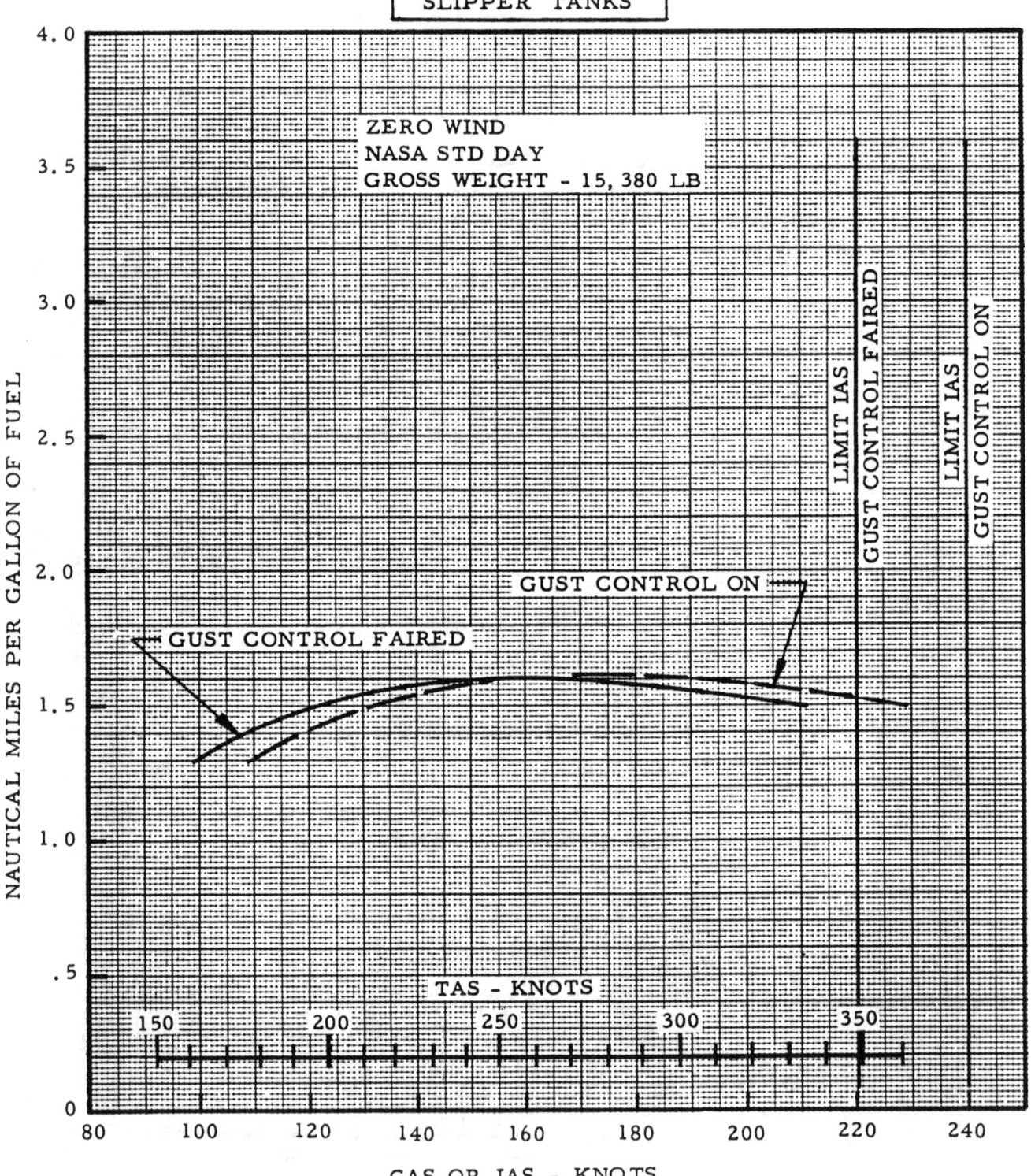

Figure A5-3

NAUTICAL MILES PER GALLON OF FUEL
ALTITUDE 45,000 FEET

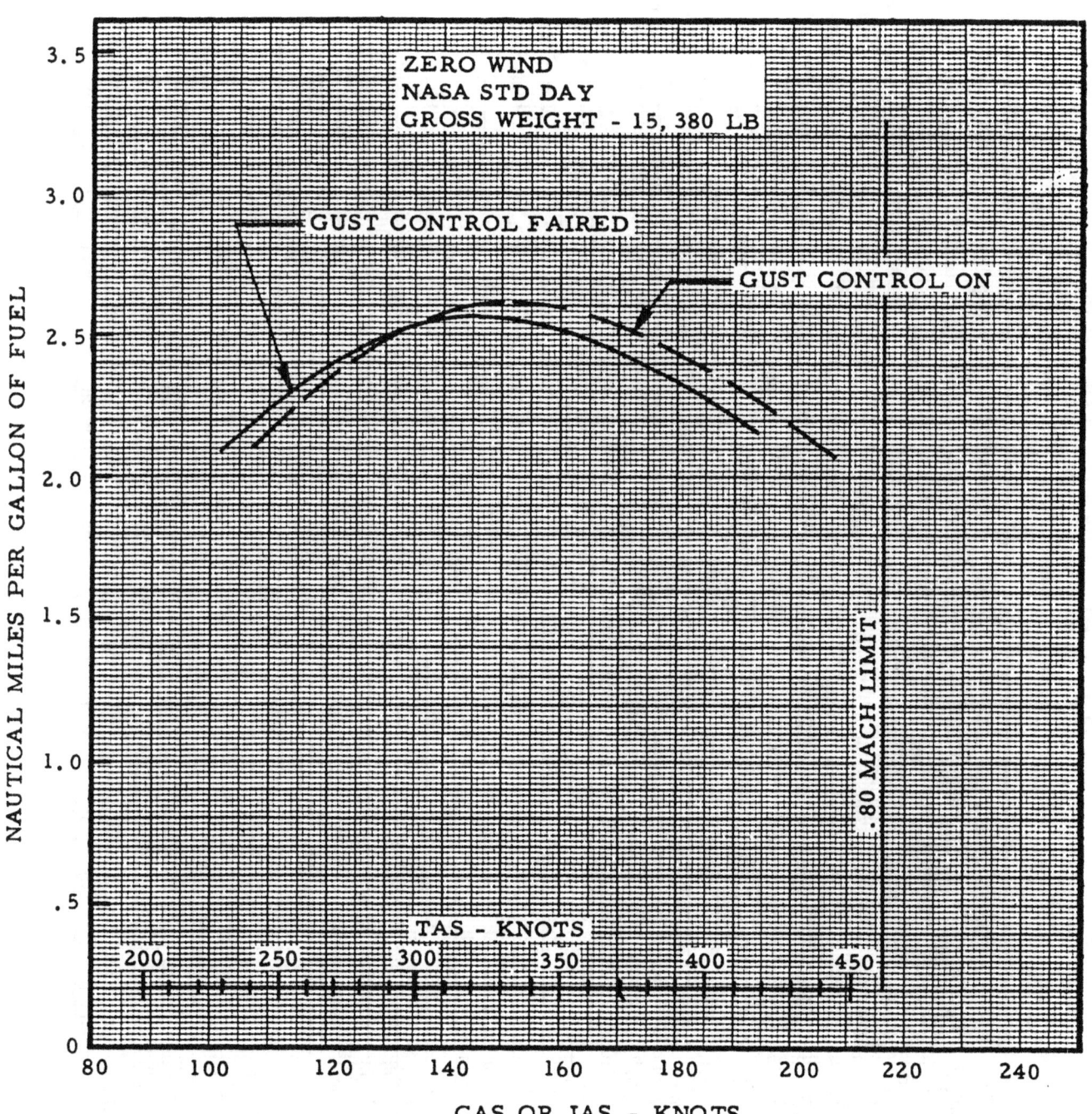

Figure A5-4

Appendix I
Part 5 - Alternate Cruise

AF (C)-1-1

NAUTICAL MILES PER GALLON OF FUEL ALTITUDE 60,000 FEET

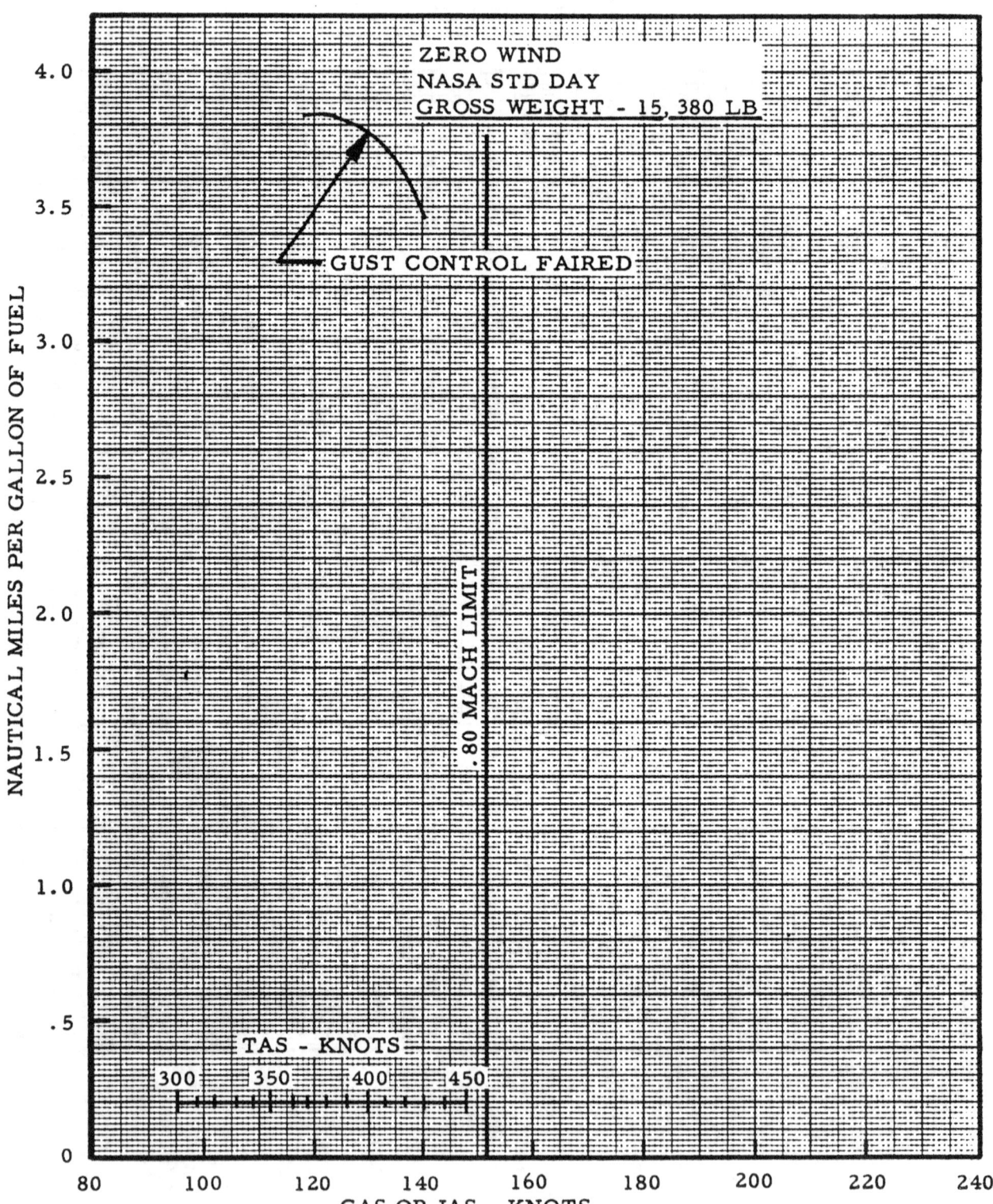

Figure A5-5

MODEL U-2

OPTIMUM RANGE FROM MISSED APPROACH
STARTING AT 10,000 FEET

NOTE:
1. AVAILABLE FUEL DOES NOT INCLUDE RESERVE FUEL. (STARTING TOTAL MINUS RESERVE FUEL = AVAILABLE FUEL.)
2. RANGE INCLUDES CLIMB, LEVEL FLIGHT, AND DESCENT DISTANCES.
3. CLIMBS AND DESCENTS SHOULD BE MADE ON NORMAL SPEED SCHEDULES.
4. ZERO WIND.

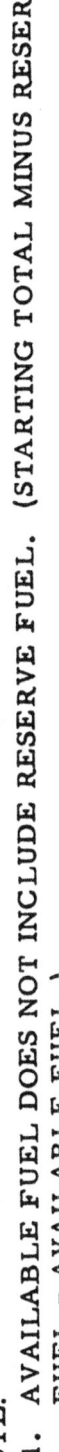

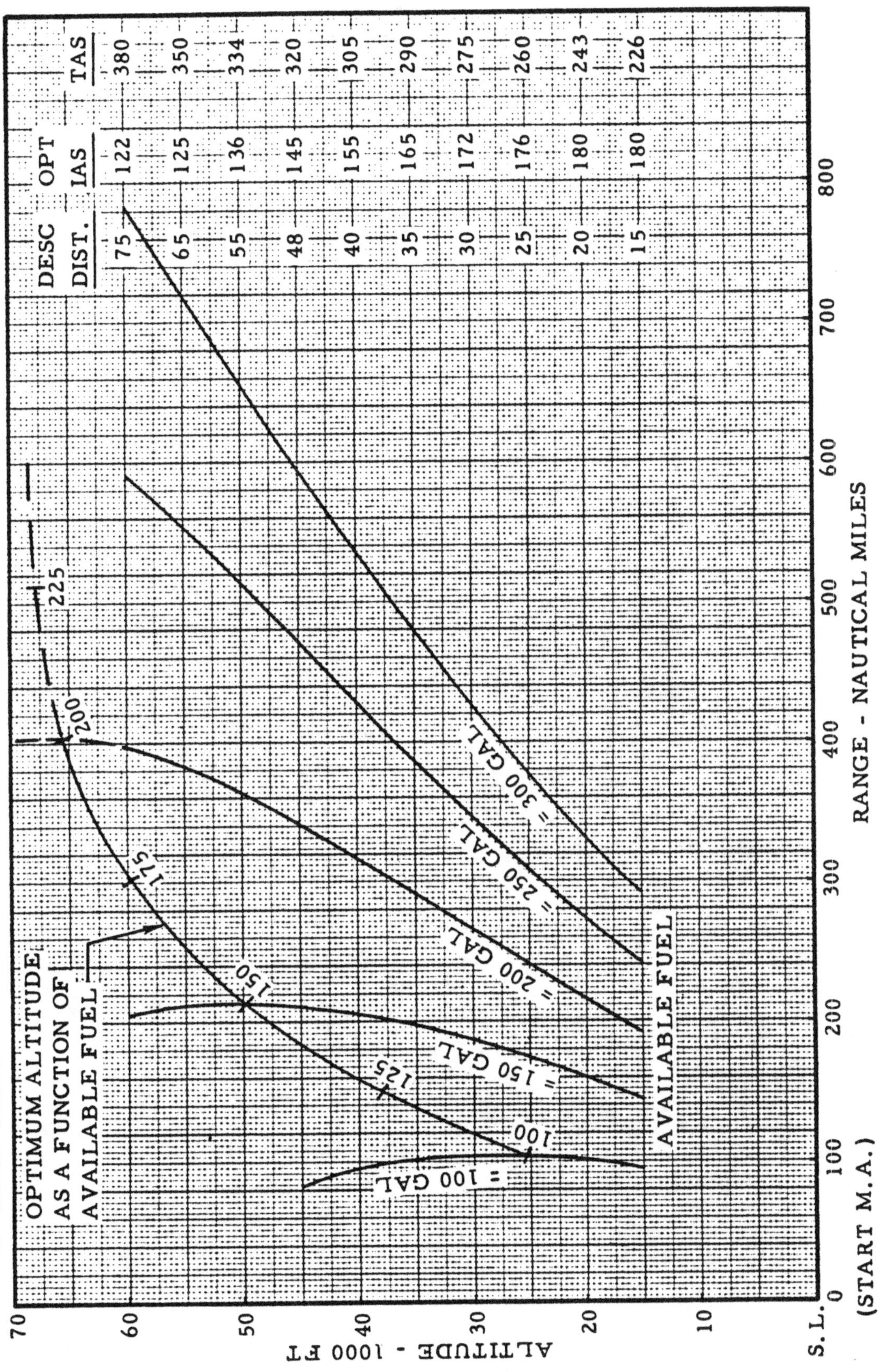

Figure A5-6

PART 6 - MAXIMUM ENDURANCE

TABLE OF CONTENTS

Subject	Page
Maximum Endurance	A6-1
Fuel Consumption	A6-1
Maximum Endurance Speed	A6-1
Power Setting	A6-1

Charts	Page
Maximum Endurance, Without Slipper Tanks	A6-2
Low Altitude Fuel Consumption, Without Slipper Tanks	A6-3

MAXIMUM ENDURANCE

FUEL CONSUMPTION

Maximum endurance is obtained by flying at the slowest practical airspeed at which the desired altitude can be maintained. The recommended airspeed (less than 600 gallons of fuel) is 115 knots IAS, as shown in the maximum endurance chart, figure A6-1. Approximate RPM settings and fuel consumption are also shown. The fuel consumption and approximate RPM settings for holding at low altitude and various airspeeds are shown in figure A6-2. The chart is based on a gross weight of 15,380 pounds and no slipper tanks. The fuel consumption will be increased slightly with slipper tanks installed.

MAXIMUM ENDURANCE SPEED

The recommended airspeed is somewhat faster than the speed for minimum power required. However, very little endurance is sacrificed because in this range there is a relatively large speed change for a very small change in power. This characteristic also makes it difficult to fly at the minimum power required speed. Therefore, 115 knots is recommended for ease of flying with very little increase in fuel consumption.

POWER SETTING

With the P-13 engine, the RPM is above idle up to 50,000 feet; idle RPM can be used at 55,000 feet.

Appendix I
Part 6 - Maximum Endurance

AF (C)-1-1

MAXIMUM ENDURANCE

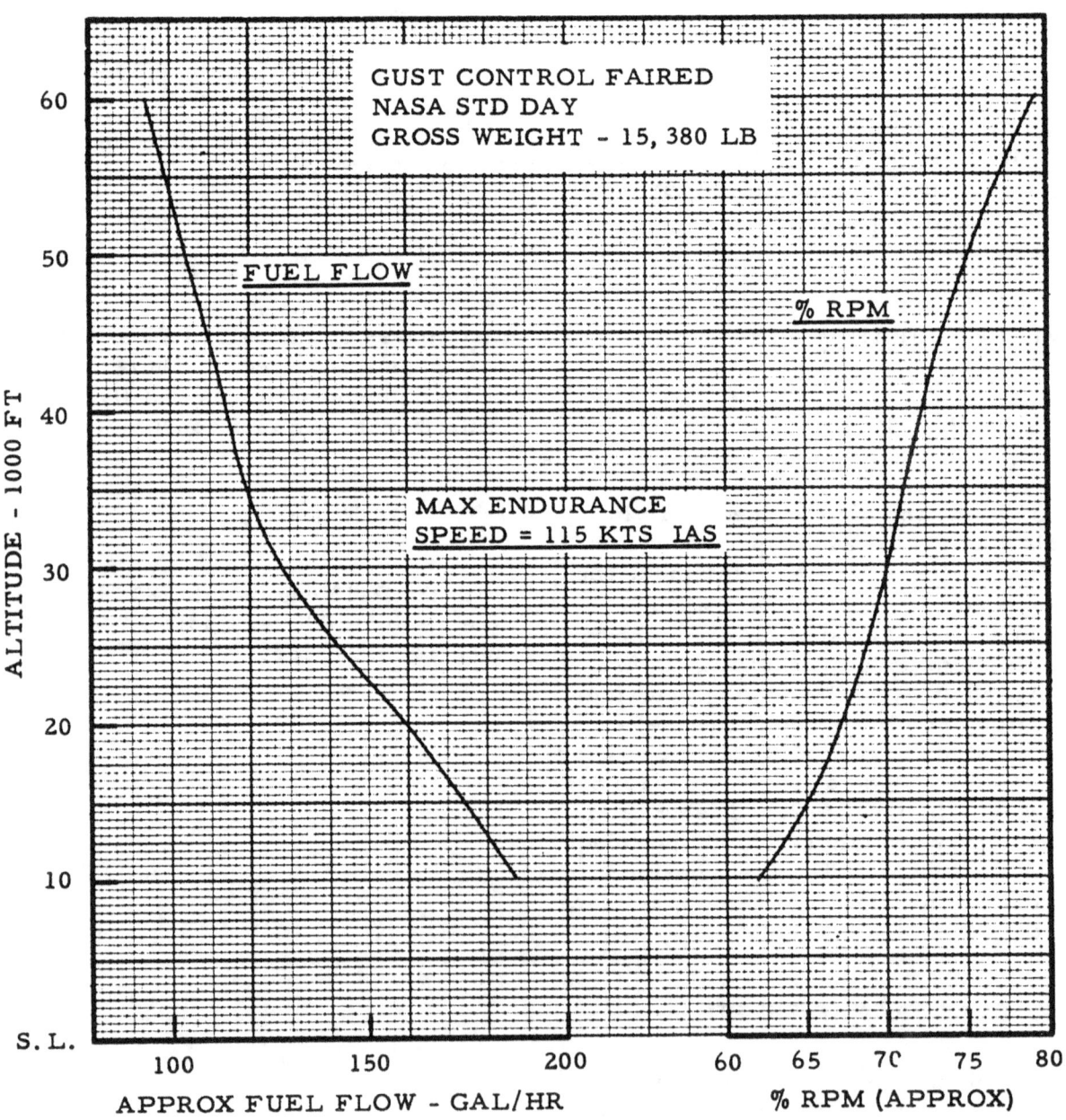

Figure A6-1

AF (C)-1-1

Appendix I
Part 6 - Maximum Endurance

LOW ALTITUDE FUEL CONSUMPTION

WITHOUT SLIPPER TANKS

GUST CONTROL FAIRED
GROSS WEIGHT - 15,050 LB

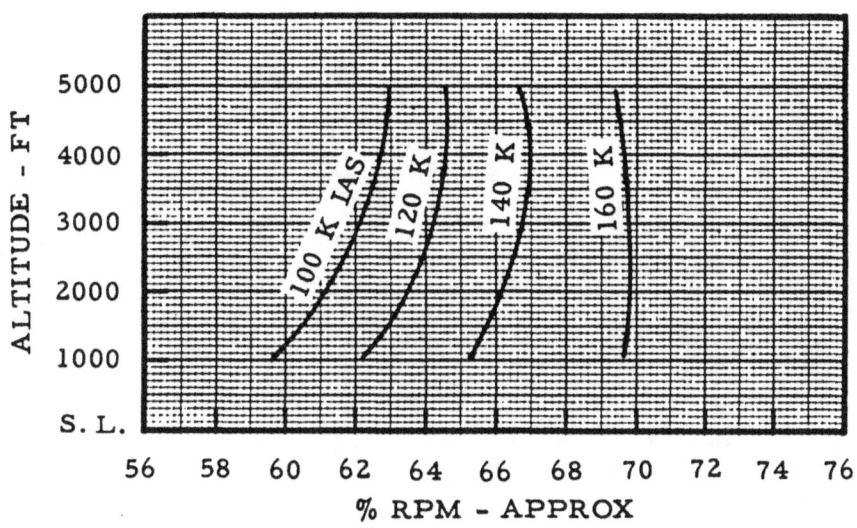

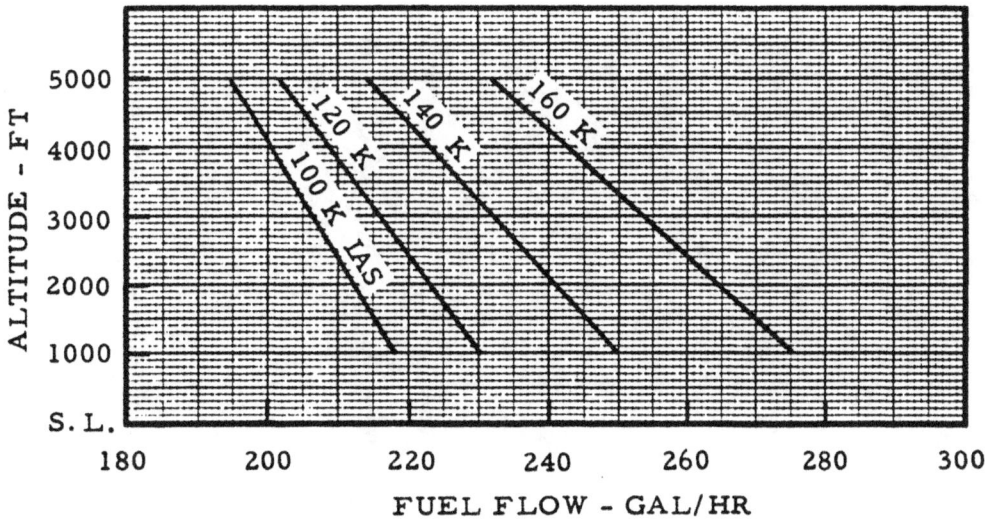

Figure A6-2

Changed 6 November 1967

AF (C)-1-1

Appendix I
Part 7 - Descent

PART 7 - DESCENT

TABLE OF CONTENTS

Subject   Page

  Descent A7-1

Charts   Page

  Descent Performance A7-2

DESCENT

The descent charts can be used to estimate zero wind distance, fuel consumption and time during fast and normal descents. Figure A7-1. Dead engine descent distances are shown in Section III.

The performance shown in the charts is based on actual flight test conditions and not on an NASA standard day. However, variations from standard day conditions will have only small effects and can be neglected. The descent performance is primarily dependent upon engine thrust, airplane configuration and airspeed. Thrust is determined by the engine speed, exhaust gas temperature and idle fuel flow. The airplane configuration and airspeed are dependent upon the descent technique and the performance can vary considerably with different techniques.

Changed 6 November 1967

A7-1

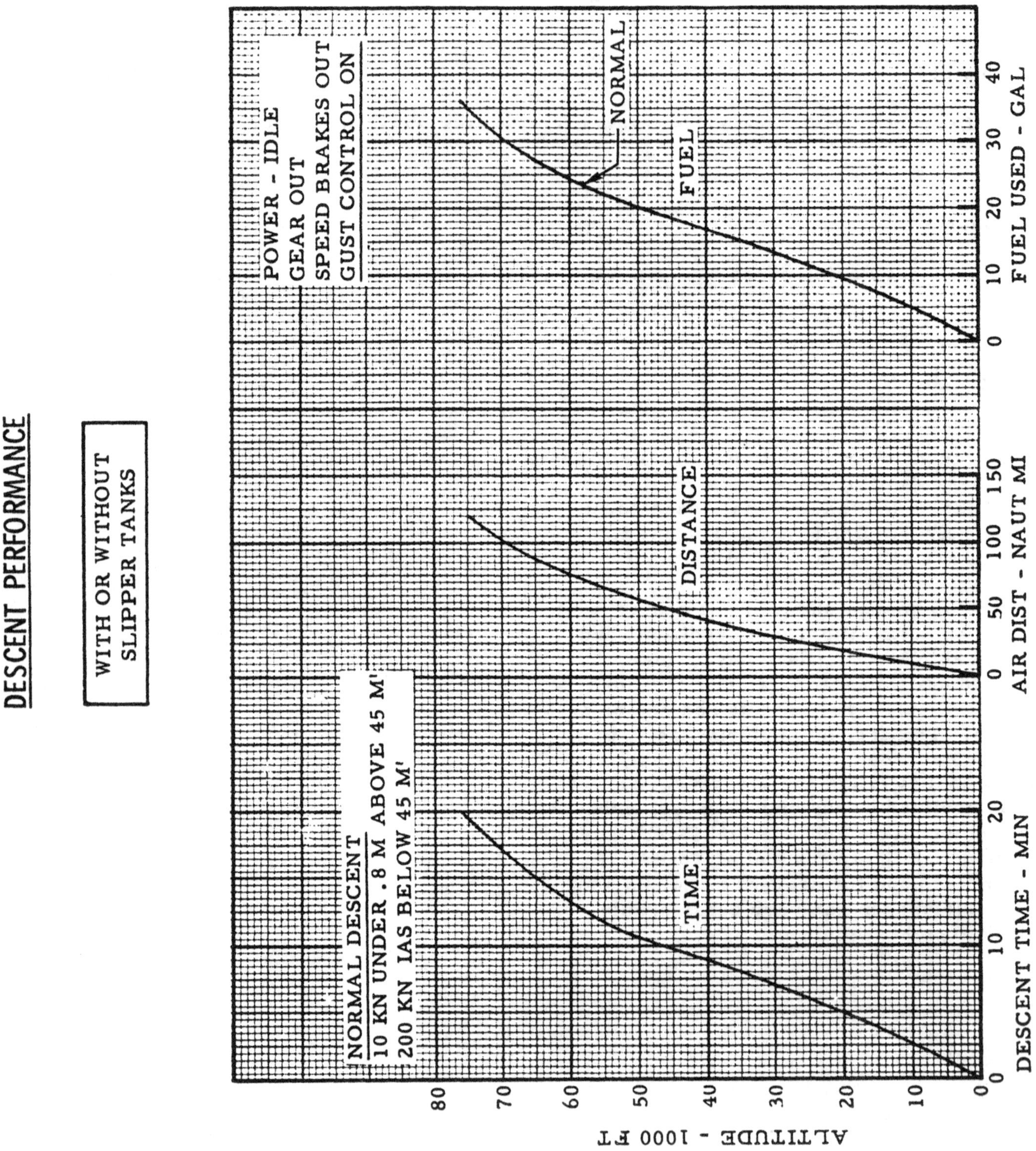

Appendix I
Part 7 - Descent

AF (C)-1-1

Figure A7-1

PART 8 - LANDING

TABLE OF CONTENTS

Subject	Page
Landing	A8-1
Landing Distance	A8-1
Effects of Thrust and Drag	A8-1
Effect of Wing Flaps	A8-2
Threshold Speed and Altitude	A8-2
Gross Weight and Braking	A8-2

Charts	Page
Landing Distance	A8-3
Threshold Speeds - Smooth Air	A8-4

LANDING

LANDING DISTANCE

Representative landing distances are tabulated in figure A8-1. The charts indicate the effects of the primary variables which influence the landing distance. The landing procedures are outlined in Section II.

The landing distances are relatively long because the airplane is very clean aerodynamically and engine idle thrust is appreciable. Major factors affecting the landing distance are as follows:

(a) Thrust.

(b) Drag.

(c) Wing Flap Position.

(d) Threshold Speed.

(e) Threshold Altitude.

(f) Braking.

(g) Gross Weight.

EFFECT OF THRUST AND DRAG

The idle thrust and low drag affect both the float distance and stopping distance. Although it is generally desirable to leave the engine running, the distance will be appreciably shorter if the engine is shut off at the threshold.

The speed brake drag is not high at landing speed, however, extending the speed brakes does shorten the distance somewhat. The drag chute is the most effective means of reducing both float distance and stopping distance.

Appendix I
Part 8 - Landing

AF (C)-1-1

EFFECT OF WING FLAPS

Wing flap setting will also affect both the float distance and stopping distance. The shortest distances are with 35 degrees of flap. Since the flap drag is appreciable, stopping distances will be shorter if the flaps are left down during the ground roll.

THRESHOLD SPEED AND ALTITUDE

The threshold speed and altitude affect the float distance, and both should be controlled closely to shorten this distance. The smooth air threshold speeds are shown in figure A8-2. The stall speed margin is 15% with less than 15° flaps, 12% with 25° flaps, and 14% with 35° flaps. For an emergency landing with flaps faired, where it is desirable to have a slow but safe speed, the stall margin is 10%.

GROSS WEIGHT AND BRAKING

The total landing distance is longer at heavier gross weight. The increased distance is in the ground roll.

Braking techniques are outlined in Section II. Differences in braking will cause variations in the landing ground roll. The braking technique during the landings is indicated in the charts.

POWER ON LANDING DISTANCE

ZERO FUEL WEIGHT = 14000 LB

SEA LEVEL - ZERO WIND

Wing Flaps	Drag Chute	Speed Brakes	Power	Fuel Gallons	Threshold Speed Knots	Threshold Altitude Feet	Braking	Float Distance (From Threshold) Feet	Stopping Distance Feet	Total Distance Feet
35°	Without	Out	Idle	300	86	5	Light at 60 knots Heavy at 40 knots	975	3360	4335
35°	With	Out	Idle	300	86	10	Light at 50 knots Heavy at 40 knots	750	2070	2820
0° *	With	In	Off after chute deploys	300	90	10	Light at 30 knots	1000	2880	3880
0° *	Without	In	Off at threshold	300	90	5	Light at 60 knots Moderate at 50 knots Heavy at 30 knots	1630	4760	6390

* Considered emergency condition

Figure A8-1

Appendix I
Part 8 - Landing

AF (C)-1-1

THRESHOLD SPEEDS-SMOOTH AIR

WITH OR WITHOUT SLIPPER TANKS

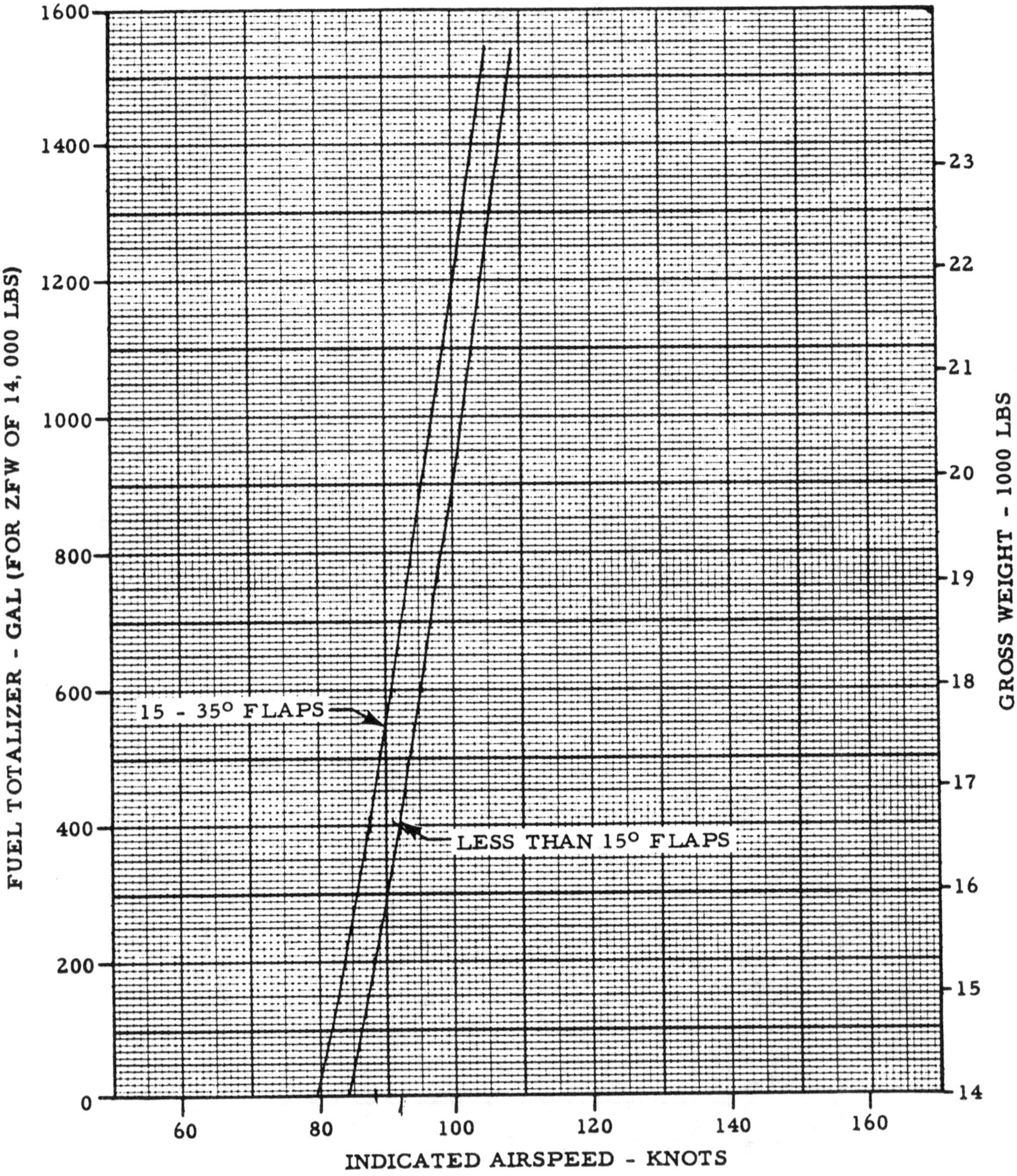

Figure A8-2

A8-4

PART 9 - STALL SPEEDS

TABLE OF CONTENTS

Subject	Page
Stall Speeds	A9-1
Effect of Flaps	A9-1

Charts	Page
Stall Speeds, P-13 Engine, With or Without Slipper Tanks	A9-2

STALL SPEEDS

Stall speed varies as a function of gross weight as shown in figure A9-1. Stall speeds are essentially the same with or without slipper tanks at a given gross weight. Figure A9-1 also shows stall speeds versus fuel totalizer for the specific case of a zero fuel weight of 14,000 pounds. The zero fuel weight can vary greatly due to differences in equipment loadings. Therefore, for cases where the zero fuel weight is significantly different from 14,000 pounds, stall speeds should be determined using the gross weight scale rather than the fuel totalizer scale.

In addition to wing root stall warning strips, the airplane has retractable stall strips which are extended in order to minimize unsymmetrical wing drop at the stall. These stall strips have a negligible effect on stall speeds.

EFFECT OF FLAPS

The stall speeds are shown for faired flaps, 15-, 25-, and 35-degree flap settings. The reduction in stall speed is only about 3 knots from 15° to 35° flaps. Therefore, a flap setting anywhere in this range can be selected for landing with little change in touchdown speed. However, as shown in Part 8 of Appendix I, the landing distances are appreciably shorter with 35° flaps due to the greater drag.

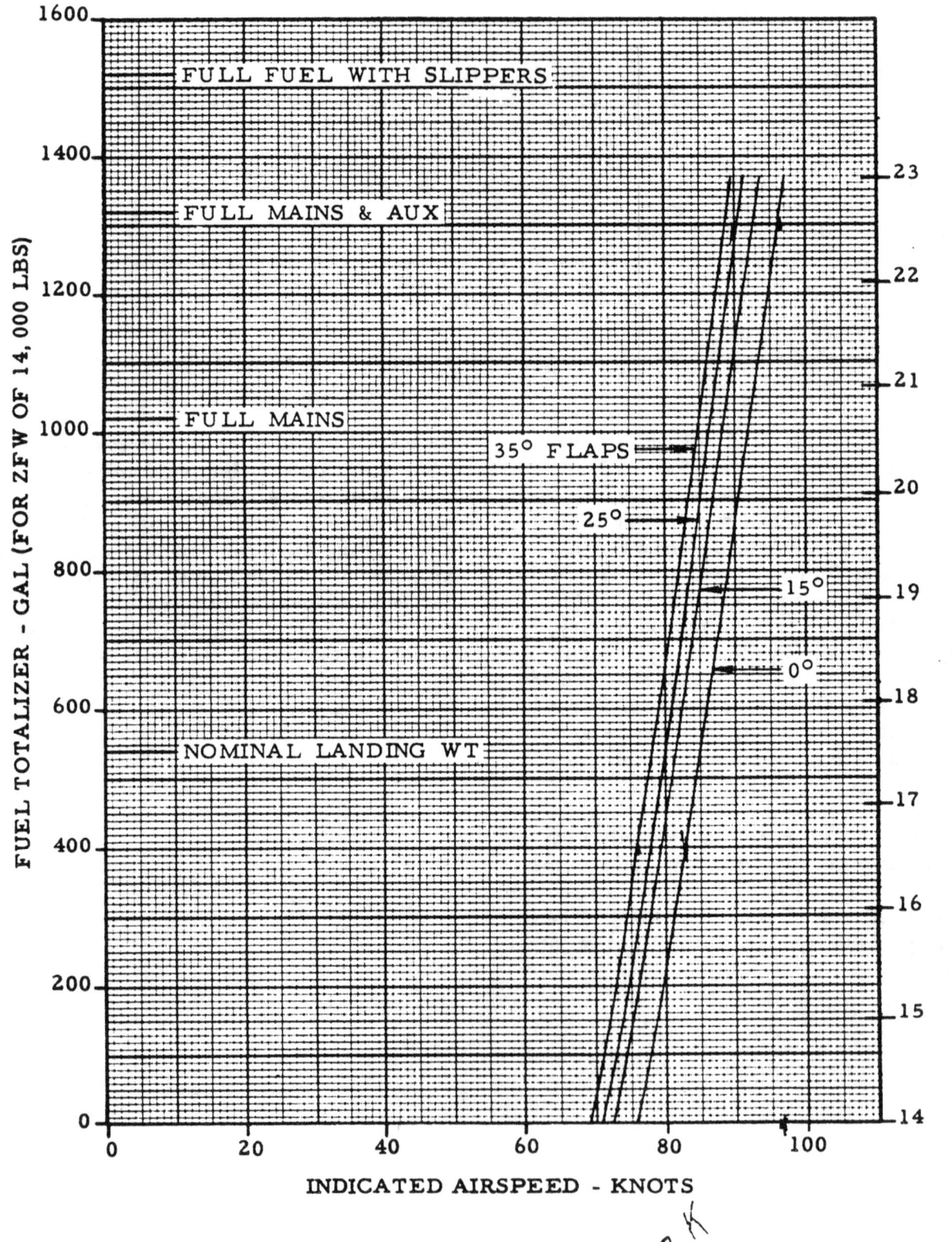

Figure A9-1

ALPHABETICAL INDEX

A

Abort, Takeoff	3-4
AC Generator	
Failure	3-34
Out Warning Light	1-24
Switch	1-24
System	1-19, 3-34
Acceleration Limitations	5-4
Adverse Yaw	6-5
Aerial Refueling (ARS)	4-29, 4-30*
Day Operation	4-34
Night Operation	4-38
Emergency	4-38
Receptacle	4-30*
Test Panel	4-31*
After Climb - Takeoff	9-8
After Landing	2-33, 9-9
After Parking	2-34
After Start	2-9
After Takeoff	2-13
After Touchdown	6-15
Aileron	6-2
Control System	1-27
Electric Tab	6-6
Trim Index	4-6, 4-6*
Air Conditioning and Pressurization	
System	4-7
Air Conditioning	
Malfunction	3-37
Automatic Temperature Control	3-37
Pressurization System Emergency	
Operation	3-38
System Schematic	4-8*
Air Ram Switch	4-10
Air Traffic Control (ATC)	4-43, 4-44*
Air Turbulence	2-13
Aircraft Attitude	2-14
Aircraft Control	2-16
Aircraft Fuel System	1-11

Aircraft Fuel System Malfunction	3-29
Aircraft Oxygen System Difficulty	3-28
Aircraft, The	1-1
Aircraft Trim	2-19
Airspeed Acceleration	6-7
Airspeed Control	2-18
Airspeed Indicator	1-32
Airspeed Limitations	5-4
Airstart	3-7
Emergency System	3-8
Low Altitude	3-10
Normal	3-8
All Weather Operation (See SECTION IX)	
Altimeter(s)	1-32
Pressurization	4-11
Altitude(s)	
Control	2-18
Level Off	2-21
Power Regulation at	5-2, 7-3
Ammeter	
Generator	1-20
High Reading	3-32
AN/APN-135 Rendezvous Beacon	4-46, 4-46
AN/APN-153(V) Control Panel	4-57*
AN/ARC-3 VHF Radio	4-47, 4-48*
AN/ARC-34 UHF Transceiver	4-40, 4-40*
AN/ARN-59 Radio Compass	4-42, 4-42*
Annunciator Warning Panel	1-20, 1-23*
Approach	6-14
Ground Controlled (GCA)	9-4, 9-5*
Final	2-28, 3-15
Initial	2-25
Missed	9-4
ARC Type 12 VHF Transceiver	4-47, 4-47*
ARC Type 15F VHF Navigation	4-42, 4-43*
Arrangement, Cockpit	1-33* thru 1-37*
ASN-66 Navigation Control	4-57*
ATC System	4-43, 4-44*
Attitude	
Aircraft	2-14
Ejection	3-25
Heading Reference System	4-1
Indicator Type J-8	1-44

Page Numbers with Asterisk (*) Denote Illustrations

INDEX AF (C)-1-1

Automatic Flight Control 4-4, 4-6*
    Failure 3-37
Automatic Observer 4-48
Automatic Pitch Trim 4-5
Automatic Temperature Control
    Cabin Heat 4-10
Autopilot 4-4, 4-6*
Auxiliary Defrosting System 4-12
Auxiliary Equipment   1-47, 4-1 thru 4-59
Auxiliary Fuel Boost Pump ... 1-17, 7-4
Auxiliary Gear 1-35
Auxiliary Tank Pressurization
    Switch 1-13
Appendix I, Performance Data A-1
    Part 1 Airspeed Corrections A1-1
    Part 2 Takeoff A2-1
    Part 3 Climb A3-1
    Part 4 Normal Cruise A4-1
    Part 5 Alternate Cruise A5-1
    Part 6 Maximum Endurance A6-1
    Part 7 Descent A7-1
    Part 8 Landing A8-1
    Part 9 Stall Speeds A9-1
Approximate Gross Weights 1-1

B

Bailout Without Ejection Seat 3-25
Basic Dimensions 1-1
Battery and Generator Switch 1-20
Battery Discharge Warning Light ... 1-23
Beacon - Rendezvous
    AN/APN-135 .... 4-46, 4-46*
Bearing-Distance-Heading Indicator
    (BDHI) 1-44, 4-55*
Before Entrance Inspection 2-2
Before Landing 2-25
Before Takeoff 2-10, 9-2, 9-8
Before Taxiing 2-9
Boost Pump Auxiliary Fuel 1-17
Boost Pump Normal Fuel 1-17

Brake(s)
    Control 1-29
    Diagram 1-31*
    Speed 1-29, 6-12
    Wheel 1-30, 1-31*
Buffet - Wing 6-7
B/400 Rate Rate Meter 4-50, 4-50*

C

Cabin Cooler
    Bypass Valve Failure 3-38
    Failure 3-38
Cabin Heat
    Automatic Temperature Control .. 4-10
    Selector 4-7
Canopy 1-46
    Defrost Control Valve 4-11
    Emergency Jettison 3-23
        Handle 1-46
    Emergency Removal 3-23
    External Opening 3-23
    Hatch Seal Controls 4-11
    Jettison 3-23
    Jettison & Seat Ejection System .. 1-48*
    Manual Release 1-46
CAUTION Definition ii
Changeover 3-11
Characteristics
    Flight 2-20, 6-1
    Flight Control 6-1
    Lateral Trim 6-6
    Mach Number 6-8
Chart
    Cruise 2-20, 2-21*
    Dead Engine Glide Distance 3-11, 3-12*
    Flight Relight   3-9*

Page Numbers with Asterisk (*) Denote Illustrations

Check(s)
   After Landing 2-33
   After Parking 2-34
   After Start 2-9
   Before Landing 2-27
   Before Takeoff 2-10
   Before Taxiing 2-9
   Climb 2-14
   Descent 2-23
      5,000 Ft (Descending) 2-25
   Fuel Dumping Ⓒ 1-16, 2-34
   Fuel Dumping Ⓕ 1-16, 1-35
   Pilot Equipment Check 2-4
      High Flight 2-4
      Low Flight 2-5
   Pilots Cockpit 2-5
   Preflight 2-3
   Preliminary Cockpit 2-3
   Prior to Boarding Aircraft 2-3
   Start Engine 2-9
   Taxiing 2-33

Chute - Drag 1-30
Climb 2-14, 3-11
   Checks 2-14
   Cruise 2-15, 6-13
   Initial 2-14
   Instrument 9-2
   Takeoff and Initial Instrument ... 9-2
   Speed 2-13
Clocks 1-45
Cockpit
   Arrangement .... 1-33* thru 1-37*
      Model U-2C ...... 1-38* thru 1-40*
      Model U-2F ...... 1-41* thru 1-43*
   Fog 3-22
   Fog or Smoke Elimination 3-22
   Lighting 4-18
   Smoke 3-22
Coding Definition iii
Cold Weather Operation 9-7
Collapsible Rudder Pedals 1-27
Comfort - Pilot 2-20, 9-8
Comm Selector Panel 4-39
   Model U-2C 4-39*
   Model U-2F 4-39*

Communication Equipment 4-39
   618T-3 HF SSB Transceiver 4-40
   AN/APN-135
      Rendezvous Beacon 4-46, 4-46*
   AN/ARC-3 VHF Radio ... 4-47, 4-48*
   AN/ARC-34 Control Panel . 4-40, 4-40*
   AN/ARN-59 Control Panel . 4-42, 4-42*
   ARC Type 12 Control Panel 4-47*
   ARC Type 15F Control Panel 4-43*
   ATC Control Panel 4-43, 4-44*
   C-832/AIC-10 Interphone 4-39
   Comm Selector Panel Ⓒ & Ⓕ 4-39*
   Failure 3-39
   Microphone Switches 4-39
Comp Mode 4-3
Compass, Magnetic Standby 1-45
Compass System 4-1
Complete Electrical Failure 3-33
Compressor Stall 7-1

Configuration, General 1-2*
Connector-Microphone and Headset 4-39
Contents, Table of................. v
Continuous Ignition 1-9
Control
   Cabin Heat Automatic
      Temperature 4-10
   Canopy and Hatch Seal 4-11
   Cruise Climb 2-15
      Aircraft 2-16
      Airspeed 2-18
      Attitude 2-18
   Directional 6-15
   Landing Gear 1-29
      Lever 1-29
   Main Fuel 1-4
   Wheel 1-26
Controls, Flight 1-26
   Aileron System 1-27
      Tab 1-27
   Elevator System 1-27
      Tab Position Indicator 1-27
   Gust 1-26, 6-12

Page Numbers with Asterisk (*) Denote Illustrations

INDEX AF (C)-1-1

Controls, Flight (Cont)
    Rudder System 1-27
    Speed Brakes 1-29
    Wing Flaps 1-28
        Position Indicator 1-28
Controllability, Fast Descent 2-24
Counter - Fuel Quantity 1-19
Course Indicator (VOR) 1-44
Cross Transfer Pump Ⓒ & Ⓕ 1-15, 7-4
Crosswind Landing 2-29
Crosswind Takeoff 2-13
Cruise 9-8
    Climb 2-15, 6-13
        Aircraft Control 2-18
        Airspeed Control 2-16
        Cruise Charts 2-20, 2-21*
        Engine Operation 2-17
        Maximum Altitude 2-18
C-823-AIC-10 Interphone Control ... 4-39

D

DC and Inverter Power Distribution . 1-21*
DC Generator Failuer 3-32
DC System 1-19, 3-32
Dead Reckoning Navigation System .. 4-54
Defroster
    Fan 4-12, 4-53
    Switch 4-12
Descent 2-22, 3-11, 6-13, 9-2, 9-8
    Check 2-23
    Controllability 2-23
    Emergency 3-20
    Maximum Range 2-23
        Maneuvers 6-7
    Normal 2-22
Descending 5,000 Ft Check 2-25
Difficulty, Oxygen 3-27
    Aircraft System 3-28
Dimension Basic 1-1
Directional Control 6-15
Directional Gyro Mode 4-3
Directional Stability 6-4
Disconnect - Seat Pack 4-15
Distribution, Electrical Power
    AC Generator Power 1-22*
    DC and Inverter Power 1-21*
Ditching Procedures 3-20

Downward Maneuvers
    Airspeed Acceleration 6-7
Downwind Leg, Before Landing ... 2-27
Drag Chute 1-30
    Landing with Deployment 2-30
Driftsight System 4-19, 4-20*
Drop Tank(s) 7-5
    Installation Ⓒ & Ⓕ 1-15
    Malfunction 3-31
        Does not Feed 3-31
        Does not Release 3-32
Dump - Fuel 2-34
    Model U-2C 1-16, 2-34
    Model U-2F 1-16, 2-35
Dynamic Stability 6-3

E

Effect of CG Position 6-4
    Of Bank Angle 6-6, 6-6*
    Of Power 6-4
Ejection 3-23
    After Ejection 3-24
    Attitude 3-25
    Prior to 3-24
    Seat 1-47
    Speed 3-26
    With Ejection Seat 3-24
Electrical Aileron Tab 6-6
Electrical System 1-19
    AC 3-34
    AC Generator 1-22*
    Complete Failure 3-33
    DC and Inverter Power
        Distribution 1-21*
    Malfunction 3-32
    DC System 3-32
Elevator 6-2
    Control System 1-27
    Trim Tab Indicator 1-27
    Trim Index 4-6, 4-6*

Page Numbers with Asterisk (*) Denote Illustrations

Emergency
   Canopy Jettison 3-23
      Handle 1-46
   Canopy Removal 3-23
   Descent 3-20
   Extension System (Landing Gear) 1-29
   Face Heat 4-17
      Operation 4-17
   Fast Descent 2-23
   Fuel Control 1-7, 3-11
   Fuel System 1-7
   Landing 3-17
      Gear Extension 3-29
   Oxygen Supply 4-16
   Procedures (See SECTION III)
   Refueling 4-38
Engine 1-4
   Banging 7-2
   Failure 3-3
      Above 45,000 Ft 3-5
      During Flight 3-5
      Turbine Bucket 3-6
   Fuel Pump 1-4
   Fuel System 1-4, 1-6*
      Malfunctions 3-10
      Fundamental 1-10A*
   Glide Distance 3-11, 3-12*
   J-75 Cutaway 1-5*
   Oil System 1-7, 1-8*
   Operation 2-14, 2-17, 7-1
   Overspeed
      At High Altitude 3-10
      At Low Altitude 3-10
   Pressure Ratio (EPR) .. 1-10, 5-1, 5-3*
   Roughness 7-2
   Shutdown 9-9
   Starting 9-8
      Just Prior to 9-8
   Vibration 3-7
Entrance to Airplane 2-2
EPR System Malfunction 3-40

Equipment
   Auxiliary .... 1-47, 4-1 thru 4-59
   Communications & Associated
      Electronic 4-39
      AN/ARC-3 VHF Radio ... 4-47, 4-48*
      AN/ARC-34 Control Panel 4-40*
      AN/ARN-59 Control Panel 4-42*
      AN/APN-135 Rendezvous
        Beacon 4-46, 4-46*
      ARC Type 12 Control Panel .. 4-47*
      ARC Type 15F Control Panel . 4-43*
      ATC Control Panel 4-43
      C-823/AIC-10 Interphone 4-39
      Comm Selector Panel ... 4-39, 4-39*
      Failure 3-39
      Microphone Switches 4-39
      618T-3 HF Transceiver . 4-40, 4-41*
   Installation - Pilot 2-2
   Instruments 1-32
      Fundamental 1-10*
   Lighting 4-18
   Miscellaneous 4-53
   Personal 4-15
   Selection 4-17
Exchanger - Fuel-Oil Heat 1-19
Exhaust Gas Temperature 1-11, 5-1, 5-3*
Extension
   Landing Gear 6-4
   Speed Brake 6-4
   Wing Flap 6-4
External Power Receptacle 1-20
External Canopy Release Handle .. 1-46
Exterior Inspection 9-8

F

F-2 Foil System 4-48, 4-49*
Face Heat 4-16
   Emergency 4-17
   Facepiece Heat Failure 3-39
   Normal Operation 4-17

Page Numbers with Asterisk (*) Denote Illustrations

INDEX

Fan
Defroster 4-12, 4-53
Switch 4-12
Fast Descent 2-23
Final Approach 2-28
Fire 3-21
    Ground 3-21
    Overheat Light on During
        Takeoff Roll 3-22
    Overheat Light on While
        Airborne 3-22
    Probable Causes - or Overheat
        Indication 3-21
    Warning System 1-45
Flameout 3-13
    Landing Pattern .... 3-11 thru 3-17*
    Landing Procedure 3-11
    Simulated 3-17
Flaps - Wing 1-28, 6-15
    Asymmetrical - Condition 3-37
    Malfunction 3-37
Flight Characteristics 2-20, 6-1
(See SECTION VI)
Flight Control System 1-26
Flight Maneuvering 6-3
Flight Plan 2-1
Flight Post 9-9
Flight Relight Chart 3-9*
Flight Restrictions 2-1
Fog 3-22
Fog or Smoke Elimination 3-22
Formation Flying 6-16
Fuel
    Auxiliary Boost 1-17, 7-4
    Boost Pump Failure 3-29
    Boost Pump - Normal 1-17
    Capacity & Totalizer Setting 4-34
    Cross Transfer 1-15, 7-4
        Transfer Pump Failure ... 3-29
    Counter Malfunction 3-30
    Dump
        Model U-2C 1-16, 2-34
        Model U-2F 1-16, 2-35
        Operation 7-5
    Emergency (Engine Fuel
        Control) 1-7, 3-11

Fuel (Cont)
Engine
    (Fuel) System 1-4, 1-6*, 1-12*, 1-14*
Feed System 1-18*
Grade 1-16
Heat Exchanger (Fuel-Oil) 1-19
Low Level Light Indication
    During Climb 3-30
Low Level Light Indication
    During Descent 3-31
Low Level Indication in Level
    Flight at Altitude 3-30
Pressure 1-11, 5-2
Pump - Auxiliary Boost 1-17
Pump - Normal Boost 1-17
Pump - Engine 1-4
Quantity Counter 1-19
Quantity Table 1-13
Sequencing System Ⓒ 1-11
Sequencing System Ⓕ 1-13
Shutoff Valve 1-17
System Operation 7-3
    Air Start Procedure 3-7, 3-9*
        Emergency 3-8
        Normal 3-8
System Malfunctions 3-10
    Aircraft 3-29
    Dump 3-37
    Low Altitude 3-10
    Overspeed - High Altitude 3-10
System Warning Lights 1-16
Transfer 1-12*, 1-14*
Flying, Formation 6-16
Flying with External Loads 6-16
Fundamental Engine Instruments .. 1-10A*

G

Gage, Hydraulic Pressure 1-26
Gas Temperature, Exhaust 1-11
GCA Pattern 9-5*
Gear, Landing 1-29
    Control Lever 1-29
    Emergency Extension 1-29, 3-29
    Warning System 1-29
Gear, Survival 4-16

Page Numbers with Asterisk (*) Denote Illustrations

General Configuration 1-2*, 1-3*
Generator, AC 1-22*
   AC Power Distribution 1-22*
   Generator Failure, AC 3-34
Generator, DC 1-21*
   Generator and Battery Switch ... 1-20
   Ammeter 1-20
   Generator Out Warning
     Light 1-23, 1-25*, 1-28
   Generator Failure, DC 3-32
Go-Around 2-32, 6-15
Gross Weights 1-1
Ground Controlled Approach (GCA) 9-4, 9-5*
Ground Egress 3-4
Ground Fire 3-21
Gust Control 1-26, 6-12
   Limitations 5-4
   Malfunction 3-36
   Operation 6-4

H

Handle, Canopy
   Emergency Jettison 1-46
   External Release 1-46
Hatch Window Heater System 4-12
Heavy Weight Landing 2-30
High Altitude Trim 6-4
High Ammeter Reading 3-32
High Mach Recovery 3-21
Higher Altitudes, Level Off at ... 2-22
Holding Pattern, Instrument Flight .. 9-2
Hot Weather and Desert Procedures .. 9-7
Hydraulic Pressure
   Gage 1-26
   Partial Loss of 3-29
Hydraulic System 1-24, 1-25*
   Emergency Operation 3-28
     Complete Pressure Loss 3-28
     Partial Pressure Loss 3-29

I

Ice and Rain 9-4
Icing 9-4
Icy Runway Landing 9-7
Ignition System 1-9
Indicator(s)
   Airspeed 1-32
   Altimeter 1-32
   Attitude Heading Reference 4-1
   Bearing-Distance-Hdg (BDHI) 1-44, 4-55*
   Course (VOR) 1-44
   Elevator Trim Position 1-27
   Landing Gear Position 1-30
   Turn and Slip 1-44
   Type J-8 Attitude 1-44
   Vertical Speed 1-32
   Wing Flap Position 1-28
Initial Approach 2-25
Inspections
   Before Entrance 2-2
   Exterior 9-8
   Preflight 2-3
Instruments ..... 1-32, 1-33* thru 1-43*
   Climb 9-2
   Cruise Flight 9-2
   Engine 1-10
   Flight Procedures 9-1
   Fundamental 1-10A*
   Markings 5-4, 5-5*
   Takeoff and Initial Climb 9-2
Installation
   Drop Tank Ⓒ & Ⓕ 1-15
   Slipper Tank 1-15
Interphone Control, C-823/AIC-10 .. 4-39
Inverter and DC Power Distribution . 1-21*
Inverter No. 1 and/or No. 2 (Main
   Inverter) Failure 3-34
Inverter, No. 1, No. 2, & Emergency
   Failure 3-34
Inverter Out Warning Light 1-24
Inverter Switch 1-24

Page Numbers with Asterisk (*) Denote Illustrations

J K

Jet Penetration Procedure	9-2, 9-3*
Junction Box, Driftsight	4-21
Just Prior to Starting Engine	9-8
JP-4 and JP-5 Fuel Limitations	5-4
J-75 Engine Cutaway	1-5*
J-75 Engine Limitations	5-1

L

Landing	2-28, 6-14, 9-9
After Landing	2-33
Before Landing	2-25
Crosswind Landing	2-29
Emergencies	3-17
Flameout Landing Pattern	3-13, 3-14*
Flameout Landing Procedure	3-11
Heavy Weight Landing	2-30
Minimum Run Landing	2-31
Night Landing	2-30
Normal Landing Pattern	2-26*
Normal Landing Run	2-29
On Unprepared Surface	3-19
Taxiing Without Pogos After Landing	2-33
Touch-and-Go-Landings	2-32
On Icy Runways	9-7
With Drag Chute Deployment	2-30
With Landing Gears Unsafe	3-18
With Wing Flaps	
Less than 15°	3-17
Extended	3-13, 3-14*
Retracted	3-15, 3-16*
Weight	6-14
Landing Gear	1-29
Control Lever	1-29
Latch Malfunction	3-29
Emergency Extension	1-29, 3-29
Extension	6-4
Position Indicator	1-30
Warning System	1-29
Lateral Trim	2-13
Characteristics	6-6
Level Off	2-22
Light(s)	
AC Generator Out Warning	1-23
Battery Discharge Warning	1-23
Cockpit	4-18
Fire or Overheat Light On During Takeoff Roll	3-22
Fire or Overheat Light On While Airborne	3-22
Fuel System Warning	1-16
Generator Out Warning	1-24
Inverter Out Warning	1-24
Landing and Taxi	4-18
Navigation	4-19
Limitations	
Acceleration	5-4
Airspeed	5-4
EGT and EPR	5-3*
Engine J-75	5-1
Gust Control	5-4
JP-4 and JP-5 Fuel	5-4
Weight and Balance	5-6
With Slipper or Drop Tanks Installed	5-6
Longitudinal Trim Changes	6-4
Effect of Power	6-4
Landing Gear Extension	6-4
Speed Brake Extension	6-4
Stability	6-2
Wing Flap Extension	6-4
Low Altitude Airstart	3-10
Low Altitude, Engine Overspeed at	3-10

Page Numbers with Asterisk (*) Denote Illustrations

M

Mach 0.72 Maximum Range Cruise..	2-19
Mach Buffet Boundary	
Maximum Cruise Power	6-9*
Stalls	6-11*
Mach Number Characteristics	6-8
Mach Recovery High	3-21
Magnetic Standby Compass	1-45
Main Fuel Control	1-4
Maneuvering Flight	6-3
Maneuvers, Prohibited	5-2
Maneuvers, Rolling	6-5
Manual Canopy Release	1-46
Map Case	4-53
Mark III Hand Control	4-21, 4-21*
Mark IIIA Hand Control	4-22, 4-22*
Markings - Instrument	5-4, 5-5*
Maximum Altitude Cruise	2-18
Maximum Range Descent	2-23
Meter, Rate Rate B/400	4-50, 4-50*
Microphone and Headset Connectors	4-39
Microphone Switches	4-39
Minimum Run Landing	2-31
Minimum Run Takeoff	3-39
Miscellaneous Equipment	4-53
Missed Approach Procedure	9-4
Mode	
Comp	4-3
Directional Gyro	4-3
Slaved	4-3

N

Navigation - Dead Reckoning System	4-54
Navigation Lights	4-19
Night Flying	9-6
Night Landing	2-30
Night Takeoff	2-13
No. 1, No. 2, and Emergency Failure	3-34
No. 1 and/or No. 2 Failure (Main Inverter)	3-34
Normal Descent	2-22
Normal Face Heat Operation	4-17
Normal Fuel Boost Pump	1-17
Normal Landing Pattern	2-26*
Normal Landing Run	2-29
Normal Procedures (See SECTION II)	
Normal Takeoff	2-11
NOTE Definition	ii

O

Oil Pressure	1-11, 5-2
Oil System	1-7, 1-8*
Malfunction	3-35
Oil Temperature	1-10, 5-2
Operating Limitations (See SECTION V)	
Operation	
Cold Weather	9-7
Engine	2-17, 7-1
Face Heat	
Emergency	4-17
Normal	4-17
Fuel System	7-3
Gust Control	6-4
Hydraulic System Emergency	3-28
Pressurization System -	
Emergency	3-38
Systems	2-20, 7-1
Outside Air Temperature System	1-44
Overspeed	5-1
Idle	5-1
Oxygen	
Emergency Supply	4-16
Excessive Consumption	3-26
Difficulty	3-27
System	4-13, 4-14*
Malfunction	3-26

Page Numbers with Asterisk (*) Denote Illustrations

INDEX AF (C)-1-1

P

P-3 Platform System	4-51, 4-52*
Panel	
AFCS Control	4-6*
AN/APN-135	
Rendezvous Beacon	4-46*
AN/APN-153(V) Control	4-57*
AN/ARC-34 Control	4-40*
AN/ARN-59 Control	4-42*
ARC Type 15F Control	4-43*
ATC Control	4-43, 4-44*
Comm Selector	4-39*
Instrument	1-2*, 1-3*, 1-33* thru 1-37*
Mark III Hand Control	4-21*
Mark IIIA Hand Control	4-22*
Navigation Control ASN-66	4-57*
Right Console Trim	4-56*
Parking Checklist, After	2-34
Partial Hydraulic Pressure Loss	3-29
Partial Power Loss After Takeoff	3-4A
Pattern, Flameout Landing	3-13, 3-14*
Pattern, Ground Controlled Approach (GCA)	9-5*
Pattern, Normal Landing	2-26*
Penetration, Jet (Typical)	9-2, 9-3*
Periscope, Driftsight	4-21
Personal Equipment	4-15
Selection	4-17
Pilots Cockpit Check	2-5
Pilots Comfort	2-20, 9-8
Hot Weather and Desert	9-7
Pilots Equipment Check	2-4
High Flight	2-4
Low Flight	2-5
Pilots Equipment Installation	2-2
Pinger	4-54
Pitch Axis and Mach Hold	4-4
Pitot Heat Switch	4-13
Pitot Static System	1-32
Pogo(s)	1-30, 6-16
Release Failure	3-39
Position Indicator - Landing Gear	1-30
Post Flight	9-9
Power Distribution	
AC Generator	1-22*
DC and Inverter	1-21*
Power Loss	
Above 45,000 Feet	3-5
Complete	3-4A
Partial	3-4A
Power Regulation at Altitude	5-2, 7-3
Preflight	2-3, 9-1
Preflight, Instrument Flight	9-1
Preliminary Cockpit Check	2-3
Pressure, Engine Oil	1-11, 5-2
Pressure, Fuel	1-11, 5-2
Pressure Gage, Hydraulic	1-26
Pressure Ratio, Engine (EPR)	1-10
Pressure Seal, Canopy	1-46
Pressurization	
Altimeter	4-11
Selector Switch	4-11
Schedule	4-9*
System Emergency Operation Operation	3-38
Primary Defrosting System	4-11
Prior to Boarding Aircraft	2-3
Procedures	
Airstart	3-7
Emergency Fuel System	3-8
Normal Fuel System	3-8
Bailout (Ejection Seat Failure)	3-25
Ditching	3-20
Ejection	3-23
Emergency	3-1
Flameout Landing	3-11
Hot Weather and Desert	9-7
Instrument Flight	9-1
Missed Approach	9-4
Normal	2-1
Prohibited Maneuvers	5-2
Pump	
Auxiliary Fuel Boost	1-17, 7-4
Cross Transfer Ⓒ & Ⓕ	1-15, 7-4
Engine Fuel	1-4
Normal Fuel Boost	1-17

Page Numbers with Asterisk (*) Denote Illustrations

Q

Quantity Counter, Fuel	1-19
Quantity Table, Fuel	1-13

R

Radio Compass, AN/ARN-59 (ADF)	4-42
Ram Air Switch	4-10
Rate Rate Meter (B/400)	4-50, 4-50*
Ratio, Pressure	1-11
Rear View Mirror, Internal	4-53
Receptacle, External Power	1-20
Release Handle	1-46
Canopy External	1-46
Canopy Manual	1-46
Relief Bottle	4-53
Rendezvous Beacon - AN/APN-135	4-46, 4-46*
Retractable Stall Strip(s)	1-28
Malfunction	3-37
Right Console Trim Panel	4-56, 4-56*
Rubber Cone	4-54
Rudder	6-2
Collapsible - Pedals	1-27
Control System	1-27
Runaway Trim in Autopilot	3-36
Runaway Trim in Manual Flight	3-35
Roll Axis - Heading Hold (Select)	4-5
Rolling Maneuvers	6-5

S

Seal Pressure, Canopy	1-46
Seal Pressure System	4-10
Seat	
Bailout Ejection Failure	3-25
Ejection	1-47, 1-48*
Pack	4-15
Disconnect	4-15
Security Information	i

Selector - Cabin Heat	4-7
Sextant	4-23
Averager	4-25
Controls	4-24
Operation	4-27
Presentation	4-25, 4-26*
Selection of Equipment	4-17
Schedule, Pressurization	4-9*
Simulated Flameout	3-17
Slave Mode	4-3
Slaving	4-3
Slipper or Drop Tanks	6-16, 7-3
Slipper Tank Installation	1-15
Speed Brakes	1-29, 6-12
Control	1-29
Extension	6-4
Speed Warning System	1-46
Spins	6-12
Stability	
Directional	6-4
Dynamic	6-3
Static	6-2
Stall(s)	6-5
Accelerated	6-6
Compressor	7-1
One-G	6-5
Speeds - KIAS	6-6*
Strips - Rectractable	1-28
Malfunction	3-37
Starting Engine	2-8, 9-8
Just Prior to	9-8
Status of the Airplane	2-1
Steep Turns	9-2
Steering System	1-30
Strainer, 200-Mesh Fuel	1-17
Sunshade	4-53
Survival Gear	4-16

Page Numbers with Asterisk (*) Denote Illustrations

Switch(es)
- AC Generator 1-24*
- Auxiliary Tank Pressurization ... 1-13
- Battery and Generator 1-20
- Defroster Fan 4-12
- Inverter 1-24
- Landing and Taxi Lights 4-18
- Light 4-19
- Main Tank Pressure 7-4
- Navigation Light 4-19
- Pitot Heat 4-13
- Pressurization-Altimeter-Selector 4-11
- Ram Air 4-10

System(s)
- Aerial Refueling (ARS) 4-29, 4-30*
- Air Conditioning 4-8*
    - Pressurization 4-7
- Aileron Control 1-27
- Aircraft Difficulty, Oxygen 3-27
- Aircraft Fuel Malfunction 3-29
- Airplane - Fuel 1-11
- Autopilot 4-4, 4-6*
- Auxiliary Defrosting 4-12
- ATC (Air Traffic Control)... 4-43, 4-44*
- Attitude Heading Reference
   (Compass) 4-1
- B/400 Rate Rate Meter 4-50, 4-50*
- Compass 4-1, 4-2*
- Canopy Jettison and Seat Ejection 1-48*
- DC Electrical Malfunction 3-32
- Dead Reckoning Navigation 4-54
- Driftsight................. 4-19, 4-20*
- Drop Tank Malfunctions 3-31
- Electrical 1-19, 1-21*, 1-22*
    - AC 1-19, 1-22*
        - Power 1-23
    - DC 1-19, 1-21*
- Electrical Malfunction 3-32
- Elevator Control 1-27
- Emergency Fuel 1-7
- Emergency Extension 1-29
- Engine Fuel 1-4, 1-6*
- Engine Oil 1-7, 1-8*, 7-2

System(s) (Cont)
- Engine Pressure Ratio
   (EPR)............... 1-9, 5-1, 5-3*
- Fire Warning 1-45
- F-2 Foil 4-48, 4-49*
- Flight Control 1-26
- Fuel 1-11, 1-14*
    - Dump Ⓒ 1-16
    - Dump Ⓕ 1-16
    - Feed 1-18*
    - Operation 7-3
        - Sequencing Ⓒ 1-11
        - Sequencing Ⓕ 1-13
    - Transfer 1-12*
    - Warning Lights 1-16
- Hatch Window Heater 4-12
- Hydraulic 1-24, 1-25*
    - Emergency Operation 3-28
- Ignition 1-9
- Intercompressor Bleed 1-10
- Landing Gear Warning 1-29
- Operation 2-20, 7-1
- Oxygen 4-13, 4-14*
    - Malfunction 3-26
- P-3 Platform 4-51, 4-51*
- Pitot Static 1-32
- Primary Defrosting 4-11
- Rudder Control 1-27
- Seal Pressure 4-10
- Sextant 4-23
- Speed Warning 1-46
- Starter 1-9
- Steering 1-30

Systems Operation   2-20, 7-1

Page Numbers with Asterisk (*) Denote Illustrations

T

Table of Contents	v
Table, Fuel Quantity	1-13
Tachometer	1-11

Takeoff
- After Climb ... 9-8
- Before ... 9-2, 9-8
- Crosswind ... 2-13
- Full Power ... 3-40
- Instrument and Initial Climb ... 9-2
- Minimum Run ... 3-39
- Night ... 2-13
- Normal ... 2-11
- Procedures After ... 2-13

Tank Vent System ... 1-19
Taxi(ing) ... 2-10, 2-33, 9-1, 9-8
- Before Takeoff ... 2-10, 9-2
- Hot Weather and Desert ... 9-7
- Without Pogos after Landing ... 2-33

Throttle ... 1-7
Touch-and-Go Landings ... 2-32
Touchdown ... 2-28
- After ... 6-15
- Before ... 6-15

Transceiver
- 618T-3 HF SSB ... 4-40, 4-41*
- AN/ARC-34 UHF ... 4-40, 4-40*
- ARC Type 12 VHF ... 4-47, 4-47*

Transfer Fuel System ... 1-12*, 1-14*
- Pump ... 1-15

Trim ... 6-2
- Aircraft ... 2-19
- Automatic Pitch ... 4-5
- Lateral ... 2-13
- Tab Malfunction ... 3-35
  - Runaway Trim, Manual ... 3-36
  - Runaway Trim, Autopilot ... 3-35

Turbine Bucket Failure ... 3-6
Turbulence and Thunderstorms ... 9-6

Turbulent Air ... 2-13
Turn and Slip Indicator ... 1-44
Turns - Steep ... 9-2
Type J-8 Attitude Indicator ... 1-44
Typical Cockpit Arrangement ... 1-38*
Typical Jet Penetration Procedure ... 9-2

U

U-2C - Oxygen Pressure Table 3-27, 4-13

V

Valve
- Auxiliary Tank Pressurization ... 7-4
- Bleed Valve ... 1-10
- Canopy Defrost Control ... 4-11
- Fuel Shutoff ... 1-17

Vent, Tank System ... 1-19
Vertical Speed Indicator ... 1-32
Vibration, Engine ... 3-7
Viewing Screen, Driftsight ... 4-21

Page Numbers with Asterisk (*) Denote Illustrations

W

WARNING Definition ii
Warning Light(s)
    AC Generator Out 1-24
    Annunciator Panel 1-20, 1-23*
    Battery Discharge 1-23
    Fire or Overheat 3-22
    Fuel System 1-16
Warning Systems
    Fire 1-45
    Landing Gear 1-29
    Speed 1-46
Weight(s)
    And Balance 2-1
        Limitations Ⓒ 5-6
    Gross - Approximate 1-1
        Limitations Ⓕ 5-6
    Landing 6-14
Wheel Brakes 1-30, 1-31*
Windshield and Canopy Defrosting
    System 4-11
Windshield Swab 4-53
Wing Buffet 6-7
Wing Flap(s) 1-28, 6-15
    Control........................ 1-28
    Extension 6-4
    Malfunction 3-37
        Asymmetrical 3-37
    Position Indicator 1-28

X Y Z

Yaw, Adverse 6-5

Zero Ammeter Reading 3-32

Page Numbers with Asterisk (*) Denote Illustrations

AF(C)-1-1

COPY NO. 21

FLIGHT MANUAL

MODELS U-2C and U-2F AIRCRAFT

NOTICE

THIS REVISION, DATED 10 MAY 1967, SUPERSEDES FLIGHT MANUAL AF (C)-1-1 DATED 15 DECEMBER 1966, CHANGED 15 FEBRUARY 1967.

THIS CHANGE, DATED 31 DECEMBER 1968, AFFECTS INFORMATION IN SECTIONS I AND II.

DESTROY SUPERSEDED DATA IN ACCORDANCE WITH AFR-205-1.

LATEST CHANGED PAGES SUPERSEDE
THE SAME PAGES OF PREVIOUS DATE

Insert changed pages into basic
publication. Destroy superseded pages.

10 MAY 1967

Changed 31 December 1968

AF (C)-1-1

LIST OF EFFECTIVE PAGES

NOTE: The portion of the text affected by the changes is indicated by a vertical line in the outer margins of the page.

Insert Latest Changed Pages, Destroy Superseded Pages

TOTAL NUMBER OF PAGES IN THIS PUBLICATION IS 310 CONSISTING OF THE FOLLOWING:

Page No.	Issue	Page No.	Issue
*Title	31 Dec 68	2-14 thru 2-17	1 Sept 68
*A	31 Dec 68	2-18 thru 2-19	6 Nov 67
*B (Added)	31 Dec 68	2-20	1 Sept 68
C (Blank)	1 Sept 68	2-21	Original
i thru v	Original	2-22	6 Nov 67
vi (Blank)	Original	2-23	1 Sept 68
1-1	1 Sept 68	2-24	22 Nov 68
1-2 thru 1-3	Original	2-25	1 Sept 68
1-4	1 Sept 68	2-26	Original
1-5	6 Nov 67	2-27	1 Sept 68
1-6 thru 1-7	Original	2-28 thru 2-29	Original
1-8	1 Sept 68	2-30	15 Oct 68
1-9 thru 1-10	6 Nov 67	2-31 thru 2-32	Original
1-10A (Added)	6 Nov 67	2-33	1 Sept 68
1-10B (Blank)	6 Nov 67	2-34 thru 2-35	6 Nov 67
1-11	6 Nov 67	2-36 (Blank)	Original
1-12	Original	2-37 thru 2-65 (Deleted)	6 Nov 67
1-13	6 Nov 67	3-1 thru 3-4	1 Sept 68
1-14	Original	3-4A (Added)	1 Sept 68
1-15	1 Sept 68	3-4B (Blank)	1 Sept 68
1-16	Original	3-5	1 Sept 68
1-17	6 Nov 67	3-6 thru 3-7	Original
1-18 thru 1-20	Original	3-8	6 Nov 67
1-21 thru 1-22	1 Sept 68	3-9 thru 3-10	Original
1-23 thru 1-26	Original	3-11	6 Nov 67
1-27	1 Sept 68	3-12 thru 3-16	Original
1-28 thru 1-29	Original	3-17	22 Nov 68
1-30	1 Sept 68	3-18 thru 3-21	1 Sept 68
1-31 thru 1-32	Original	3-22 thru 3-26	Original
1-33 thru 1-34	1 Sept 68	3-27	1 Sept 68
1-35	Original	3-28	Original
1-36	1 Sept 68	3-29	1 Sept 68
1-36A (Added)	1 Sept 68	3-30 thru 3-34	Original
1-36B (Added)	1 Sept 68	3-35	15 Aug 68
*1-37	31 Dec 68	3-36	22 Nov 68
1-38 thru 1-46	Original	3-37	15 Aug 68
1-47	6 Nov 67	3-38 thru 3-39	Original
1-48 thru 1-49	Original	3-40	6 Nov 67
1-50 (Blank)	Original	3-41 thru 3-66 (Deleted)	6 Nov 67
2-1	6 Nov 67	4-1	Original
2-2	Original	4-2	6 Nov 67
2-3	1 Sept 68	4-3	5 July 67
2-4 thru 2-5	Original	4-4 thru 4-6	1 Sept 68
2-6 thru 2-10	1 Sept 68	4-6A thru 4-6B (Added)	1 Sept 68
2-11	6 Nov 67	4-7 thru 4-20	Original
2-12 thru 2-13	Original	4-21	6 Nov 67
		4-22 thru 4-28	Original

* The asterisk indicates pages changed, added, or deleted by the current change.

A Changed 31 December 1968

AF (C)-1-1

NOTE: The portion of the text affected by the changes is indicated by a vertical line in the outer margins of the page.

LIST OF EFFECTIVE PAGES

Insert Latest Changed Pages, Destroy Superseded Pages

Page No.	Issue	Page No.	Issue
4-29	1 Sept 68	A5-1	6 Nov 67
4-30 thru 4-33	Original	A5-2	Original
4-34	6 Nov 67	A5-3 thru A5-8	6 Nov 67
4-35 thru 4-38	Original	A5-9 (Added)	6 Nov 67
*4-39	31 Dec 68	A5-10 (Blank)	6 Nov 67
4-40 thru 4-48	Original	A6-1 thru A6-3	6 Nov 67
4-49	1 Sept 68	A6-4 (Blank)	Original
4-50 thru 4-52	Original	A7-1 thru A7-2	6 Nov 67
4-53	6 Nov 67	A8-1 thru A8-4	Original
4-54 thru 4-55	Original	A9-1 thru A9-2	Original
4-56 thru 4-57	12 Dec 67	X-1	1 Sept 68
4-58 thru 4-59	Original	X-2 thru X-2	6 Nov 67
4-60 (Blank)	Original	X-4	1 Sept 68
5-1	6 Nov 67	X-5	6 Nov 67
5-2	1 Sept 68	X-6	15 Aug 68
5-3	6 Nov 67	X-7	1 Sept 68
5-4	Original	X-8 thru X-9	6 Nov 67
5-5 thru 5-6	1 Sept 68	X-10	1 Sept 68
5-7 thru 5-16	Original	X-11	15 Aug 68
6-1 thru 6-15	Original	X-12	6 Nov 67
6-16	1 Sept 68	X-13 thru X-14	15 Aug 68
7-1	Original		
7-2 thru 7-3	6 Nov 67		
7-4 thru 7-5	Original		
7-6 (Blank)	Original		
8-1	Original		
8-2 (Blank)	Original		
9-1 thru 9-9	Original		
9-10 (Blank)	Original		
A-1	Original		
A-2 (Blank)	Original		
A1-1 thru A1-2	Original		
A1-3	6 Nov 67		
A1-4 thru A1-7	Original		
A1-8 (Blank)	Original		
A2-1	6 Nov 67		
A2-2	Original		
A3-1	6 Nov 67		
A3-2 thru A3-5	Original		
A3-6 (Blank)	Original		
A4-1	6 Nov 67		
A4-2	Original		
A4-3 thru A4-17	6 Nov 67		
A4-18 (Blank)	Original		

*The asterisk indicates pages changed, added, or deleted by the current change.

Changed 31 December 1968

B/C

COCKPIT ARRANGEMENT

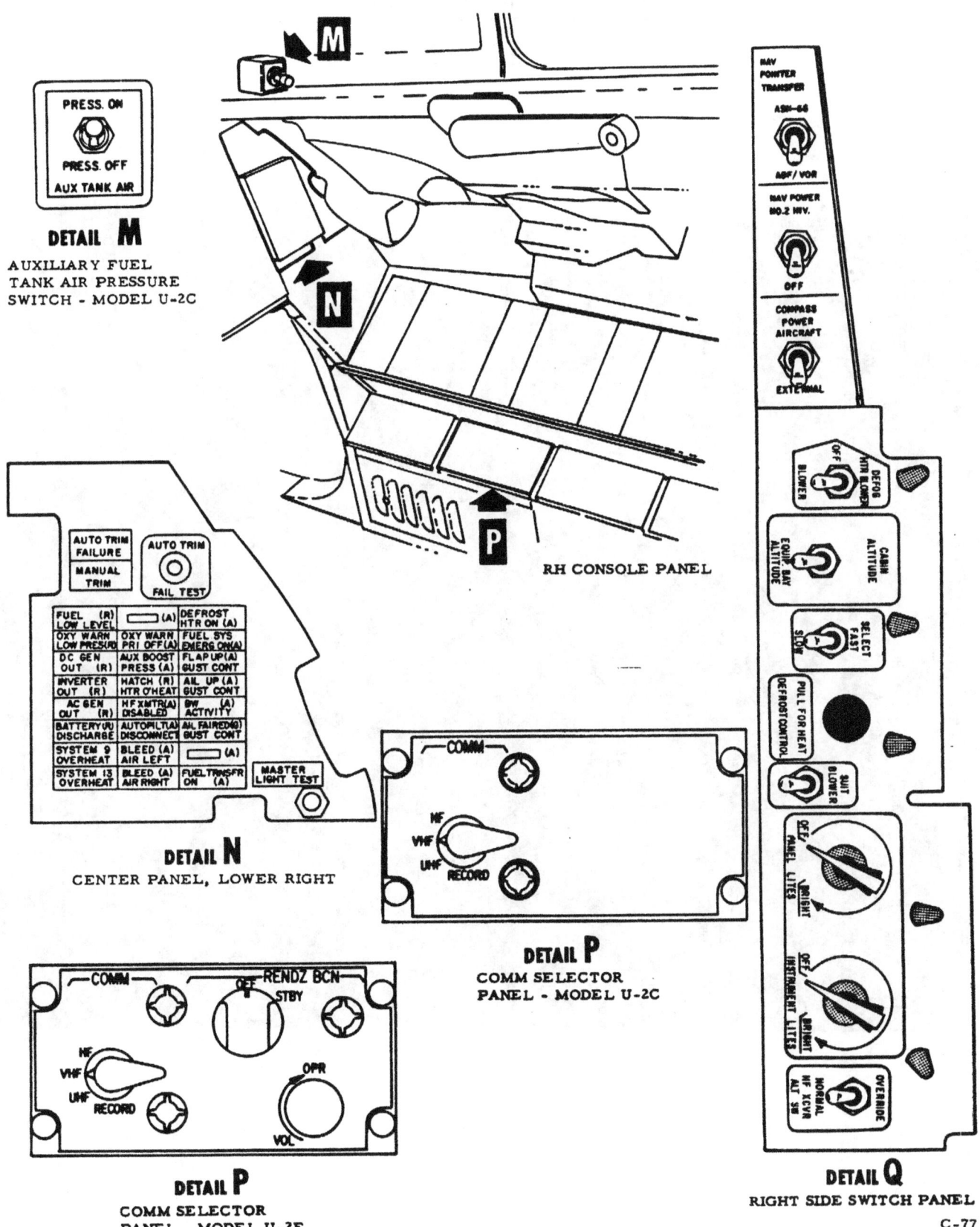

Figure 1-15 (Sheet 7)

SECTION I AF (C)-1-1

TYPICAL COCKPIT - MODEL U-2C
LEFT SIDE

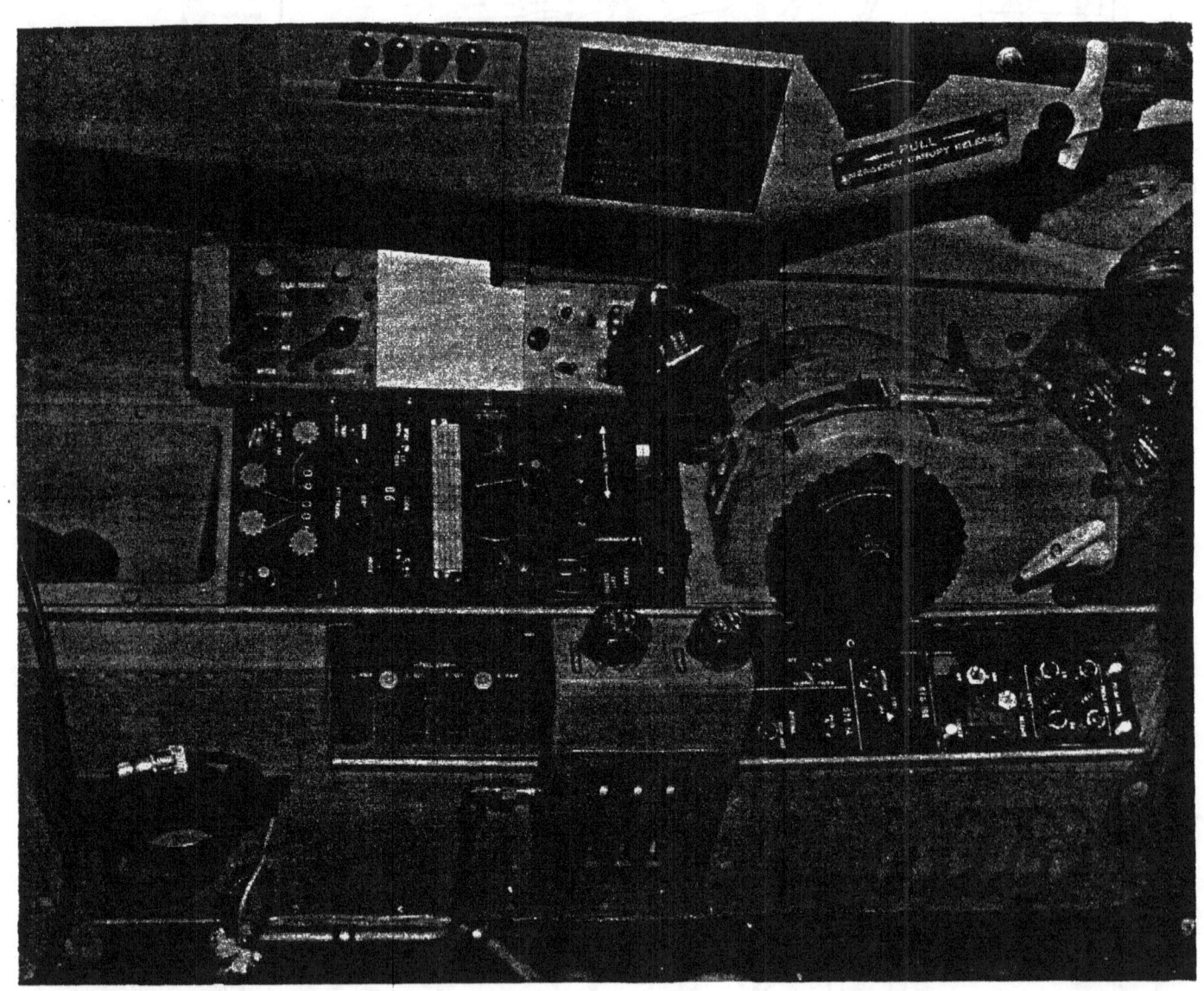

Figure 1-16 (Sheet 1)

COMMUNICATIONS AND ASSOCIATED ELECTRONIC EQUIPMENT

MICROPHONE AND HEADSET CONNECTORS

A quick disconnect fitting, located on the left-hand side of the seat near the floor, provides a connection for the pilot's microphone and headset. The connection will separate in the event of ejection.

A cable which bypasses the quick disconnect is stowed behind the seat. This cable can be used to connect a microphone and headset directly to the communications system. This cable is normally used by the pilot's assistant during the preflight aircraft and equipment checkouts. The pilot's use of this cable will often improve the quality of communications. However, automatic disconnection of the cable is not provided for in the event that ejection is necessary. It must be disconnected prior to ejection.

MICROPHONE SWITCHES

Two microphones thumb-switches are provided. They are both of the momentary button type. One is located on the throttle grip and the other is on the left control wheel grip.

C-823A/AIC-10 INTERPHONE CONTROL

The Interphone Control is used in this aircraft primarily to improve intelligibility of audio communication. The unit consists basically of an audio frequency amplifier which preamplifies the pilots transmissions and boost-amplifies radio receiver outputs.

COMM SELECTOR PANEL

The COMM selector panel, figures 4-13 and 4-14, is used to select the communication system. This panel is located in the No. 2 position in the right console step panel.

COMM SELECTOR PANEL - MODEL U-2C

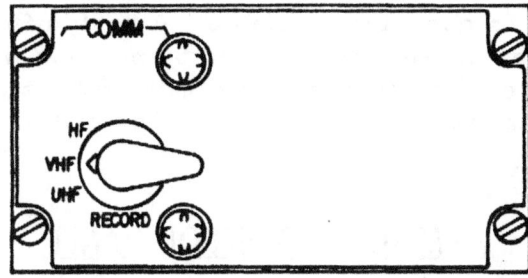

Figure 4-13

COMM SELECTOR PANEL - MODEL U-2F

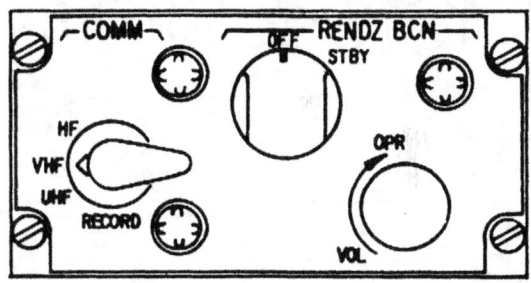

Figure 4-14

The COMM selector has RECORD, UHF, VHF, and HF positions which determine whether voice is recorded or transmitted over the radio(s) selected.

Receiving is not affected by this switch, except that in the RECORD position, no receiving is possible. Receiving is controlled by the individual receiver switches and their respective volume controls.

AN/ARC-34 UHF TRANSCEIVER

A standard AN/ARC-34 UHF receiver-transmitter having a 10 watt output is provided for communications. This unit is located in the aft end of the equipment bay. The control unit is located on the left-hand console. (See figure 4-15.) This radio provides for voice communications in the frequency band range of 225 to 399.9 megacycles. In this range there are 1750 separate frequencies available. Twenty of these can be preset and selected by number with the selector switch. Any of the others can be manually selected in the cockpit.

AN/ARC-34 CONTROL PANEL

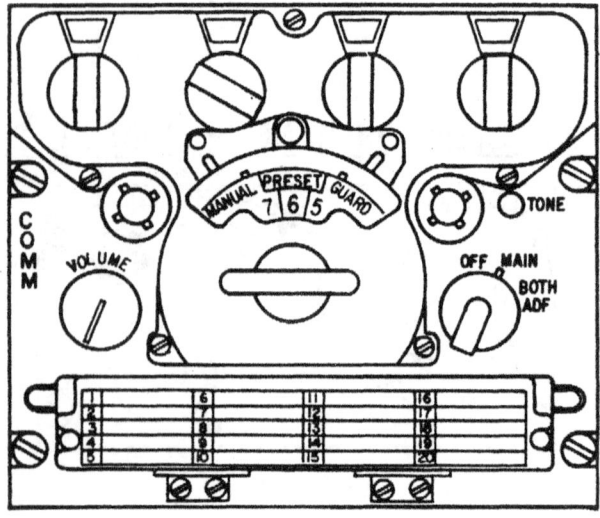

Figure 4-15

Operation

AN/ARC-34 circuit breaker in equipment bay must be closed.

1. Check the channel preset frequencies as indicated on the plastic write-in card. Change preset frequencies as required for the mission.

2. Close the interphone circuit breaker.

3. Rotate the function switch to BOTH if simultaneous monitoring of the preset channel and the guard channel is desired.

4. Set the mode switch so that PRESET is visible through the clear window.

5. Select the preset channel using the channel selector so that the channel number appears in the clear window.

6. Select UHF on COM selector switch.

7. Before transmission of a message, check for operation and warm up of the transmitter by using either the microphone button or tone button and listening for side tone.

8. If it is desired to transmit and receive on a frequency not preset on the channel selector, place the mode switch in the manual position and set up the new frequency with the manual frequency selector knobs.

9. Turn the function switch to OFF to deenergize the set.

618T-3 HF TRANSCEIVER

The Collins 618T is an airborne HF transceiver capable of operating in single band or AM modes in the frequency band of 2.5 to 24 megacycles. The major components and their locations are as follows:

1. Transceiver - Pressurized box in nose.

2. Antenna Tuner - Pressurized box in the top fairing.

3. Antenna - External wire between the antenna tuner and the vertical fin.

COPY NO. 25

FLIGHT MANUAL SUPPLEMENT

AF (C)-1-1 S-3

MODEL U-2   AIRCRAFT

SERIAL NO. 56-6681, 56-6714 AND 56-6718

SUMP TANK FUEL LEVEL GAGE
INSTRUMENT AND SWITCH LOCATIONS
LANDING GEAR
STALL STRIP WARNING LIGHTS
AUTOPILOT PITCH TRIM

29 OCTOBER 1968

INTRODUCTION

This document supplements U-2 flight manual AF (C)-1-1 and is applicable to peculiar configurations of U-2 aircraft Serial Numbers 56-6681, 56-6714 and 56-6718.

DESCRIPTION

The following description deals in greater part with the relocation of instruments and switches described in the basic flight manual. Exceptions are the description and operation of a sump tank fuel quantity gage incorporated in these airplanes, and a brief description of a modified landing gear.

The operating procedures for these airplanes are the same as for the U-2C except for the minor differences outlined in this supplement.

INSTRUMENTS

SUMP TANK FUEL LEVEL GAGE

The sump tank fuel level gage is on the lower left-hand instrument panel. It is powered by either main inverter. The gage is calibrated in gallons and covers the range from 21 to 92 gallons. Green markings show the primary float levels (77 gallons and 57 gallons for the auxiliary and main tanks, respectively); and a yellow marking shows the secondary float level (25 gallons). Other yellow markings indicate areas of concern. The level will dwell at 77 gallons until the auxiliary tanks are empty and gradually lower to 57 gallons and dwell there until the main tanks are empty.

When the sump tank level drops below the low level light (50 gallons), follow instructions in Section III of flight manual.

OUTSIDE AIR TEMPERATURE INDICATOR

The outside air temperature indicator is located aft in the right-hand console.

WING FLAP POSITION INDICATOR

The wing flap position indicator has a range from minus 4 to plus 60 degrees. However, the wing flap down limit is set at 35 degrees.

DC AMMETER

The DC ammeter is on the center instrument panel below the 8-day clock.

A clip bracket is provided on the control column, suitable for retaining a hack watch if desired.

SWITCHES

INTERPHONE SWITCH

The record switch on the communications selector panel must be held in the RECORD VOICE position to transmit on the interphone.

VHF NAV CONTROL PANEL

The VHF navigation control panel is on the left-hand console.

UHF CONTROL PANEL

The UHF control panel is located on the right-hand console.

THROTTLE SWITCHES

A microphone switch, speed brake and spoiler control switches are on the throttle handle. The spoilers are deactivated and the spoiler switch can not be energized.

LANDING GEAR

The landing gear on these airplanes is strengthened to accomodate higher sink rates at touchdown. The main gear extends 2 1/2 inches lower and the tail gear extends 1 1/2 inch lower. A shrink bar on the main gear strut automatically compresses the strut during retraction sufficiently for the gear to fit in the well. The tail gear retracts normally.

STALL STRIP WARNING LIGHTS

There are two amber lights (left and right stall strip) located on the annunciator panel which illuminate whenever the stall strips are retracted and the landing gear is down and locked. The lights go out when the stall strips are extended. When the landing gear is not down and locked, the lights will not illuminate regardless of the position of the stall strips.

AUTOPILOT PITCH TRIM

An Autopilot Pitch Trim switch with a red guard is located above the annunciator panel on the lower right side of the instrument panel. This is a two position switch: AUTO (guard down) and MANUAL (guard up). In the AUTO position, with the autopilot on, the autopilot controls the pitch trim automatically to relieve control forces. If the autopilot is not on, pitch trim is manual even with the A/P Pitch Trim switch in the AUTO position. Selecting the MANUAL position transfers the pitch trim control from automatic to manual. In the MANUAL position, pitch trim is manual, on or off autopilot.

The Auto Trim Failure light on the annunciator panel along with the MASTER CAUTION light will illuminate whenever the autopilot attempts to trim the tab to increase control forces. The MASTER CAUTION light does not illuminate in manual trim operation.

The AUTO TRIM FAIL TEST switch is located to the right of the autopilot pitch trim switch. It is used for a system check. Pressing the button simulates a failure in the automatic pitch trim control circuit. If the AUTO TRIM FAILURE and MASTER CAUTION lights illuminate, proper operation of the auto trim failure indicating system is indicated. Switching the autopilot trim switch (guard up) activates manual trim and illuminates the MANUAL TRIM light. Switching the autopilot pitch trim switch to AUTO (guard down) restores the automatic pitch trim function.

END OF SUPPLEMENT

www.ingramcontent.com/pod-product-compliance
Lightning Source LLC
Chambersburg PA
CBHW082029300426
44117CB00015B/2408